Contemporary Exercises a[...]
practice a concept in a real-w[...]

Cooperative Learning Exercises provide an opportunity to work with others to solve a problem.

26. **Lotteries** The PowerBall lottery commission chooses 5 white balls from a drum containing 59 balls marked with the numbers 1 through 59, and 1 red ball from a separate drum containing 39 balls. The following table shows the approximate probability of winning a certain prize if the numbers you choose match those chosen by the lottery commission.

Match	Prize	Approximate probability
⚪⚪⚪⚪⚪ + 🔴	Jackpot	1:195,249,054
⚪⚪⚪⚪⚪	$200,000	1:5,138,133

A variety of **End-of-Chapter** features help you prepare for a test.

The **Chapter Summary** reviews the major concepts discussed in the chapter. For each concept, there is a reference to a worked example illustrating how the concept is used and at least one exercise in the Chapter Review Exercises relating to that concept.

Chapter Review Exercises help you review all of the concepts in the chapter. Answers to all the Chapter Review Exercises are in the answer section, along with a reference to the section from which the exercise was taken. If you miss an exercise, use that reference to review the concept.

CHAPTER 11 SUMMARY

The following table summarizes essential concepts in this chapter. The references given in the right-hand column list Examples and Exercises that can be used to test your understanding of a concept.

11.1 Simple Interest

Simple Interest Formula The simple interest formula is $I = Prt$, where I is the interest, P is the principal, r is the interest rate, and t is the time period.	See Examples 2, 4, and 5 on pages 661 and 662, and then try Exercises 1, 2, and 3 on page 729.

CHAPTER 11 REVIEW EXERCISES

1. Simple Interest Calculate the simple interest due on a 4-month loan of $2750 if the interest rate is 6.75%.

2. Simple Interest Find the simple interest due on an 8-month loan of $8500 if the interest rate is 1.15% per month.

3. Simple Interest What is the simple interest earned in 120 days on a deposit of $4000 if the interest rate is 6.75%?

4. Maturity Value Calculate the maturity value of a simple interest, 108-day loan of $7000 if the interest rate is 10.4%.

CHAPTER 11 REVIEW EXERCISES *page 729*

1. $61.88 [Sec. 11.1] 2. $782 [Sec. 11.1] 3. $90 [Sec. 11.1] 4. $7218.40 [Sec. 11.1] 5. 7.5% [
6. $3654.90 [Sec. 11.2] 7. $11,609.72 [Sec. 11.2] 8. $7859.52 [Sec. 11.2] 9. $200.23 [Sec. 11.2]
11. a. $11,318.23 b. $3318.23 [Sec. 11.2] 12. $19,225.50 [Sec. 11.2] 13. 1.1% [Sec. 11.4] 14.
15. $1.59 [Sec. 11.2] 16. $43,650.68 [Sec. 11.2] 17. 6.06% [Sec. 11.2] 18. 5.4% compounded semia
19. $431.16 [Sec. 11.3] 20. $6.12 [Sec. 11.3] 21. a. $259.38 b. 12.9% [Sec. 11.3] 22. a. $36
23. $45.41 [Sec. 11.3] 24. a. $10,092.69 b. $2018.54 c. $253.01 [Sec. 11.3] 25. $664.40 [Se
b. $12,196.80 [Sec. 11.3] 27. a. Profit of $5325 b. $256.10 [Sec. 11.4] 28. 200 shares [Sec. 11.4]
30. a. $1659.11 b. $597,279.60 c. $341,479.60 [Sec. 11.5] 31. a. $1396.69 b. $150,665.74 [S
32. $2658.53 [Sec. 11.5]

The **Chapter Test** gives you a chance to practice a possible test for the chapter. Answers to all Chapter Test questions are in the answer section, along with a section reference for the question.

For the Chapter Test, besides a reference to the section from which an exercise was taken, there is a reference to an example that is similar to the exercise.

CHAPTER 11 TEST

1. Simple Interest Calculate the simple interest due on a 3-month loan of $5250 if the interest rate is 8.25%.

2. Simple Interest Find the simple interest earned in 180 days on a deposit of $6000 if the interest rate is 6.75%.

3. Maturity Value Calculate the maturity value of a simple interest, 200-day loan of $8000 if the interest rate is 9.2%.

4. Simple Interest Rate The simple interest charged on a 2-month loan of $7600 is $114. Find the simple interest

10. Bonds Suppose you purchase a $5000 bond that has a 3.8% coupon and a 10-year maturity. Calculate the total of the interest payments that you will receive.

11. **Inflation** In 2009 the median value of a single-family house was $169,000. Use an ann[...] inflation rate of 7% to calculate the median val[...] single family house in 2029. (*Source:* money.cn[...]

12. Effective Interest Rate Calculate the effectiv[...] rate of 6.25% compounded quarterly. Round to [...] est hundredth of a percent.

CHAPTER 11 TEST *page 731*

1. $108.28 [Sec. 11.1, Example 2] 2. $202.50 [Sec. 11.1, Example 1] 3. $8408.89 [Sec. 11.1, Example 6]
4. 9% [Sec. 11.1, Example 5] 5. $7340.87 [Sec. 11.2, Check Your Progress 2] 6. $312.03 [Sec. 11.2, Example
7. a. $15,331.03 b. $4831.03 [Sec. 11.1, Example 6] 8. $21,949.06 [Sec. 11.2, Example 6] 9. 1.2% [Sec.
10. $1900 [Sec. 11.4, Example 4] 11. $653,976.67 [Sec. 11.2, Check Your Progress 8] 12. 6.40% [Sec. 11.2, C
13. 4.6% compounded semiannually [Sec. 11.2, Example 11] 14. $7.79 [Sec. 11.3, Example 1] 15. a. $48.56
b. 16.6% [Sec. 11.3, Example 4] 16. $56.49 [Sec. 11.3, Example 3] 17. a. loss of $4896 b. $226.16 [Se
18. 208 shares [Sec. 11.4, Example 5] 19. a. $6985.94 b. $1397.19 c. $174.62 [Sec. 11.3, Example 4]
20. $60,083.50 [Sec. 11.5, Example 1] 21. a. $1530.69 [Sec. 11.5, Example 2a] b. $221,546.46 [Sec. 11.5, Exa
22. $2595.97 [Sec. 11.5, Example 2a, Example 5]

Mathematical Excursions

Enhanced Edition

Third Edition

Aufmann | Lockwood | Nation | Clegg

CENGAGE
Learning·

Australia • Brazil • Japan • Korea • Mexico • Singapore • Spain • United Kingdom • United States

Mathematical Excursions: Enhanced Edition, Third Edition

Mathematical Excursions, Enhanced Edition, 3rd Edition
Richard N. Aufmann | Joanne S. Lockwood | Richard D. Nation | Daniel K. Clegg

For product information and technology assistance, contact us at
Cengage Learning Customer & Sales Support, 1-800-354-9706

For permission to use material from this text or product,
submit all requests online at **cengage.com/permissions**
Further permissions questions can be emailed to
permissionrequest@cengage.com

This book contains select works from existing Cengage Learning resources and was produced by Cengage Learning Custom Solutions for collegiate use. As such, those adopting and/or contributing to this work are responsible for editorial content accuracy, continuity and completeness.

Compilation © 2017 Cengage Learning

ISBN: 978-1-337-32684-1

Cengage Learning
20 Channel Center Street
Boston, MA 02210
USA

Cengage Learning is a leading provider of customized learning solutions with office locations around the globe, including Singapore, the United Kingdom, Australia, Mexico, Brazil, and Japan. Locate your local office at: www.international.cengage.com/region.

Cengage Learning products are represented in Canada by Nelson Education, Ltd.

For your lifelong learning solutions, visit www.cengage.com/custom.

Visit our corporate website at www.cengage.com.

Brief Contents

Mathematical Excursions is about mathematics as a system of knowing or understanding our surroundings. It is similar to an English literature textbook, an introduction to philosophy textbook, or perhaps an introductory psychology textbook. Each of those books provides glimpses into the thoughts and perceptions of some of the world's greatest writers, philosophers, and psychologists. Reading and studying their thoughts enables us to better understand the world we inhabit.

In a similar way, *Mathematical Excursions* provides glimpses into the nature of mathematics and how it is used to understand our world. This understanding, in conjunction with other disciplines, contributes to a more complete portrait of the world. Our contention is that:

- Planning a shopping trip to several local stores, or several cities scattered across Europe, is more interesting when one has knowledge of efficient routes, which is a concept from the field of graph theory.
- Problem solving is more enjoyable after you have studied a variety of problem-solving techniques and have practiced using George Polya's four-step, problem-solving strategy.
- The challenges of sending information across the Internet are better understood by examining prime numbers.
- The perils of radioactive waste take on new meaning with knowledge of exponential functions.
- Generally, knowledge of mathematics strengthens the way we know, perceive, and understand our surroundings.

The central purpose of *Mathematical Excursions* is to explore those facets of mathematics that will strengthen your quantitative understandings of our environs. We hope you enjoy the journey.

Updates to This Edition

- NEW! The Excursion activities and exercises that appear at the end of every section in the textbook now appear in Enhanced WebAssign. Many are accompanied by videos and interactive simulations to reinforce conceptual understanding and to more fully engage students with the mathematics of that section.
- NEW! Chapter-level reviews are now available in Enhanced WebAssign. These preloaded and assignable reviews present students with questions that focus on prerequisite/co-requisite algebra skills. The reviews will help students refresh their knowledge and fill in any gaps so they may advance more smoothly through new concepts or topics.
- NEW! The table of contents has been reorganized and rearranged to group the chapters by broad topics that can be covered sequentially.
- NEW! In the News exercises, based on media sources, have been added to this edition, providing another way to engage students by demonstrating the contemporary use of mathematics.
- NEW! Chapter Summaries now appear in an easy-to-use grid format organized by section. Each summary point is now paired with page numbers of an example that illustrates the concept and exercises that students can use to test their understanding.
- NEW! In the Answer Section, answers to Chapter Test exercises now include a reference to a similar example in the text, making it easy for students to review relevant material for exercises that they have answered incorrectly.
- Application Examples, Exercises, and Excursions have been updated to reflect recent data and trends.
- Definitions are now boxed and highlighted for greater prominence throughout the text, facilitating study and review.

Interactive Method

The AIM FOR SUCCESS STUDENT PREFACE explains what is required of a student to be successful and how this text has been designed to foster student success. This "how to use this text" preface can be used as a lesson on the first day of class or as a project for students to complete to strengthen their study skills. ————

AIM for Success

Welcome to *Mathematical Excursions*, Third Edition. As you begin this course, we know two important facts: (1) You want to succeed. (2) We want you to succeed. In order to accomplish these goals, an effort is required from each of us. For the next few pages, we are going to show you what is required of you to achieve your goal and how we have designed this text to help you succeed.

Motivation

One of the most important keys to success is motivation. We can try to motivate you by offering interesting or important ways that you can benefit from mathematics. But, in the end, the motivation must come from you. On the first day of class it is easy to be motivated. Eight weeks into the term, it is harder to keep that motivation.

To stay motivated, there must be outcomes from this course that are worth your time, money, and energy. List some reasons you are taking this course. Do not make a mental list—actually write them out. Do this now.

Although we hope that one of the reasons you listed was an interest in mathematics, we know that many of you are taking this course because it is required to graduate, it is a prerequisite for a course you must take, or because it is required for your major. If you are motivated to graduate or complete the requirements for your major, then use that motivation to succeed in this course. Do not become distracted from your goal to complete your education!

Commitment

To be successful, you must make a commitment to succeed. This means devoting time to math so that you achieve a better understanding of the subject.

List some activities (sports, hobbies, talents such as dance, art, or music) that you enjoy and at which you would like to become better. Do this now.

Next to these activities, put the number of hours each week that you spend practicing these activities.

Whether you listed surfing or sailing, aerobics or restoring cars, or any other activity you enjoy, note how many hours a week you spend on each activity. To succeed in math, you must be willing to commit the same amount of time. Success requires some sacrifice.

The "I Can't Do Math" Syndrome

There may be things you cannot do, for instance, lift a two-ton boulder. You can, however, do math. It is much easier than lifting the two-ton boulder. When you first learned the activities you listed above, you probably could not do them well. With practice, you got better. With practice, you will be better at math. Stay focused, motivated, and committed to success.

It is difficult for us to emphasize how important it is to overcome the "I Can't Do Math Syndrome." If you listen to interviews of very successful athletes after a particularly bad performance, you will note that they focus on the positive aspect of what they did, not the negative. Sports psychologists encourage athletes to always be positive—to have a "Can Do" attitude. You need to develop this attitude toward math.

TAKE NOTE

Motivation alone will not lead to success. For instance, suppose a person who cannot swim is placed in a boat, taken out to the middle of a lake, and then thrown overboard. That person has a lot of motivation to swim but there is a high likelihood the person will drown without some help. Motivation gives us the desire to learn but is not the same as learning.

Each CHAPTER OPENER includes a list of sections that can be found within the chapter and includes an anecdote, description, or explanation that introduces the student to a topic in the chapter. ————

1

1.1 Inductive and Deductive Reasoning

1.2 Problem Solving with Patterns

1.3 Problem-Solving Strategies

Problem Solving

Most occupations require good problem-solving skills. For instance, architects and engineers must solve many complicated problems as they design and construct modern buildings that are aesthetically pleasing, functional, and that meet stringent safety requirements. Two goals of this chapter are to help you become a better problem solver and to demonstrate that problem solving can be an enjoyable experience.

One problem that many have enjoyed is the Monty Hall (host of the game show *Let's Make a Deal*) problem, which is stated as follows. The grand prize in *Let's Make a Deal* is behind one of three doors. Less desirable prizes (for instance, a goat and a box of candy) are behind the other two doors. You select one of the doors, say door 1. Monty Hall reveals one of the less desirable prizes behind one of the other doors. You are then given the opportunity either to stay with your original choice or to choose the remaining closed door.

Example: You choose door 1. Monty Hall reveals a goat behind door 3. You can stay with door 1 or switch to door 2.

Marilyn vos Savant, author of the "Ask Marilyn" column featured in *Parade Magazine*, analyzed this problem,[1] claiming that you *double* your chances of winning the grand prize by switching to the other closed door. Many readers, including some mathematicians, responded with arguments that contradicted Marilyn's analysis.

What do you think? Do you have a better chance of winning the grand prize by switching to the other closed door or staying with your original choice?

Of course there is also the possibility that it does not matter, if the chances of winning are the same with either strategy.

Discuss the Monty Hall problem with some of your friends and classmates. Is everyone in agreement? Additional information on this problem is given in Exploration Exercise 56 on page 15.

1. "Ask Marilyn," *Parade Magazine*, September 9, 1990, p. 15.

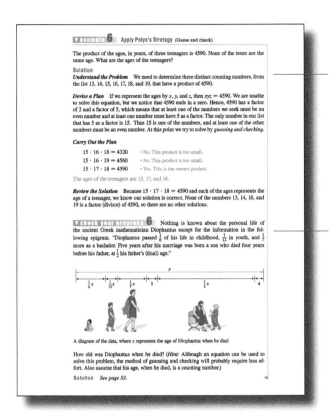

Each section contains a variety of WORKED EXAMPLES. Each example is given a title so that the student can see at a glance the type of problem that is being solved. Most examples include annotations that assist the student in moving from step to step, and the final answer is in color in order to be readily identifiable.

Following each worked example is a CHECK YOUR PROGRESS exercise for the student to work. By solving this exercise, the student actively practices concepts as they are presented in the text. For each Check Your Progress exercise, there is a detailed solution in the Solutions appendix.

At various places throughout the text, a QUESTION is posed about the topic that is being discussed. This question encourages students to pause, think about the current discussion, and answer the question. Students can immediately check their understanding by referring to the ANSWER to the question provided in a footnote on the same page. This feature creates another opportunity for the student to interact with the textbook.

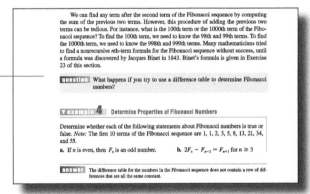

Each section ends with an EXCURSION along with corresponding EXCURSION EXERCISES. These activities engage students in the mathematics of the section and encourage them to take an active part in the learning process. Some are designed as in-class cooperative learning activities that lend themselves to a hands-on approach. They can also be assigned as projects or extra credit. In addition, the Excursions appear in Enhanced WebAssign for assignment as online homework. Many are accompanied by videos and interactive simulations to more fully engage students with the underlying mathematics.

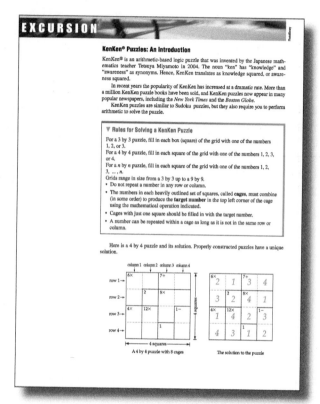

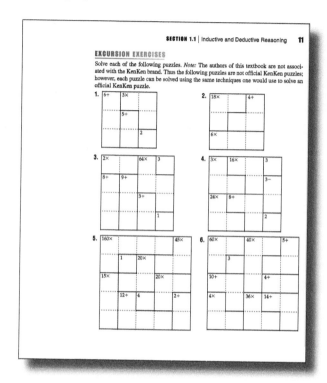

The EXERCISE SETS were carefully written to provide a wide variety of exercises that range from drill and practice to interesting challenges. Exercise sets emphasize skill building, skill maintenance, concepts, and applications. Icons are used to identify various types of exercises.

Writing exercises

Data analysis exercises

Graphing calculator exercises

Exercises that require the Internet

EXTENSION EXERCISES are placed at the end of each exercise set. These exercises are designed to extend concepts. In most cases these exercises are more challenging and require more time and effort than the preceding exercises. The Extension exercises always include at least two of the following types of exercises:

- Critical Thinking
- Cooperative Learning
- Exploration

Some Critical Thinking exercises require the application of two or more procedures or concepts.

The Cooperative Learning exercises are designed for small groups of two to four students.

Many of the Exploration exercises require students to search the Internet or through reference materials in a library.

41. Sally likes 225 but not 224; she likes 900 but not 800; she likes 144 but not 145. Which of the following does she like?
   a. 1600   b. 1700

42. There are 1200 elephants in a herd. Some have pink and green stripes, some are all pink, and some are all blue. One third are pure pink. Is it true that 400 elephants are definitely blue?
   a. Yes   b. No

43. Following the pattern shown in the number sequence below, what is the missing number?

   1 8 27 ? 125 216

   a. 36   b. 45   c. 46
   d. 64   e. 99

EXTENSIONS

Critical Thinking

44. **Compare Exponential Expressions**
   a. How many times as large is $3^{(3^3)}$ than $(3^3)^3$?
   b. How many times as large is $4^{(4^4)}$ than $(4^4)^4$? *Note:* Most calculators will not display the answer to this problem because it is too large. However, the answer can be determined in exponential form by applying the following properties of exponents.

   $$(a^m)^n = a^{mn} \text{ and } \frac{a^m}{a^n} = a^{m-n}$$

45. **A Famous Puzzle** The mathematician Augustus De Morgan once wrote that he had the distinction of being x years old in the year x^2. He was 43 in the year 1849.
   a. Explain why people born in the year 1980 might share the distinction of being x years old in the year x^2. *Note:* Assume x is a natural number.
   b. What is the next year after 1980 for which people born in that year might be x years old in the year x^2?

46. **Verify a Procedure** Select a two-digit number between 50 and 100. Add 83 to your number. From this number form a new number by adding the digit in the hundreds place to the number formed by the other two digits (the digits in the tens place and the ones place). Now subtract this newly formed number from your original number. Your final result is 16. Use a deductive approach to show that the final result is always 16 regardless of which number you start with.

47. **Numbering Pages** How many digits does it take in total to number a book from page 1 to page 240?

48. **Mini Sudoku** Sudoku is a deductive reasoning, number-placement puzzle. The object in a 6 by 6 mini-Sudoku puzzle is to fill all empty squares so that the counting numbers 1 to 6 appear exactly once in each row, each

column, and each of the 2 by 3 regions, which are delineated by the thick line segments. Solve the following 6 by 6 mini-Sudoku puzzle.

6	2			5	
		4	3		
	6	5			4
		1		3	
1		6	2		5
	4		1	6	

Cooperative Learning

49. **The Four 4s Problem** The object of this exercise is to create mathematical expressions that use exactly four 4s and that simplify to a counting number from 1 to 20, inclusive. You are allowed to use the following mathematical symbols: +, −, ×, ÷, √, (, and). For example,

   $$\frac{4}{4} + \frac{4}{4} = 2, \ 4^{(4-4)} + 4 = 5, \text{ and}$$

   $$4 - \sqrt{4} + 4 \times 4 = 18$$

50. **A Cryptarithm** The following puzzle is a famous cryptarithm.

```
  S E N D
+ M O R E
---------
M O N E Y
```

Each letter in the cryptarithm represents one of the digits 0 through 9. The leading digits, represented by S and M, are not zero. Determine which digit is represented by each of the letters so that the addition is correct. *Note:* A letter that is used more than once, such as M, represents the same digit in each position in which it appears.

CHAPTER 2 SUMMARY

The following table summarizes essential concepts in this chapter. The references given in the right-hand column list Examples and Exercises that can be used to test your understanding of a concept.

2.1 Basic Properties of Sets

The Roster Method The roster method is used to represent a set by listing each element of the set inside a pair of braces. Commas are used to separate the elements.	See **Example 1** on page 52, and then try Exercises 1 and 2 on page 108.
Basic Number Sets Natural Numbers or Counting Numbers $N = \{1, 2, 3, 4, 5, \dots\}$ Whole Numbers $W = \{0, 1, 2, 3, 4, 5, \dots\}$ Integers $I = \{\dots, -4, -3, -2, -1, 0, 1, 2, 3, 4, \dots\}$ Rational Numbers $Q =$ the set of all terminating or repeating decimals Irrational Numbers $\mathcal{I} =$ the set of all nonterminating, nonrepeating decimals Real Numbers $R =$ the set of all rational or irrational numbers	See **Example 3** and **Check Your Progress 3** on pages 53 and 54, and then try Exercises 3 to 6 on page 108.
Set-Builder Notation Set-builder notation is used to represent a set, by describing its elements.	See **Example 5** on page 55, and then try Exercises 7 to 10 on page 108.
Cardinal Number of a Finite Set The cardinal number of a finite set is the number of elements in the set. The cardinal number of a finite set A is denoted by the notation $n(A)$.	See **Example 6** on page 55, and then try Exercises 63 to 67 on page 109.
Equal Sets and Equivalent Sets Two sets are equal if and only if they have exactly the same elements. Two sets are equivalent if and only if	See **Example 7** on page 56, and then try Exercises 11 and 12 on page 108.

At the end of each chapter is a CHAPTER SUMMARY that describes the concepts presented in each section of the chapter. Each concept is paired with page numbers of examples that illustrate the concept and exercises that students can use to test their understanding of a concept.

CHAPTER REVIEW EXERCISES are found near the end of each chapter. These exercises were selected to help the student integrate the major topics presented in the chapter. The answers to all the Chapter Review exercises appear in the answer section along with a section reference for each exercise. These section references indicate the section or sections where a student can locate the concepts needed to solve the exercise.

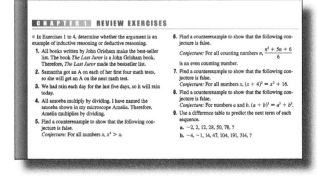

CHAPTER 1 REVIEW EXERCISES

▪ In Exercises 1 to 4, determine whether the argument is an example of inductive reasoning or deductive reasoning.

1. All books written by John Grisham make the best-seller list. The book *The Last Juror* is a John Grisham book. Therefore, *The Last Juror* made the bestseller list.

2. Samantha got an A on each of her first four math tests, so she will get an A on the next math test.

3. We had rain each day for the last five days, so it will rain today.

4. All amoeba multiply by dividing. I have named the amoeba shown in my microscope Amelia. Therefore, Amelia multiplies by dividing.

5. Find a counterexample to show that the following conjecture is false.
   Conjecture: For all numbers x, $x^4 > x$.

6. Find a counterexample to show that the following conjecture is false.
   Conjecture: For all counting numbers n, $\dfrac{n^3 + 5n + 6}{6}$ is an even counting number.

7. Find a counterexample to show that the following conjecture is false.
   Conjecture: For all numbers x, $(x + 4)^2 = x^2 + 16$.

8. Find a counterexample to show that the following conjecture is false.
   Conjecture: For numbers a and b, $(a + b)^3 = a^3 + b^3$.

9. Use a difference table to predict the next term of each sequence.
   a. $-2, 2, 12, 28, 50, 78, ?$
   b. $-4, -1, 14, 47, 104, 191, 314, ?$

The CHAPTER TEST exercises are designed to emulate a possible test of the material in the chapter. The answers to all the Chapter Test exercises appear in the answer section along with a section reference and an example reference for each exercise. The section references indicate the section or sections where a student can locate the concepts needed to solve the exercise, and the example references allow students to readily find an example that is similar to a given test exercise.

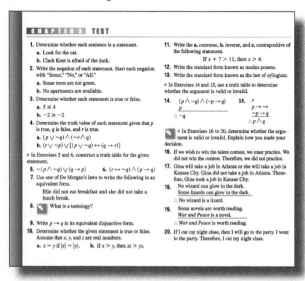

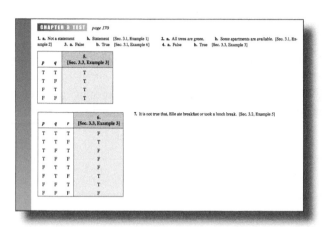

Other Key Features

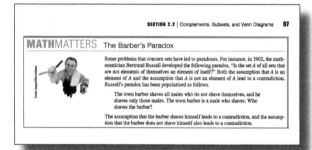

Math Matters

This feature of the text typically contains an interesting sidelight about mathematics, its history, or its applications.

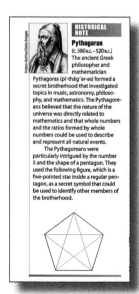

Historical Note

These margin notes provide historical background information related to the concept under discussion or vignettes of individuals who were responsible for major advancements in their fields of expertise.

POINT OF INTEREST

Waterfall by M.C. Escher

M. C. Escher (1898–1972) created many works of art that defy logic. In this lithograph, the water completes a full cycle even though the water is always traveling downward.

Point of Interest

These short margin notes provide interesting information related to the mathematical topics under discussion. Many of these are of a contemporary nature and, as such, they help students understand that math is an interesting and dynamic discipline that plays an important role in their daily lives.

TAKE NOTE

The alternative procedure for constructing a truth table, as described to the right, generally requires less writing, less time, and less effort than the truth table procedure that was used in Examples 1 and 2.

Take Note

These notes alert students to a point requiring special attention, or they are used to amplify the concepts currently being developed.

CALCULATOR NOTE

Some calculators display $\frac{7}{27}$ as 0.25925925926. However, the last digit 6 is not correct. It is a result of the rounding process. The actual decimal representation of $\frac{7}{27}$ is the decimal 0.259259... or $0.\overline{259}$, in which the digits continue to repeat the 259 pattern forever.

Calculator Note

These notes provide information about how to use the various features of a calculator.

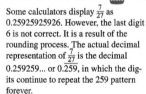

IN THE NEWS

Guidelines for Saving for Your Child's College Education

Fidelity Investments has released guidelines for parents outlining how much they should save in a 529 college savings plan. Fidelity estimated expenses for a 4-year college education beginning 18 years from now by using a 5.4% annual increase in costs. From that amount, Fidelity subtracted the amount that a family could expect to receive in scholarships, grants, and family gifts. According to Fidelity, a family with an annual income of $55,000 needs to save $48,000 for a public college education, while a family with an annual income of $75,000 would need to save $51,000.

SOURCE: Fidelity Investments

NEW! In the News

These application exercises help students master the utility of mathematics in our everyday world. They are based on information found in popular media sources, including newspapers, magazines, and the Web.

Instructor Resources

PRINT SUPPLEMENTS

Annotated Instructor's Edition (ISBN: 978-1-305-07789-8)
The Annotated Instructor's Edition features answers to all problems in the book as well as an appendix denoting which problems can be found in Enhanced WebAssign.

Instructor's Solutions Manual (ISBN: 978-1-133-11220-4)
Author: Kathryn Pearson, Hudson Valley Community College
This manual contains complete solutions to all the problems in the text.

ELECTRONIC SUPPLEMENTS

Enhanced WebAssign® (ISBN: 978-0-538-73810-1)
Exclusively from Cengage Learning, Enhanced WebAssign combines the exceptional mathematics content that you know and love with the most powerful online homework solution, WebAssign. Enhanced WebAssign engages students with immediate feedback, rich tutorial content, and interactive eBooks helping students to develop a deeper conceptual understanding of the subject matter. Online assignments can be built by selecting from thousands of text-specific problems or supplemented with problems from any Cengage Learning textbook.

Enhanced WebAssign: Start Smart Guide for Students (ISBN: 978-0-495-38479-3)
Author: Brooks/Cole
The Enhanced WebAssign Student Start Smart Guide helps students get up and running quickly with Enhanced WebAssign so they can study smarter and improve their performance in class.

Text-Specific Videos
Author: Dana Mosely
Hosted by Dana Mosely, these text-specific instructional videos provide students with visual reinforcement of concepts and explanations, conveyed in easy-to-understand terms with detailed examples and sample problems. A flexible format offers versatility for quickly accessing topics or catering lectures to self-paced, online, or hybrid courses. Closed captioning is provided for the hearing impaired. These videos are available through Enhanced WebAssign and CourseMate.

PowerLecture with Diploma® (ISBN: 978-1-133-11218-1)
This CD-ROM provides the instructor with dynamic media tools for teaching. Create, deliver, and customize tests (both print and online) in minutes with Diploma's Computerized Testing featuring algorithmic equations. Easily build solution sets for homework or exams using Solution Builder's online solutions manual. Quickly and easily update your syllabus with the new Syllabus Creator, which was created by the authors and contains the new edition's table of contents. Practice Sheets, First Day of Class PowerPoint® lecture slides, art and figures from the book, and a test bank in electronic format are also included on this CD-ROM.

Syllabus Creator (Included on the PowerLecture)
Authors: Richard N. Aufmann and Joanne S. Lockwood
NEW! Easily write, edit, and update your syllabus with the Aufmann/Lockwood Syllabus Creator. This software program allows you to create your new syllabus in six easy steps: select the required course objectives, add your contact information, course information, student expectations, the grading policy, dates and location, and your course outline. And now you have your syllabus!

Solution Builder
This online instructor database offers complete worked-out solutions to all exercises in the text, allowing you to create customized, secure solutions printouts (in PDF format) matched exactly to the problems you assign in class. For more information, visit www.cengage.com/solutionbuilder.

Printed Access Card for CourseMate with eBook (ISBN: 978-1-133-50998-1)

Instant Access Card for CourseMate with eBook
(See your local sales representative for details.)
Complement your text and course content with study and practice materials. Cengage Learning's Developmental Mathematics CourseMate brings course concepts to life with interactive learning, study, and exam preparation tools that support the printed textbook. Watch student comprehension soar as your class works with the printed textbook and the textbook-specific website. Liberal Arts Mathematics CourseMate goes beyond the book to deliver what you need!

Student Resources

PRINT SUPPLEMENTS

Student Solutions Manual (ISBN: 978-1-133-11221-1)
Author: Kathryn Pearson, Hudson Valley Community College
Go beyond the answers—see what it takes to get there and improve your grade! This manual provides worked-out, step-by-step solutions to the odd-numbered problems in the text. You'll have the information you need to truly understand how the problems are solved.

ELECTRONIC SUPPLEMENTS

Enhanced WebAssign (ISBN: 978-0-538-73810-1)
Enhanced WebAssign (assigned by the instructor) provides instant feedback on homework assignments to students. This online homework system is easy to use and includes helpful links to textbook sections, video examples, and problem-specific tutorials.

Enhanced WebAssign: Start Smart Guide for Students (ISBN: 978-0-495-38479-3)
Author: Brooks/Cole
If your instructor has chosen to package Enhanced WebAssign with your text, this manual will help you get up and running quickly with the Enhanced WebAssign system so you can study smarter and improve your performance in class.

Text-Specific Videos
Author: Dana Mosely
Hosted by Dana Mosely, these text-specific instructional videos provide you with visual reinforcement of concepts and explanations. Easy-to-understand descriptions are illustrated with detailed examples and sample problems. A flexible format lets you access topics quickly. Closed captioning is provided for the hearing impaired. These videos are available through Enhanced WebAssign and CourseMate.

Printed Access Card for CourseMate with eBook
(ISBN: 978-1-133-50998-1)

Instant Access Card for CourseMate with eBook
(See your local sales representative for details.)
The more students study, the better the results. Students can make the most of their study time by accessing everything they need to succeed in one place: read the textbook, take notes, review flashcards, watch videos, and take practice quizzes—online with CourseMate.

Acknowledgments

The authors would like to thank the people who have reviewed this manuscript and provided many valuable suggestions.

Brenda Alberico, *College of DuPage*

Beverly R. Broomell, *Suffolk County Community College*

Donald Cater, *Monroe Community College*

Henjin Chi, *Indiana State University*

Ivette Chuca, *El Paso Community College*

Marcella Cremer, *Richland Community College*

Margaret Finster, *Erie Community College*

Kenny Fister, *Murray State University*

Luke Foster, *Northeastern State University*

Rita Fox, *Kalamazoo Valley Community College*

Sue Grapevine, *Northwest Iowa Community College*

Shane Griffith, *Lee University*

Elizabeth Henkle, *Longview Community College*

Robert Jajcay, *Indiana State University*

Dr. Nancy R. Johnson, *Manatee Community College*

Brian Karasek, *South Mountain Community College*

Dr. Vernon Kays, *Richland Community College*

Dr. Suda Kunyosying, *Shepherd College*

Kathryn Lavelle, *Westchester Community College*

Roger Marty, *Cleveland State University*

Eric Matsuoka, *Leeward Community College*

Beverly Meyers, *Jefferson College*

Dr. Alec Mihailovs, *Shepherd University*

Leona Mirza, *North Park University*

Bette Nelson, *Alvin Community College*

Kathleen Offenholley, *Brookdale Community College*

Kathy Pinchback, *University of Memphis*

Michael Polley, *Southeastern Community College*

Dr. Anne Quinn, *Edinboro University of Pennsylvania*

Brenda Reed, *Navarro College*

Marc Renault, *Shippensburg University*

Chistopher Rider, *North Greenville College*

Cynthia Roemer, *Union County College*

Sharon M. Saxton, *Cascadia Community College*

Mary Lee Seitz, *Erie Community College–City Campus*

Dr. Sue Stokley, *Spartanburg Technical College*

Dr. Julie M. Theoret, *Lyndon State College*

Walter Jacob Theurer, *Fulton Montgomery Community College*

Jamie Thomas, *University of Wisconsin Colleges–Manitowoc*

William Twentyman, *ECPI College of Technology*

Denise A. Widup, *University of Wisconsin–Parkside*

Nancy Wilson, *Marshall University*

Jane-Marie Wright, *Suffolk Community College*

Diane Zych, *Erie Community College*

INSTRUCTOR NOTE

See the *PowerLecture CD* for teaching tools and resources for this lesson.

TAKE NOTE

Motivation alone will not lead to success. For instance, suppose a person who cannot swim is placed in a boat, taken out to the middle of a lake, and then thrown overboard. That person has a lot of motivation to swim but there is a high likelihood the person will drown without some help. Motivation gives us the desire to learn but is not the same as learning.

Welcome to *Mathematical Excursions,* Third Edition. As you begin this course, we know two important facts: (1) You want to succeed. (2) We want you to succeed. In order to accomplish these goals, an effort is required from each of us. For the next few pages, we are going to show you what is required of you to achieve your goal and how we have designed this text to help you succeed.

Motivation

One of the most important keys to success is motivation. We can try to motivate you by offering interesting or important ways that you can benefit from mathematics. But, in the end, the motivation must come from you. On the first day of class it is easy to be motivated. Eight weeks into the term, it is harder to keep that motivation.

To stay motivated, there must be outcomes from this course that are worth your time, money, and energy. List some reasons you are taking this course. Do not make a mental list—actually write them out. Do this now.

Although we hope that one of the reasons you listed was an interest in mathematics, we know that many of you are taking this course because it is required to graduate, it is a prerequisite for a course you must take, or because it is required for your major. If you are motivated to graduate or complete the requirements for your major, then use that motivation to succeed in this course. Do not become distracted from your goal to complete your education!

Commitment

To be successful, you must make a commitment to succeed. This means devoting time to math so that you achieve a better understanding of the subject.

List some activities (sports, hobbies, talents such as dance, art, or music) that you enjoy and at which you would like to become better. Do this now.

Next to these activities, put the number of hours each week that you spend practicing these activities.

Whether you listed surfing or sailing, aerobics or restoring cars, or any other activity you enjoy, note how many hours a week you spend on each activity. To succeed in math, you must be willing to commit the same amount of time. Success requires some sacrifice.

The "I Can't Do Math" Syndrome

There may be things you cannot do, for instance, lift a two-ton boulder. You can, however, do math. It is much easier than lifting the two-ton boulder. When you first learned the activities you listed above, you probably could not do them well. With practice, you got better. With practice, you will be better at math. Stay focused, motivated, and committed to success.

It is difficult for us to emphasize how important it is to overcome the "I Can't Do Math Syndrome." If you listen to interviews of very successful athletes after a particularly bad performance, you will note that they focus on the positive aspect of what they did, not the negative. Sports psychologists encourage athletes to always be positive—to have a "Can Do" attitude. You need to develop this attitude toward math.

Strategies for Success

Know the Course Requirements   To do your best in this course, you must know exactly what your instructor requires. Course requirements may be stated in a *syllabus*, which is a printed outline of the main topics of the course, or they may be presented orally. When they are listed in a syllabus or on other printed pages, keep them in a safe place. When they are presented orally, make sure to take complete notes. In either case, it is important that you understand them completely and follow them exactly. Be sure you know the answer to each of the following questions.

1. What is your instructor's name?
2. Where is your instructor's office?
3. At what times does your instructor hold office hours?
4. Besides the textbook, what other materials does your instructor require?
5. What is your instructor's attendance policy?
6. If you must be absent from a class meeting, what should you do before returning to class? What should you do when you return to class?
7. What is the instructor's policy regarding collection or grading of homework assignments?
8. What options are available if you are having difficulty with an assignment? Is there a math tutoring center?
9. If there is a math lab at your school, where is it located? What hours is it open?
10. What is the instructor's policy if you miss a quiz?
11. What is the instructor's policy if you miss an exam?
12. Where can you get help when studying for an exam?

Remember: Your instructor wants to see you succeed. If you need help, ask! Do not fall behind. If you were running a race and fell behind by 100 yards, you may be able to catch up, but it will require more effort than had you not fallen behind.

TAKE NOTE

Besides time management, there must be realistic ideas of how much time is available. There are very few people who can *successfully* work full-time and go to school full-time. If you work 40 hours a week, take 15 units, spend the recommended study time given at the right, and sleep 8 hours a day, you use over 80% of the available hours in a week. That leaves less than 20% of the hours in a week for family, friends, eating, recreation, and other activities.

Time Management   We know that there are demands on your time. Family, work, friends, and entertainment all compete for your time. We do not want to see you receive poor job evaluations because you are studying math. However, it is also true that we do not want to see you receive poor math test scores because you devoted too much time to work. When several competing and important tasks require your time and energy, the only way to manage the stress of being successful at both is to manage your time efficiently.

Instructors often advise students to spend twice the amount of time outside of class studying as they spend in the classroom. Time management is important if you are to accomplish this goal and succeed in school. The following activity is intended to help you structure your time more efficiently.

Take out a sheet of paper and list the names of each course you are taking this term, the number of class hours each course meets, and the number of hours you should spend outside of class studying course materials. Now create a weekly calendar with the days of the week across the top and each hour of the day in a vertical column. Fill in the calendar with the hours you are in class, the hours you spend at work, and other commitments such as sports practice, music lessons, or committee meetings. Then fill in the hours that are more flexible, for example, study time, recreation, and meal times.

	Monday	Tuesday	Wednesday	Thursday	Friday	Saturday	Sunday
10–11 a.m.	History	Rev Spanish	History	Rev Span Vocab	History	Jazz Band	
11–12 p.m.	Rev History	Spanish	Study group	Spanish	Math tutor	Jazz Band	
12–1 p.m.	Math		Math		Math		Soccer

We know that many of you must work. If that is the case, realize that working 10 hours a week at a part-time job is equivalent to taking a three-unit class. If you must work, consider letting your education progress at a slower rate to allow you to be successful at both work and school. There is no rule that says you must finish school in a certain time frame.

Schedule Study Time   As we encouraged you to do by filling out the time management form, schedule a certain time to study. You should think of this time like being at work or class. Reasons for "missing study time" should be as compelling as reasons for missing work or class. "I just didn't feel like it" is not a good reason to miss your scheduled study time. Although this may seem like an obvious exercise, list a few reasons you might want to study. Do this now.

Of course we have no way of knowing the reasons you listed, but from our experience one reason given quite frequently is "To pass the course." There is nothing wrong with that reason. If that is the most important reason for you to study, then use it to stay focused.

One method of keeping to a study schedule is to form a ***study group***. Look for people who are committed to learning, who pay attention in class, and who are punctual. Ask them to join your group. Choose people with similar educational goals but different methods of learning. You can gain from seeing the material from a new perspective. Limit groups to four or five people; larger groups are unwieldy.

There are many ways to conduct a study group. Begin with the following suggestions and see what works best for your group.

1. Test each other by asking questions. Each group member might bring two or three sample test questions to each meeting.
2. Practice teaching each other. Many of us who are teachers learned a lot about our subject when we had to explain it to someone else.
3. Compare class notes. You might ask other students about material in your notes that is difficult for you to understand.
4. Brainstorm test questions.
5. Set an agenda for each meeting. Set approximate time limits for each agenda item and determine a quitting time.

And now, probably the most important aspect of studying is that it should be done in relatively small chunks. If you can study only three hours a week for this course (probably not enough for most people), do it in blocks of one hour on three separate days, preferably after class. Three hours of studying on a Sunday is not as productive as three hours of paced study.

Features of This Text That Promote Success

Preparing for Class   Before the class meeting in which your professor begins a new chapter, you should read the title of each section. Next, browse through the chapter material, being sure to note each word in bold type. These words indicate important concepts that you must know to learn the material. Do not worry about trying to understand all the material. Your professor is there to assist you with that endeavor. The purpose of browsing through the material is so that your brain will be prepared to accept and organize the new information when it is presented to you.

Math Is Not a Spectator Sport   To learn mathematics you must be an active participant. Listening and watching your professor do mathematics is not enough. Mathematics requires that you interact with the lesson you are studying. If you have been writing down the things we have asked you to do, you were being interactive. There are other ways this textbook has been designed so that you can be an active learner.

Check Your Progress   One of the key instructional features of this text is a completely worked-out example followed by a *Check Your Progress*.

▼ **example** 8   **Applications of the Blood Transfusion Table**

Use the blood transfusion table and Figures 2.3 and 2.4 to answer the following questions.

a. Can Sue safely be given a type O+ blood transfusion?

b. Why is a person with type O− blood called a *universal donor?*

Solution

a. Sue's blood type is A−. The blood transfusion table shows that she can safely receive blood only if it is type A− or type O−. Thus it is not safe for Sue to receive type O+ blood in a blood transfusion.

b. The blood transfusion table shows that all eight blood types can safely receive type O− blood. Thus a person with type O− blood is said to be a universal donor.

page 79

Note that each Example is completely worked out and the *Check Your Progress* following the example is not. Study the worked-out example carefully by working through each step. You should do this with paper and pencil.

Now work the *Check Your Progress*. If you get stuck, refer to the page number following the word *Solution,* which directs you to the page on which the *Check Your Progress* is solved—a complete worked-out solution is provided. Try to use the given solution to get a hint for the step you are stuck on. Then try to complete your solution.

When you have completed the solution, check your work against the solution we provide.

▼ **check your progress** 8   Use the blood transfusion table and Figures 2.3 and 2.4 to answer the following questions.

a. Is it safe for Alex to receive type A− blood in a blood transfusion?

b. What blood type do you have if you are classified as a *universal recipient?*

Solution   *See page S6.*   ◄

page 79

Be aware that frequently there is more than one way to solve a problem. Your answer, however, should be the same as the given answer. If you have any question as to whether your method will "always work," check with your instructor or with someone in the math center.

Remember: Be an active participant in your learning process. When you are sitting in class watching and listening to an explanation, you may think that you understand. However, until you actually try to do it, you will have no confirmation of the new knowledge or skill. Most of us have had the experience of sitting in class thinking we knew how to do something only to get home and realize we didn't.

Rule Boxes   Pay special attention to definitions, theorems, formulas, and procedures that are presented in a rectangular box, because they generally contain the most important concepts in each section.

> ### ▼ Simple Interest Formula
>
> The simple interest formula is
>
> $$I = Prt$$
>
> where I is the interest, P is the principal, r is the interest rate, and t is the time period.

page 660

Chapter Exercises When you have completed studying a section, do the section exercises. Math is a subject that needs to be learned in small sections and practiced continually in order to be mastered. Doing the exercises in each exercise set will help you master the problem-solving techniques necessary for success. As you work through the exercises, check your answers to the odd-numbered exercises against those in the back of the book.

Preparing for a Test There are important features of this text that can be used to prepare for a test.

- Chapter Summary
- Chapter Review Exercises
- Chapter Test

After completing a chapter, read the Chapter Summary. (See page 106 for the Chapter 2 Summary.) This summary highlights the important topics covered in each section of the chapter. Each concept is paired with page numbers of examples that illustrate the concept and exercises that will provide you with practice on the skill or technique.

Following the Chapter Summary are Chapter Review Exercises (see page 108). Doing the review exercises is an important way of testing your understanding of the chapter. The answer to each review exercise is given at the back of the book, along with, in brackets, the section reference from which the question was taken (see page A5). After checking your answers, restudy any section from which a question you missed was taken. It may be helpful to retry some of the exercises for that section to reinforce your problem-solving techniques.

Each chapter ends with a Chapter Test (see page 110). This test should be used to prepare for an exam. We suggest that you try the Chapter Test a few days before your actual exam. Take the test in a quiet place and try to complete the test in the same amount of time you will be allowed for your exam. When taking the Chapter Test, practice the strategies of successful test takers: (1) scan the entire test to get a feel for the questions; (2) read the directions carefully; (3) work the problems that are easiest for you first; and perhaps most importantly, (4) try to stay calm.

When you have completed the Chapter Test, check your answers for each exercise (see page A6). Next to each answer is, in brackets, the reference to the section from which the question was taken and an example reference for each exercise. The section references indicate the section or sections where you can locate the concepts needed to solve a given exercise, and the example reference allows you to easily find an example that is similar to the given test exercise. If you missed a question, review the material in that section and rework some of the exercises from that section. This will strengthen your ability to perform the skills in that section.

Is it difficult to be successful? YES! Successful music groups, artists, professional
Your career goal goes here. ⟶ athletes, teachers, sociologists, chefs, and _____ have to work very hard to achieve their goals. They focus on their goals and ignore distractions. The things we ask you to do to achieve success take time and commitment. We are confident that if you follow our suggestions, you will succeed.

2

Sets

In mathematics, any group or collection of objects is called a set. A simple application of sets occurs when you use a search engine (such as Google or Bing) to find a topic on the Internet. You enter a few words describing what you are searching for and click the Search button. The search engine then creates a list (set) of websites that contain a match for the words you submitted.

For instance, suppose you wish to make a cake. You search the Internet for a cake recipe and you obtain a set containing over 30 million matches. This is a very large number, so you narrow your search. One method of narrowing your search is to use the AND option found in the Advanced Search link of some search engines. An AND search is an all-words search. That is, an AND search finds only those sites that contain all of the words submitted. An AND search for "flourless chocolate cake recipe" produces a set containing 210,400 matches. This is a more reasonable number, but it is still quite large.

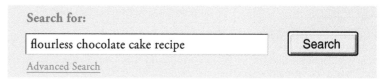

Search for:

flourless chocolate cake recipe   Search

Advanced Search

You narrow the search even further by using an AND search for "foolproof flourless chocolate cake recipe," which returns a few hundred matches. One of these sites provides you with a recipe that provides clear directions and has several good reviews.

Sometimes it is helpful to perform a search using the OR option. An OR search is an any-words search. That is, an OR search finds all those sites that contain any of the words you submitted.

Many additional applications of sets are given in this chapter.

Fuse/Getty Images

Basic Properties of Sets

Sets

In an attempt to better understand the universe, ancient astronomers classified certain groups of stars as constellations. Today we still find it extremely helpful to classify items into groups that enable us to find order and meaning in our complicated world.

Any group or collection of objects is called a **set**. The objects that belong in a set are the **elements**, or **members**, of the set. For example, the set consisting of the four seasons has spring, summer, fall, and winter as its elements.

The following two methods are often used to designate a set.

- Describe the set using words.
- List the elements of the set inside a pair of braces, { }. This method is called the **roster method**. Commas are used to separate the elements.

For instance, let's use S to represent the set consisting of the four seasons. Using the roster method, we would write

$$S = \{\text{spring, summer, fall, winter}\}$$

The order in which the elements of a set are listed is not important. Thus the set consisting of the four seasons can also be written as

$$S = \{\text{winter, spring, fall, summer}\}$$

The following table gives two examples of sets, where each set is designated by a word description and also by using the roster method.

TABLE 2.1 Define Sets by Using a Word Description and the Roster Method

Description	Roster method
The set of denominations of U.S. paper currency in production at this time	{$1, $2, $5, $10, $20, $50, $100}
The set of states in the United States that border the Pacific Ocean	{California, Oregon, Washington, Alaska, Hawaii}

▼ **example** **1** Use The Roster Method to Represent a Set

Use the roster method to represent the set of the days in a week.

Solution {Sunday, Monday, Tuesday, Wednesday, Thursday, Friday, Saturday}

▼ **check your progress** **1** Use the roster method to represent the set of months that start with the letter A.

Solution *See page S4.*

▼ **example** **2** Use a Word Description to Represent a Set

Write a word description for the set

$$A = \{\text{a, b, c, d, e, f, g, h, i, j, k, l, m, n, o, p, q, r, s, t, u, v, w, x, y, z}\}$$

Solution Set A is the set of letters of the English alphabet.

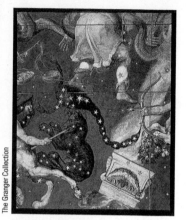

The constellation Scorpius is a set of stars.

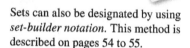

TAKE NOTE

Sets can also be designated by using *set-builder notation*. This method is described on pages 54 to 55.

POINT OF INTEREST

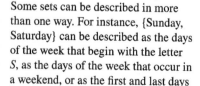

Paper currency in denominations of $500, $1000, $5000, and $10,000 has been in circulation, but production of these bills ended in 1945. If you just happen to have some of these bills, you can still cash them for their face value.

TAKE NOTE

Some sets can be described in more than one way. For instance, {Sunday, Saturday} can be described as the days of the week that begin with the letter *S*, as the days of the week that occur in a weekend, or as the first and last days of a week.

2 Write a word description for the set {March, May}.

Solution *See page S4.*

The following sets of numbers are used extensively in many areas of mathematics.

▼ Basic Number Sets

Natural Numbers or Counting Numbers $N = \{1, 2, 3, 4, 5, ...\}$

Whole Numbers $W = \{0, 1, 2, 3, 4, 5, ...\}$

Integers $I = \{..., -4, -3, -2, -1, 0, 1, 2, 3, 4, ...\}$

Rational Numbers Q = the set of all terminating or repeating decimals

Irrational Numbers $\mathcal{I}$ = the set of all nonterminating, nonrepeating decimals

Real Numbers R = the set of all rational or irrational numbers

---

TAKE NOTE

In this chapter, the letters N, W, I, Q, $\mathcal{I}$, and R will often be used to represent the basic number sets defined at the right.

The set of natural numbers is also called the set of counting numbers. The three dots ... are called an **ellipsis** and indicate that the elements of the set continue in a manner suggested by the elements that are listed.

The integers ..., -4, -3, -2, -1 are **negative integers**. The integers 1, 2, 3, 4, ... are **positive integers**. Note that the natural numbers and the positive integers are the same set of numbers. The integer zero is neither a positive nor a negative integer.

If a number in decimal form terminates or repeats a block of digits, then the number is a rational number. Rational numbers can also be written in the form $\frac{p}{q}$, where p and q are integers and $q \neq 0$. For example,

$$\frac{1}{4} = 0.25 \quad \text{and} \quad \frac{3}{11} = 0.\overline{27}$$

are rational numbers. The bar over the 27 means that the block of digits 27 repeats without end; that is, $0.\overline{27} = 0.27272727....$

A decimal that neither terminates nor repeats is an **irrational number**. For instance, 0.35335333533335... is a nonterminating, nonrepeating decimal and thus is an irrational number.

Every real number is either a rational number or an irrational number.

▼ example **3**   Use The Roster Method to Represent a Set of Numbers

Use the roster method to write each of the given sets.

a. The set of natural numbers less than 5

b. The solution set of $x + 5 = -1$

c. The set of negative integers greater than -4

Solution

a. The set of natural numbers is given by $\{1, 2, 3, 4, 5, 6, 7, ...\}$. The natural numbers less than 5 are 1, 2, 3, and 4. Using the roster method, we write this set as $\{1, 2, 3, 4\}$.

b. Adding -5 to each side of the equation produces $x = -6$. The solution set of $x + 5 = -1$ is $\{-6\}$.

c. The set of negative integers greater than -4 is $\{-3, -2, -1\}$.

INSTRUCTOR NOTE

Students should be reminded that a listing such as 1, 2, 3 is not a set. With the roster method, elements need to be placed inside braces to form a set.

 **3** Use the roster method to write each of the given sets.

a. The set of whole numbers less than 4

b. The set of counting numbers larger than 11 and less than or equal to 19

c. The set of negative integers between −5 and 7

Solution *See page S4.*

Definitions Regarding Sets

A set is **well defined** if it is possible to determine whether any given item is an element of the set. For instance, the set of letters of the English alphabet is well defined. The set of *great songs* is not a well-defined set. It is not possible to determine whether any given song is an element of the set or is not an element of the set because there is no standard method for making such a judgment.

The statement "4 is an element of the set of natural numbers" can be written using mathematical notation as $4 \in N$. The symbol $\in$ is read "is an element of." To state that "−3 is not an element of the set of natural numbers," we use the "is not an element of" symbol, $\notin$, and write $-3 \notin N$.

TAKE NOTE ✓

Recall that N denotes the set of natural numbers, I denotes the set of integers, and W denotes the set of whole numbers.

▼ **example 4** **Apply Definitions Regarding Sets**

Determine whether each statement is true or false.

a. $4 \in \{2, 3, 4, 7\}$ **b.** $-5 \in N$ **c.** $\frac{1}{2} \notin I$

d. The set of nice cars is a well-defined set.

Solution

a. Since 4 is an element of the given set, the statement is true.

b. There are no negative natural numbers, so the statement is false.

c. Since $\frac{1}{2}$ is not an integer, the statement is true.

d. The word *nice* is not precise, so the statement is false.

▼ **check your progress 4** Determine whether each statement is true or false.

a. $5.2 \in \{1, 2, 3, 4, 5, 6\}$ **b.** $-101 \in I$ **c.** $2.5 \notin W$

d. The set of all integers larger than π is a well-defined set.

Solution *See page S4.*

The **empty set**, or **null set**, is the set that contains no elements. The symbol $\varnothing$ or $\{\ \}$ is used to represent the empty set. As an example of the empty set, consider the set of natural numbers that are negative integers.

Another method of representing a set is **set-builder notation**. Set-builder notation is especially useful when describing infinite sets. For instance, in set-builder notation, the set of natural numbers greater than 7 is written as follows:

TAKE NOTE ✓

Neither the set $\{0\}$ nor the set $\{\varnothing\}$ represents the empty set because each set has one element.

membership
conditions

$$\{x \mid x \in N \text{ and } x > 7\}$$

| the set | of all elements x | such that | x is an element of the set of natural numbers | and x is greater than 7 |

The preceding set-builder notation is read as "the set of all elements x such that x is an element of the set of natural numbers and x is greater than 7." It is impossible to list all the elements of the set, but set-builder notation defines the set by describing its elements.

▼ **example 5**   Use Set-Builder Notation to Represent a Set

Use set-builder notation to write the following sets.

a. The set of integers greater than -3

b. The set of whole numbers less than 1000

Solution

a. $\{x \mid x \in I \text{ and } x > -3\}$ **b.** $\{x \mid x \in W \text{ and } x < 1000\}$

▼ **check your progress 5**   Use set-builder notation to write the following sets.

a. The set of integers less than 9

b. The set of natural numbers greater than 4

Solution   *See page S4.*   ◄

A set is **finite** if the number of elements in the set is a whole number. The **cardinal number** of a finite set is the number of elements in the set. The cardinal number of a finite set A is denoted by the notation $n(A)$. For instance, if $A = \{1, 4, 6, 9\}$, then $n(A) = 4$. In this case, A has a cardinal number of 4, which is sometimes stated as "A has a *cardinality* of 4."

▼ **example 6**   The Cardinality of a Set

Find the cardinality of each of the following sets.

a. $J = \{2, 5\}$ **b.** $S = \{3, 4, 5, 6, 7, ..., 31\}$ **c.** $T = \{3, 3, 7, 51\}$

Solution

a. Set J contains exactly two elements, so J has a cardinality of 2. Using mathematical notation, we state this as $n(J) = 2$.

b. Only a few elements are actually listed. The number of natural numbers from 1 to 31 is 31. If we omit the numbers 1 and 2, then the number of natural numbers from 3 to 31 must be $31 - 2 = 29$. Thus $n(S) = 29$.

c. Elements that are listed more than once are counted only once. Thus $n(T) = 3$.

▼ **check your progress 6**   Find the cardinality of the following sets.

a. $C = \{-1, 5, 4, 11, 13\}$ **b.** $D = \{0\}$ **c.** $E = \varnothing$

Solution   *See page S4.*   ◄

The following definitions play an important role in our work with sets.

▼ **Equal Sets**

Set A is **equal** to set B, denoted by $A = B$, if and only if A and B have exactly the same elements.

For instance $\{d, e, f\} = \{e, f, d\}$.

▼ **Equivalent Sets**

Set A is **equivalent** to set B, denoted by $A \sim B$, if and only if A and B have the same number of elements.

---

question If two sets are equal, must they also be equivalent?

▼ **example 7** Equal Sets and Equivalent Sets

State whether each of the following pairs of sets are equal, equivalent, both, or neither.

a. {a, e, i, o, u}, {3, 7, 11, 15, 19}   **b.** {4, −2, 7}, {3, 4, 7, 9}

Solution

a. The sets are not equal. However, each set has exactly five elements, so the sets are equivalent.

b. The first set has three elements and the second set has four elements, so the sets are not equal and are not equivalent.

▼ **check your progress 7** State whether each of the following pairs of sets are equal, equivalent, both, or neither.

a. $\{x \mid x \in W \text{ and } x \le 5\}$, $\{\alpha, \beta, \Gamma, \Delta, \delta, \varepsilon\}$

b. $\{5, 10, 15, 20, 25, 30, \ldots, 80\}$, $\{x \mid x \in N \text{ and } x < 17\}$

Solution *See page S4.*

MATHMATTERS Georg Cantor

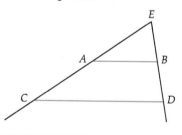

Georg Cantor

Georg Cantor (kăn't ər) (1845–1918) was a German mathematician who developed many new concepts regarding the theory of sets. Cantor studied under the famous mathematicians Karl Weierstrass and Leopold Kronecker at the University of Berlin. Although Cantor demonstrated a talent for mathematics, his professors were unaware that Cantor would produce extraordinary results that would cause a major stir in the mathematical community.

Cantor never achieved his lifelong goal of a professorship at the University of Berlin. Instead he spent his active career at the undistinguished University of Halle. It was during this period, when Cantor was between the ages of 29 and 39, that he produced his best work. Much of this work was of a controversial nature. One of the simplest of the controversial concepts concerned points on a line segment. For instance, consider the line segment $\overline{AB}$ and the line segment $\overline{CD}$ in the figure at the left. Which of these two line segments do you think contains the most points? Cantor was able to prove that they both contain the same number of points. In fact, he was able to prove that any line segment, no matter how short, contains the same number of points as a line, or a plane, or all of three-dimensional space. We will take a closer look at some of the mathematics developed by Cantor in the last section of this chapter.

---

answer Yes. If the sets are equal, then they have exactly the same elements; therefore, they also have the same number of elements.

EXCURSION

Fuzzy Sets

Lotfi Zadeh Lotfi Zadeh's work in the area of fuzzy sets has led to a new area of mathematics called *soft computing*. On the topic of soft computing, Zadeh stated, "The essence of soft computing is that unlike the traditional, hard computing, soft computing is aimed at an accommodation with the pervasive imprecision of the real world. Thus the guiding principle of soft computing is: Exploit the tolerance for imprecision, uncertainty and partial truth to achieve tractability, robustness, low solution cost and better rapport with reality. In the final analysis, the role model for soft computing is the human mind."

In traditional set theory, an element either belongs to a set or does not belong to the set. For instance, let $A = \{x \mid x$ is an even integer$\}$. Given $x = 8$, we have $x \in A$. However, if $x = 11$, then $x \notin A$. For any given integer, we can decide whether x belongs to A.

Now consider the set $B = \{x \mid x$ is a number close to 10$\}$. Does 8 belong to this set? Does 9.9 belong to the set? Does 10.001 belong to the set? Does 10 belong to the set? Does -50 belong to the set? Given the imprecision of the words "close to," it is impossible to know which numbers belong to set B.

In 1965, Lotfi A. Zadeh of the University of California, Berkeley, published a paper titled *Fuzzy Sets* in which he described the mathematics of fuzzy set theory. This theory proposed that "to some degree," many of the numbers 8, 9.9, 10.001, 10, and -50 belong to set B defined in the previous paragraph. Zadeh proposed giving each element of a set a *membership grade* or *membership value*. This value is a number from 0 to 1. The closer the membership value is to 1, the greater the certainty that an element belongs to the set. The closer the membership value is to 0, the less the certainty that an element belongs to the set. Elements of fuzzy sets are written in the form (element, membership value). Here is an example of a fuzzy set.

$$C = \{(8, 0.4), (9.9, 0.9), (10.001, 0.999), (10, 1), (-50, 0)\}$$

An examination of the membership values suggests that we are certain that 10 belongs to C (membership value is 1) and we are certain that -50 does not belong to C (membership value is 0). Every other element belongs to the set "to some degree."

The concept of a fuzzy set has been used in many real-world applications. Here are a few examples.

- Control of heating and air-conditioning systems
- Compensation against vibrations in camcorders
- Voice recognition by computers
- Control of valves and dam gates at power plants
- Control of robots
- Control of subway trains
- Automatic camera focusing

▼ A Fuzzy Heating System

Typical heating systems are controlled by a thermostat that turns a furnace on when the room temperature drops below a set point and turns the furnace off when the room temperature exceeds the set point. The furnace either runs at full force or it shuts down completely. This type of heating system is inefficient, and the frequent off and on changes can be annoying. A fuzzy heating system makes use of "fuzzy" definitions such as cold, warm, and hot to direct the furnace to run at low, medium, or full force. This results in a more efficient heating system and fewer temperature fluctuations.

EXCURSION EXERCISES

1. Mark, Erica, Larry, and Jennifer have each defined a fuzzy set to describe what they feel is a "good" grade. Each person paired the letter grades A, B, C, D, and F with a membership value. The results are as follows.

   Mark: $M = \{(A, 1), (B, 0.75), (C, 0.5), (D, 0.5), (F, 0)\}$

   Erica: $E = \{(A, 1), (B, 0), (C, 0), (D, 0), (F, 0)\}$

   Larry: $L = \{(A, 1), (B, 1), (C, 1), (D, 1), (F, 0)\}$

   Jennifer: $J = \{(A, 1), (B, 0.8), (C, 0.6), (D, 0.1), (F, 0)\}$

   a. Which of the four people considers an A grade to be the only good grade?

   b. Which of the four people is most likely to be satisfied with a grade of D or better?

   c. Write a fuzzy set that you would use to describe the set of good grades. Consider only the letter grades A, B, C, D, and F.

2. In some fuzzy sets, membership values are given by a *membership graph* or by a *formula*. For instance, the following figure is a graph of the membership values of the fuzzy set *OLD*.

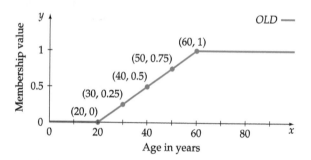

   Use the membership graph of *OLD* to determine the membership value of each of the following.

   a. $x = 15$ b. $x = 50$ c. $x = 65$

   d. Use the graph of *OLD* to determine the age x with a membership value of 0.25.

   An ordered pair (x, y) of a fuzzy set is a **crossover point** if its membership value is 0.5.

   e. Find the crossover point for *OLD*.

3. The following membership graph provides a definition of real numbers x that are "about" 4.

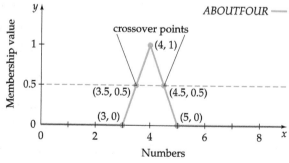

   Use the graph of *ABOUTFOUR* to determine the membership value of:

   a. $x = 2$ b. $x = 3.5$ c. $x = 7$

   d. Use the graph of *ABOUTFOUR* to determine its crossover points.

4. The membership graphs in the following figure provide definitions of the fuzzy sets *COLD* and *WARM*.

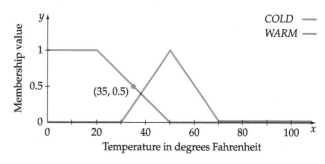

The point (35, 0.5) on the membership graph of *COLD* indicates that the membership value for $x = 35$ is 0.5. Thus, by this definition, 35°F is 50% cold. Use the above graphs to estimate

a. the *WARM* membership value for $x = 40$.

b. the *WARM* membership value for $x = 50$.

c. the crossover points of *WARM*.

5. The membership graph in Excursion Exercise 2 shows one person's idea of what ages are "old." Use a grid similar to the following to draw a membership graph that you feel defines the concept of being "young" in terms of a person's age in years.

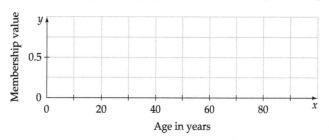

Show your membership graph to a few of your friends. Do they concur with your definition of "young"?

INSTRUCTOR NOTE

The process of creating a fuzzy definition is called **fuzzification.** The graphs in Excursion Exercise 4 are fuzzifications of the terms *cold* and *warm*. Fuzzification is not a unique process. For instance, "cold" can be defined in many different ways. Have your students create membership graphs that provide fuzzifications of terms such as *middle aged, short, tall, rich, poor, good test scores,* and *poor test scores.* It can be fun to compare their fuzzifications.

EXERCISE SET **2.1**   (Suggested Assignment: The Enhanced WebAssign Exercises and Exercises 2, 6, 13, 27, 37, 45, 47, 60, 78, 81, and 87)

■ In Exercises 1 to 12, use the roster method to write each of the given sets. For some exercises you may need to consult a reference, such as the Internet or an encyclopedia.

1. The set of U.S. coins with a value of less than 50¢

2. The set of months of the year with a name that ends with the letter y

3. The set of planets in our solar system with a name that starts with the letter M

4. The set of the seven dwarfs

5. The set of U.S. presidents who have served after Bill Clinton

6. The set of months with exactly 30 days

7. The set of negative integers greater than −6

8. The set of whole numbers less than 8

9. The set of integers x that satisfy $x - 4 = 3$

10. The set of integers x that satisfy $2x - 1 = -11$

11. The set of natural numbers x that satisfy $x + 4 = 1$

12. The set of whole numbers x that satisfy $x - 1 < 4$

■ In Exercises 13 to 20, write a description of each set. There may be more than one correct description.

13. {Tuesday, Thursday}

14. {Libra, Leo}

15. {Mercury, Venus}

16. {penny, nickel, dime}

17. {1, 2, 3, 4, 5, 6, 7, 8, 9}

18. {2, 4, 6, 8}

19. $\{x \mid x \in N \text{ and } x \leq 7\}$

20. $\{x \mid x \in W \text{ and } x < 5\}$

■ In Exercises 21 to 30, determine whether each statement is true or false. If the statement is false, give a reason.

21. b ∈ {a, b, c}

22. 0 ∉ N

23. {b} ∈ {a, b, c}

24. {1, 5, 9} ~ {Ψ, Π, Σ}

25. {0} ~ ∅

26. The set of large numbers is a well-defined set.

27. The set of good teachers is a well-defined set.

28. The set $\{x \mid 2 \le x \le 3\}$ is a well-defined set.

29. $\{x^2 \mid x \in I\} = \{x^2 \mid x \in N\}$

30. 0 ∈ ∅

■ In Exercises 31 to 40, use set-builder notation to write each of the following sets.

31. {1, 2, 3, 4, 5, 6, 7, 8, 9, 10, 11, 12}

32. {45, 55, 65, 75}

33. {5, 10, 15}

34. {1, 4, 9, 16, 25, 36, 49, 64, 81}

35. {January, March, May, July, August, October, December}

36. {Iowa, Ohio, Utah}

37. {Arizona, Alabama, Arkansas, Alaska}

38. {Mexico, Canada}

39. {spring, summer}

40. {1900, 1901, 1902, 1903, 1904, ... , 1999}

Charter Schools The following horizontal bar graph shows the eight states with the greatest number of charter schools in the fall of 2010.

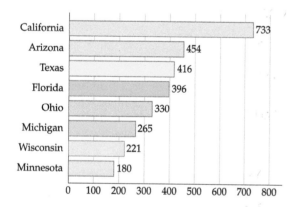

States with the Greatest Number of Charter Schools, 2010
SOURCE: U.S. Charter Schools webpage, http://www.uscharterschools.org

■ Use the data in the above graph and the roster method to represent each of the sets in Exercises 41 to 44.

41. The set of states that have more than 200 but less than 300 charter schools.

42. The set of states with more than 350 charter schools

43. $\{x \mid x$ is a state that has between 325 and 450 charter schools$\}$

44. $\{x \mid x$ is a state with at least 425 charter schools$\}$

Affordability of Housing The following bar graph shows the monthly principal and interest payment needed to purchase an average-priced existing home in the United States for the years from 2002 to 2009.

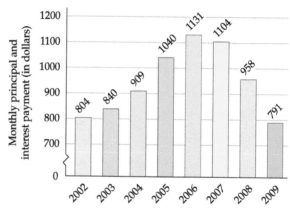

Monthly Principal and Interest Payment for an Average-Priced Existing Home
SOURCE: National Association of REALTORS® as reported in the World Almanac, 2010, p. 79

■ Use the data in the graph and the roster method to represent each of the sets in Exercises 45 to 48.

45. The set of years in which the monthly principal and interest payment, for an average-priced existing home, exceeded $1000.

46. The set of years in which the monthly principal and interest payment, for an average-priced existing home, was between $900 and $1000.

47. $\{x \mid x$ is a year for which the monthly principal and interest payment, for an average-priced existing home, was between $700 and 800\}$

48. $\{x \mid x$ is a year for which the monthly principal and interest payment, for an average-priced existing home, was less than 950\}$

Gasoline Prices The following graph shows the U.S. city average unleaded regular gasoline retail prices for the years from 2000 to 2009.

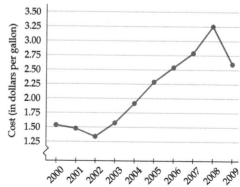

U.S. City Average Unleaded Regular Gasoline Retail Prices
SOURCE: The World Almanac and Book of Facts, 2010, p.118

■ Use the information in the graph and the roster method to represent each of the sets in Exercises 49 to 52.

49. The set of years in which the average unleaded regular gasoline retail price was more than $2.00 but less than $3.00 per gallon

50. The set of years in which the average unleaded regular gasoline retail price was more than $2.40 per gallon

51. {x | x is a year for which the average unleaded regular gasoline retail prices were more than $1.75 but less than $2.50 per gallon}

52. {x | x is a year for which the average unleaded regular gasoline retail prices were more than $3.00 per gallon}

Ticket Prices   The following table shows the average U.S. movie theatre ticket price for each year from 1994 to 2009.

Average U.S. Movie Theatre Ticket Prices (in dollars)

Year	Price	Year	Price
1994	4.08	2002	5.80
1995	4.35	2003	6.03
1996	4.42	2004	6.21
1997	4.59	2005	6.41
1998	4.69	2006	6.55
1999	5.06	2007	6.88
2000	5.39	2008	7.18
2001	5.65	2009	7.50

SOURCE: National Association of Theatre Owners, http://www.natoonline.org/statisticstickets.htm

■ Use the information in the table and the roster method to represent each of the sets in Exercises 53 to 56.

53. The set of years in which the average ticket price was less than $5.00

54. The set of years in which the average ticket price was greater than $7.00

55. {x | x is a year for which the average ticket price was greater than $6.15 but less than $6.75}

56. {x | x is a year for which the average ticket price was less than $7.00 but greater than $6.00}

■ In Exercises 57 to 66, find the cardinality of each of the following sets. For some exercises you may need to consult a reference, such as the Internet or an encyclopedia.

57. A = {2, 4, 6, 8, 10, 12, 14, 16, 18, 20, 22}

58. B = {7, 14, 21, 28, 35, 42, 49, 56}

59. D = the set of all dogs that can spell "elephant"

60. S = the set of all states in the United States

61. J = the set of all states of the United States that border Minnesota

62. T = the set of all stripes on the U.S. flag

63. N = the set of all baseball teams in the National League

64. C = the set of all chess pieces on a chess board at the start of a chess game

65. {3, 6, 9, 12, 15, ..., 363}

66. {7, 11, 15, 19, 23, 27, ..., 407}

■ In Exercises 67 to 74, state whether each of the given pairs of sets are equal, equivalent, both, or neither.

67. The set of U.S. senators; the set of U.S. representatives

68. The set of single-digit natural numbers; the set of pins used in a regulation bowling game

69. The set of positive whole numbers; the set of counting numbers

70. The set of single-digit natural numbers; the set of single-digit integers

71. {1, 2, 3}; {I, II, III}

72. {6, 8, 10, 12}; {1, 2, 3, 4}

73. {2, 5}; {0, 1}

74. { }; {0}

■ In Exercises 75 to 86, determine whether each of the sets is a well-defined set.

75. The set of good foods

76. The set of the six most heavily populated cities in the United States

77. The set of tall buildings in the city of Chicago

78. The set of states that border Colorado

79. The set of even integers

80. The set of rational numbers of the form $\frac{1}{p}$, where p is a counting number

81. The set of former presidents of the United States who are alive at the present time

82. The set of real numbers larger than 89,000

83. The set of small countries

84. The set of great cities in which to live

85. The set consisting of the best soda drinks

86. The set of fine wines

EXTENSIONS

Critical Thinking

■ In Exercises 87 to 90, determine whether the given sets are equal. Recall that W represents the set of whole numbers and N represents the set of natural numbers.

87. $A = \{2n + 1 \,|\, n \in W\}$
$B = \{2n - 1 \,|\, n \in N\}$

88. $A = \left\{16\left(\dfrac{1}{2}\right)^{n-1} \middle|\, n \in N\right\}$
$B = \left\{16\left(\dfrac{1}{2}\right)^{n} \middle|\, n \in W\right\}$

89. $A = \{2n - 1 \,|\, n \in N\}$
$B = \left\{\dfrac{n(n + 1)}{2} \middle|\, n \in N\right\}$

90. $A = \{3n + 1 \,|\, n \in W\}$
$B = \{3n - 2 \,|\, n \in N\}$

91. Give an example of a set that cannot be written using the roster method.

Explorations

92. In this section, we have introduced the concept of *cardinal numbers*. Use the Internet or a mathematical textbook to find information about *ordinal numbers* and *nominal numbers*. Write a few sentences that explain the differences among these three types of numbers.

---

section **2.2** Complements, Subsets, and Venn Diagrams

The Universal Set and the Complement of a Set

In complex problem-solving situations and even in routine daily activities, we need to understand the set of all elements that are under consideration. For instance, when an instructor assigns letter grades, the possible choices may include A, B, C, D, F, and I. In this case the letter H is not a consideration. When you place a telephone call, you know that the area code is given by a natural number with three digits. In this instance a rational number such as $\frac{2}{3}$ is not a consideration. The set of all elements that are being considered is called the **universal set**. We will use the letter U to denote the universal set.

> ▼ **The Complement of a Set**
>
> The **complement** of a set A, denoted by A', is the set of all elements of the universal set U that are not elements of A.

TAKE NOTE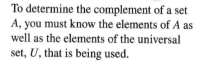

To determine the complement of a set A, you must know the elements of A as well as the elements of the universal set, U, that is being used.

▼ **example 1** Find the Complement of a Set

Let $U = \{1, 2, 3, 4, 5, 6, 7, 8, 9, 10\}$, $S = \{2, 4, 6, 7\}$, and $T = \{x \,|\, x < 10 \text{ and } x \in \text{ the odd counting numbers}\}$. Find

a. S' **b.** T'

Solution

a. The elements of the universal set are 1, 2, 3, 4, 5, 6, 7, 8, 9, and 10. From these elements we wish to exclude the elements of S, which are 2, 4, 6, and 7. Therefore $S' = \{1, 3, 5, 8, 9, 10\}$.

b. $T = \{1, 3, 5, 7, 9\}$. Excluding the elements of T from U gives us $T' = \{2, 4, 6, 8, 10\}$.

▼**check your progress** **1** Let $U = \{0, 2, 3, 4, 6, 7, 17\}$, $M = \{0, 4, 6, 17\}$, and $P = \{x \mid x < 7$ and $x \in$ the even natural numbers$\}$. Find

a. M' **b.** P'

Solution *See page S4.*

There are two fundamental results concerning the universal set and the empty set. Because the universal set contains all elements under consideration, the complement of the universal set is the empty set. Conversely, the complement of the empty set is the universal set, because the empty set has no elements and the universal set contains all the elements under consideration. Using mathematical notation, we state these fundamental results as follows:

▼ **The Complement of the Universal Set and the Complement of the Empty Set**

$U' = \varnothing$ and $\varnothing' = U$

Subsets

Consider the set of letters in the alphabet and the set of vowels $\{a, e, i, o, u\}$. Every element of the set of vowels is an element of the set of letters in the alphabet. The set of vowels is said to be a *subset* of the set of letters in the alphabet. We will often find it useful to examine subsets of a given set.

▼ **A Subset of a Set**

Set A is a **subset** of set B, denoted by $A \subseteq B$, if and only if every element of A is also an element of B.

Here are two fundamental subset relationships.

▼ **Subset Relationships**

$A \subseteq A$, for any set A
$\varnothing \subseteq A$, for any set A

To convince yourself that the empty set is a subset of any set, consider the following. We know that a set is a subset of a second set provided every element of the first set is an element of the second set. Pick an arbitrary set A. Because every element of the empty set (*there are none*) is an element of A, we know that $\varnothing \subseteq A$.

The notation $A \not\subseteq B$ is used to denote that A is *not* a subset of B. To show that A is not a subset of B, it is necessary to find at least one element of A that is not an element of B.

▼**example** **2** Apply the Definition of a Subset

Determine whether each statement is true or false.

a. $\{5, 10, 15, 20\} \subseteq \{0, 5, 10, 15, 20, 25, 30\}$

b. $W \subseteq N$

TAKE NOTE

Recall that W represents the set of whole numbers and N represents the set of natural numbers.

c. $\{2, 4, 6\} \subseteq \{2, 4, 6\}$

d. $\varnothing \subseteq \{1, 2, 3\}$

Solution

a. True; every element of the first set is an element of the second set.

b. False; 0 is a whole number, but 0 is not a natural number.

c. True; every set is a subset of itself.

d. True; the empty set is a subset of every set.

▼ **check your progress** **2**   Determine whether each statement is true or false.

a. $\{1, 3, 5\} \subseteq \{1, 5, 9\}$

b. The set of counting numbers is a subset of the set of natural numbers.

c. $\varnothing \subseteq U$

d. $\{-6, 0, 11\} \subseteq I$

Solution   *See page S4.*

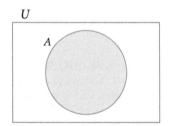
U
A

A Venn diagram

The English logician John Venn (1834–1923) developed diagrams, which we now refer to as *Venn diagrams,* that can be used to illustrate sets and relationships between sets. In a **Venn diagram**, the universal set is represented by a rectangular region and subsets of the universal set are generally represented by oval or circular regions drawn inside the rectangle. The Venn diagram at the left shows a universal set and one of its subsets, labeled as set *A*. The size of the circle is not a concern. The region outside of the circle, but inside of the rectangle, represents the set A'.

question   What set is represented by $(A')'$?

Proper Subsets of a Set

> ▼ **Proper Subset**
>
> Set *A* is a **proper subset** of set *B*, denoted by $A \subset B$, if every element of *A* is an element of *B*, and $A \neq B$.

To illustrate the difference between subsets and proper subsets, consider the following two examples.

1. Let $R = \{$Mars, Venus$\}$ and $S = \{$Mars, Venus, Mercury$\}$. The first set, *R*, is a subset of the second set, *S*, because every element of *R* is an element of *S*. In addition, *R* is also a proper subset of *S*, because $R \neq S$.

2. Let $T = \{$Europe, Africa$\}$ and $V = \{$Africa, Europe$\}$. The first set, *T*, is a subset of the second set, *V*; however, *T* is *not* a proper subset of *V* because $T = V$.

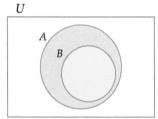
U
A
B

B is a proper subset of *A*.

Venn diagrams can be used to represent proper subset relationships. For instance, if a set *B* is a proper subset of a set *A*, then we illustrate this relationship in a Venn diagram by drawing a circle labeled *B* inside of a circle labeled *A*. See the Venn diagram at the left.

answer   The set A' contains the elements of *U* that are not in *A*. By definition, the set $(A')'$ contains only the elements of *U* that are elements of *A*. Thus $(A')' = A$.

▼ example **3**   **Proper Subsets**

For each of the following, determine whether the first set is a proper subset of the second set.

a. {a, e, i, o, u}, {e, i, o, u, a}   **b.** *N*, *I*

Solution

a. Because the sets are equal, the first set is not a proper subset of the second set.

b. Every natural number is an integer, so the set of natural numbers is a subset of the set of integers. The set of integers contains elements that are not natural numbers, such as −3. Thus the set of natural numbers is a proper subset of the set of integers.

▼ check your progress **3**   For each of the following, determine whether the first set is a proper subset of the second set.

a. *N*, *W*   **b.** {1, 4, 5}, {5, 1, 4}

Solution   *See page S4.*

Some counting problems in the study of probability require that we find all of the subsets of a given set. One way to find all the subsets of a given set is to use the method of making an organized list. First list the empty set, which has no elements. Next list all the sets that have exactly one element, followed by all the sets that contain exactly two elements, followed by all the sets that contain exactly three elements, and so on. This process is illustrated in the following example.

▼ example **4**   **List All the Subsets of a Set**

Set *C* shows the four condiments that a hot dog stand offers on its hot dogs.

$$C = \{\text{mustard, ketchup, onions, relish}\}$$

List all the subsets of *C*.

Solution
An organized list shows the following subsets.

{ }	• **Subsets with 0 elements**
{mustard}, {ketchup}, {onions}, {relish}	• **Subsets with 1 element**
{mustard, ketchup}, {mustard, onions}, {mustard, relish}, {ketchup, onions}, {ketchup, relish}, {onions, relish}	• **Subsets with 2 elements**
{mustard, ketchup, onions}, {mustard, ketchup, relish}, {mustard, onions, relish}, {ketchup, onions, relish}	• **Subsets with 3 elements**
{mustard, ketchup, onion, relish}	• **Subsets with 4 elements**

▼ check your progress **4**   List all of the subsets of {a, b, c, d, e}.

Solution   *See page S4.*

svry/Shutterstock.com

Number of Subsets of a Set

In Example 4, we found that a set with 4 elements has 16 subsets. The following formula can be used to find the number of subsets of a set with n elements, where n is a natural number.

▼ The Number of Subsets of a Set

A set with n elements has 2^n subsets.

INSTRUCTOR NOTE

Ask your students the following question:

If a set has n elements, then how many proper subsets does it have?

Answer: $2^n - 1$

Examples

- $\{1, 2, 3, 4, 5, 6\}$ has 6 elements, so it has $2^6 = 64$ subsets.
- $\{4, 5, 6, 7, 8, ..., 15\}$ has 12 elements, so it has $2^{12} = 4096$ subsets.
- The empty set has 0 elements, so it has $2^0 = 1$ subset.

In Example 5, we apply the formula for the number of subsets of a set to determine the number of different variations of pizzas that a restaurant can serve.

▼ **example 5**　Pizza Variations

A restaurant sells pizzas for which you can choose from seven toppings.
a. How many different variations of pizzas can the restaurant serve?
b. What is the minimum number of toppings the restaurant must provide if it wishes to advertise that it offers over 1000 variations of its pizzas?

Solution

a. The restaurant can serve a pizza with no topping, one topping, two toppings, three toppings, and so forth, up to all seven toppings.

　Let T be the set consisting of the seven toppings. The elements in each subset of T describe exactly one of the variations of toppings that the restaurant can serve. Consequently, the number of different variations of pizzas that the restaurant can serve is the same as the number of subsets of T.

　Thus the restaurant can serve $2^7 = 128$ different variations of its pizzas.

b. Use the method of guessing and checking to find the smallest natural number n for which $2^n > 1000$.

$$2^8 = 256$$

$$2^9 = 512$$

$$2^{10} = 1024$$

The restaurant must provide a minimum of 10 toppings if it wishes to offer over 1000 variations of its pizzas.

▼ **check your progress 5**　A company makes a car with 11 upgrade options.
a. How many different versions of this car can the company produce? Assume that each upgrade option is independent of the other options.
b. What is the minimum number of upgrade options the company must provide if it wishes to offer at least 8000 different versions of this car?

Solution　*See page S4.*

MATHMATTERS   The Barber's Paradox

Some problems that concern sets have led to paradoxes. For instance, in 1902, the mathematician Bertrand Russell developed the following paradox. "Is the set A of all sets that are not elements of themselves an element of itself?" Both the assumption that A is an element of A and the assumption that A is not an element of A lead to a contradiction. Russell's paradox has been popularized as follows.

> The town barber shaves all males who do not shave themselves, and he shaves only those males. The town barber is a male who shaves. Who shaves the barber?

The assumption that the barber shaves himself leads to a contradiction, and the assumption that the barber does not shave himself also leads to a contradiction.

EXCURSION

Subsets and Complements of Fuzzy Sets

This excursion extends the concept of fuzzy sets that was developed in the Excursion in Section 2.1. Recall that the elements of a fuzzy set are ordered pairs. For any ordered pair (x, y) of a fuzzy set, the membership value y is a real number such that $0 \leq y \leq 1$.

The set of all x-values that are being considered is called the **universal set for the fuzzy set** and it is denoted by X.

TAKE NOTE

A set such as {3, 5, 9} is called a *crisp set*, to distinguish it from a fuzzy set.

▼ A Fuzzy Subset

If the fuzzy sets $A = \{(x_1, a_1), (x_2, a_2), (x_3, a_3), ...\}$ and
$B = \{(x_1, b_1), (x_2, b_2), (x_3, b_3), ...\}$ are both defined on the universal set
$X = \{x_1, x_2, x_3, ...\}$, then $A \subseteq B$ if and only if $a_i \leq b_i$ for all i.

A fuzzy set A is a **subset** of a fuzzy set B if and only if the membership value of each element of A *is less than or equal to* its corresponding membership value in set B. For instance, in Excursion Exercise 1 in Section 2.1, Mark and Erica used fuzzy sets to describe the set of good grades as follows:

Mark: $M = \{(A, 1), (B, 0.75), (C, 0.5), (D, 0.5), (F, 0)\}$

Erica: $E = \{(A, 1), (B, 0), (C, 0), (D, 0), (F, 0)\}$

In this case, fuzzy set E is a subset of fuzzy set M because each membership value of set E *is less than or equal to* its corresponding membership value in set M.

▼ The Complement of a Fuzzy Set

Let A be the fuzzy set $\{(x_1, a_1), (x_2, a_2), (x_3, a_3), ...\}$ defined on the universal set $X = \{x_1, x_2, x_3, ...\}$. Then the **complement** of A is the fuzzy set
$A' = \{(x_1, b_1), (x_2, b_2), (x_3, b_3), ...\}$, where each $b_i = 1 - a_i$.

Each element of the fuzzy set A' has a membership value that is 1 minus its membership value in the fuzzy set A. For example, the complement of

$$S = \{(\text{math}, 0.8), (\text{history}, 0.4), (\text{biology}, 0.3), (\text{art}, 0.1), (\text{music}, 0.7)\}$$

is the fuzzy set

$$S' = \{(\text{math}, 0.2), (\text{history}, 0.6), (\text{biology}, 0.7), (\text{art}, 0.9), (\text{music}, 0.3)\}.$$

The membership values in S' were calculated by subtracting the corresponding membership values in S from 1. For instance, the membership value of math in set S is 0.8. Thus the membership value of math in set S' is $1 - 0.8 = 0.2$.

EXCURSION EXERCISES

1. Let $K = \{(1, 0.4), (2, 0.6), (3, 0.8), (4, 1)\}$ and
$J = \{(1, 0.3), (2, 0.6), (3, 0.5), (4, 0.1)\}$ be fuzzy sets defined on $X = \{1, 2, 3, 4\}$.
Is $J \subseteq K$? Explain.

2. Consider the following membership graphs of *YOUNG* and *ADOLESCENT* defined on $X = \{x \mid 0 \le x \le 50\}$, where x is age in years.

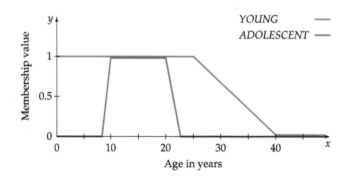

Is the fuzzy set *ADOLESCENT* a subset of the fuzzy set *YOUNG*? Explain.

3. Let the universal set be $\{A, B, C, D, F\}$ and let

$$G = \{(A, 1), (B, 0.7), (C, 0.4), (D, 0.1), (F, 0)\}$$

be a fuzzy set defined by Greg to describe what he feels is a good grade. Determine G'.

4. Let $C = \{(\text{Ferrari}, 0.9), (\text{Ford Mustang}, 0.6), (\text{Dodge Neon}, 0.5), (\text{Hummer}, 0.7)\}$
be a fuzzy set defined on the universal set {Ferrari, Ford Mustang, Dodge Neon, Hummer}. Determine C'.

Consider the following membership graph.

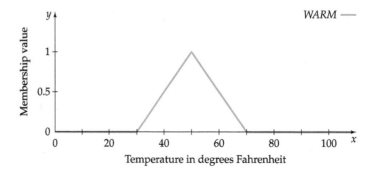

The membership graph of *WARM'* can be drawn by *reflecting* the graph of *WARM* about the graph of the line $y = 0.5$, as shown in the following figure.

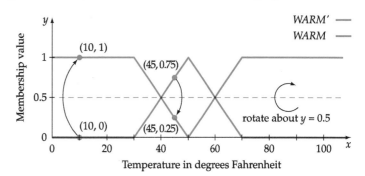

Note that when the membership graph of *WARM* is at a height of 0, the membership graph of *WARM'* is at a height of 1, and vice versa. In general, for any point (x, a) on the graph of *WARM*, there is a corresponding point $(x, 1 - a)$ on the graph of *WARM'*.

5. Use the following membership graph of *COLD* to draw the membership graph of *COLD'*.

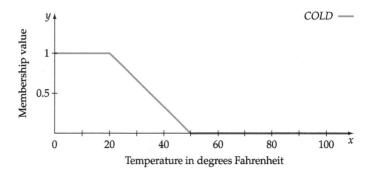

EXERCISE SET 2.2

(Suggested Assignment: The Enhanced WebAssign Exercises and Exercises 6, 8, 10, 15, 25, 26, and 38)

■ In Exercises 1 to 8, find the complement of the set given that $U = \{0, 1, 2, 3, 4, 5, 6, 7, 8\}$.

1. $\{2, 4, 6, 7\}$ **2.** $\{3, 6\}$

3. $\varnothing$ **4.** $\{4, 5, 6, 7, 8\}$

5. $\{x \mid x < 7 \text{ and } x \in N\}$

6. $\{x \mid x < 6 \text{ and } x \in W\}$

7. The set of odd counting numbers less than 8

8. The set of even counting numbers less than 10

■ In Exercises 9 to 18, insert either $\subseteq$ or $\nsubseteq$ in the blank space between the sets to make a true statement.

9. $\{a, b, c, d\}$ $\{a, b, c, d, e, f, g\}$

10. $\{3, 5, 7\}$ $\{3, 4, 5, 6\}$

11. $\{big, small, little\}$ $\{large, petite, short\}$

12. $\{red, white, blue\}$ $\{the colors in the American flag\}$

13. the set of integers   the set of rational numbers

14. the set of real numbers   the set of integers

15. $\varnothing$ $\{a, e, i, o, u\}$

16. $\{all sandwiches\}$ $\{all hamburgers\}$

17. $\{2, 4, 6, \ldots, 5000\}$ the set of even whole numbers

18. $\{x \mid x < 10 \text{ and } x \in Q\}$ the set of integers

■ In Exercises 19 to 36, let $U = \{p, q, r, s, t\}$, $D = \{p, r, s, t\}$, $E = \{q, s\}$, $F = \{p, t\}$, and $G = \{s\}$. Determine whether each statement is true or false.

19. $F \subseteq D$ **20.** $D \subseteq F$

21. $F \subset D$ **22.** $E \subset F$

23. $G \subset E$ **24.** $E \subset D$

25. $G' \subset D$ **26.** $E = F'$

27. $\varnothing \subset D$ **28.** $\varnothing \subset \varnothing$

29. $D' \subset E$ **30.** $G \in E$

31. $F \in D$ **32.** $G \nsubseteq F$

33. D has exactly eight subsets and seven proper subsets.

34. U has exactly 32 subsets.

35. F and F' each have exactly four subsets.

36. $\{0\} = \varnothing$

37. A class of 16 students has 2^{16} subsets. Use a calculator to determine how long (to the nearest hour) it would take you to write all the subsets, assuming you can write each subset in 1 second.

38. A class of 32 students has 2^{32} subsets. Use a calculator to determine how long (to the nearest year) it would take you to write all the subsets, assuming you can write each subset in 1 second.

■ In Exercises 39 to 42, list all subsets of the given set.

39. $\{\alpha, \beta\}$

40. $\{\alpha, \beta, \Gamma, \Delta\}$

41. $\{I, II, III\}$

42. $\varnothing$

■ In Exercises 43 to 50, find the number of subsets of the given set.

43. $\{2, 5\}$

44. $\{1, 7, 11\}$

45. $\{x \mid x$ is an even counting number between 7 and 21$\}$

46. $\{x \mid x$ is an odd integer between -4 and 8$\}$

47. The set of eleven players on a football team

48. The set of all letters of our alphabet

49. The set of all negative whole numbers

50. The set of all single-digit natural numbers

51. Suppose you have a nickel, two dimes, and a quarter. One of the dimes was minted in 1976, and the other one was minted in 1992.

   a. Assuming you choose at least one coin, how many different sets of coins can you form?

   b. Assuming you choose at least one coin, how many different sums of money can you produce?

   c. Explain why the answers in part a and part b are not the same.

52. The number of subsets of a set with n elements is 2^n.

   a. Use a calculator to find the exact value of 2^{18}, 2^{19}, and 2^{20}.

   b. What is the largest integer power of 2 for which your calculator will display the exact value?

53. Attribute Pieces   Elementary school teachers use plastic pieces called *attribute pieces* to illustrate subset concepts and to determine whether a student has learned to distinguish among different shapes, sizes, and colors. The following figure shows 12 attribute pieces.

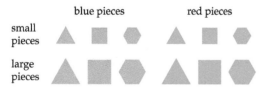

A set of 12 attribute pieces

A student has been asked to form the following sets. Determine the number of elements the student should have in each set.

   a. The set of red attribute pieces

   b. The set of red squares

   c. The set of hexagons

   d. The set of large blue triangles

54. Sandwich Choices   A delicatessen makes a roast-beef-on-sourdough sandwich for which you can choose from eight condiments.

   a. How many different types of roast-beef-on-sourdough sandwiches can the delicatessen prepare?

   b. What is the minimum number of condiments the delicatessen must have available if it wishes to offer at least 2000 different types of roast-beef-on-sourdough sandwiches?

55. Omelet Choices   A restaurant provides a brunch where the omelets are individually prepared. Each guest is allowed to choose from 10 different ingredients.

   a. How many different types of omelets can the restaurant prepare?

   b. What is the minimum number of ingredients that must be available if the restaurant wants to advertise that it offers over 4000 different omelets?

56. Truck Options   A truck company makes a pickup truck with 12 upgrade options. Some of the options are air conditioning, chrome wheels, and a satellite radio.

   a. How many different versions of this truck can the company produce?

   b. What is the minimum number of upgrade options the company must be able to provide if it wishes to offer at least 14,000 different versions of this truck?

EXTENSIONS

Critical Thinking

57. **a.** Explain why $\{2\} \notin \{1, 2, 3\}$.

b. Explain why $1 \not\subseteq \{1, 2, 3\}$.

c. Consider the set $\{1, \{1\}\}$. Does this set have one or two elements? Explain.

58. a. A set has 1024 subsets. How many elements are in the set?

b. A set has 255 proper subsets. How many elements are in the set?

c. Is it possible for a set to have an odd number of subsets? Explain.

59. Voting Coalitions Five people, designated A, B, C, D, and E, serve on a committee. To pass a motion, at least three of the committee members must vote for the motion. In such a situation, any set of three or more voters is called a **winning coalition** because if this set of people votes for a motion, the motion will pass. Any nonempty set of two or fewer voters is called a **losing coalition**.

a. List all the winning coalitions.

b. List all the losing coalitions.

Explorations

60. Subsets and Pascal's Triangle Following is a list of all the subsets of $\{a, b, c, d\}$.
Subsets with

0 elements: $\{\ \}$

1 element: $\{a\}, \{b\}, \{c\}, \{d\}$

2 elements: $\{a, b\}, \{a, c\}, \{a, d\}, \{b, c\}, \{b, d\}, \{c, d\}$

3 elements: $\{a, b, c\}, \{a, b, d\}, \{a, c, d\}, \{b, c, d\}$

4 elements: $\{a, b, c, d\}$

There is 1 subset with zero elements, and there are 4 subsets with exactly one element, 6 subsets with exactly two elements, 4 subsets with exactly three elements, and 1 subset with exactly four elements. Note that the numbers 1, 4, 6, 4, 1 are the numbers in row 4 of Pascal's triangle, which is shown below. Recall that the numbers in Pascal's triangle are created in the following manner. Each row begins and ends with the number 1. Any other number in a row is the sum of the two closest numbers above it. For instance, the first 10 in row 5 is the sum of the first 4 and the 6 in row 4.

```
            1                 row 0
          1   1               row 1
        1   2   1             row 2
      1   3   3   1           row 3
    1   4   6   4   1         row 4
  1   5  10  10   5   1       row 5
```

Pascal's triangle

a. Use Pascal's triangle to make a conjecture about the numbers of subsets of $\{a, b, c, d, e\}$ that have: zero elements, exactly one element, exactly two elements, exactly three elements, exactly four elements, and exactly five elements. Use your work from Check Your Progress 4 on page 65 to verify that your conjecture is correct.

b. Extend Pascal's triangle to show row 6. Use row 6 of Pascal's triangle to make a conjecture about the number of subsets of $\{a, b, c, d, e, f\}$ that have exactly three elements. Make a list of all the subsets of $\{a, b, c, d, e, f\}$ that have exactly three elements to verify that your conjecture is correct.

---

section **2.3** Set Operations

Intersection and Union of Sets

In Section 2.2 we defined the operation of finding the complement of a set. In this section we define the set operations *intersection* and *union*. In everyday usage, the word "intersection" refers to the *common region* where two streets cross. The intersection of two sets is defined in a similar manner.

▼ Intersection of Sets

The **intersection** of sets A and B, denoted by $A \cap B$, is the set of elements common to both A and B.

$$A \cap B = \{x \,|\, x \in A \quad \text{and} \quad x \in B\}$$

U

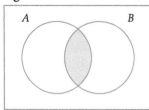

$A \cap B$

In the figure at the left, the region shown in blue represents the intersection of sets *A* and *B*.

▼ **example 1**   **Find Intersections**

Let $A = \{1, 4, 5, 7\}$, $B = \{2, 3, 4, 5, 6\}$, and $C = \{3, 6, 9\}$. Find

a. $A \cap B$ **b.** $A \cap C$

Solution

a. The elements common to *A* and *B* are 4 and 5.

$$A \cap B = \{1, 4, 5, 7\} \cap \{2, 3, 4, 5, 6\}$$
$$= \{4, 5\}$$

b. Sets *A* and *C* have no common elements. Thus $A \cap C = \varnothing$.

TAKE NOTE

It is a mistake to write

$$\{1, 5, 9\} \cap \{3, 5, 9\} = 5, 9$$

The intersection of two sets is a set. Thus

$$\{1, 5, 9\} \cap \{3, 5, 9\} = \{5, 9\}$$

▼ **check your progress 1**   Let $D = \{0, 3, 8, 9\}$, $E = \{3, 4, 8, 9, 11\}$, and $F = \{0, 2, 6, 8\}$. Find

a. $D \cap E$ **b.** $D \cap F$

Solution   *See page S4.* ◄

Two sets are **disjoint** if their intersection is the empty set. The sets *A* and *C* in Example 1b are disjoint. The Venn diagram at the left illustrates two disjoint sets.

In everyday usage, the word "union" refers to the act of uniting or joining together. The union of two sets has a similar meaning.

U

$A \cap C = \varnothing$

▼ **Union of Sets**

The **union** of sets *A* and *B*, denoted by $A \cup B$, is the set that contains all the elements that belong to *A* or to *B* or to both.

$$A \cup B = \{x \mid x \in A \quad \text{or} \quad x \in B\}$$

U

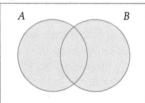

$A \cup B$

In the figure at the left, the region shown in blue represents the union of sets *A* and *B*.

▼ **example 2**   **Find Unions**

Let $A = \{1, 4, 5, 7\}$, $B = \{2, 3, 4, 5, 6\}$, and $C = \{3, 6, 9\}$. Find

a. $A \cup B$ **b.** $A \cup C$

Solution

a. List all the elements of set *A*, which are 1, 4, 5, and 7. Then add to your list the elements of set *B* that have not already been listed—in this case 2, 3, and 6. Enclose all elements with a pair of braces. Thus

$$A \cup B = \{1, 4, 5, 7\} \cup \{2, 3, 4, 5, 6\}$$
$$= \{1, 2, 3, 4, 5, 6, 7\}$$

b. $A \cup C = \{1, 4, 5, 7\} \cup \{3, 6, 9\}$
$$= \{1, 3, 4, 5, 6, 7, 9\}$$

▼ **check your progress 2**   Let $D = \{0, 4, 8, 9\}$, $E = \{1, 4, 5, 7\}$, and $F = \{2, 6, 8\}$. Find

a. $D \cup E$ **b.** $D \cup F$

Solution   *See page S4.* ◄

In mathematical problems that involve sets, the word "and" is interpreted to mean *intersection*. For instance, the phrase "the elements of A and B" means the elements of $A \cap B$. Similarly, the word "or" is interpreted to mean *union*. The phrase "the elements of A or B" means the elements of $A \cup B$.

▼ **example 3**   Describe Sets

Write a sentence that describes the set.

a. $A \cup (B \cap C)$ **b.** $J \cap K'$

Solution

a. The set $A \cup (B \cap C)$ can be described as "the set of all elements that are in A, or are in B and C."

b. The set $J \cap K'$ can be described as "the set of all elements that are in J and are not in K."

▼ **check your progress 3**   Write a sentence that describes the set.

a. $D \cap (E' \cup F)$ **b.** $L' \cup M$

Solution   *See page S5.*

Venn Diagrams and Equality of Sets

The Venn diagram in Figure 2.1 shows the four regions formed by two intersecting sets in a universal set U. It shows the four possible relationships that can exist between an element of a universal set U and two sets A and B.

An element of U:

- may be an element of both A and B.   Region i
- may be an element of A, but not B.   Region ii
- may be an element of B, but not A.   Region iii
- may not be an element of either A or B.   Region iv

We can use Figure 2.1 to determine whether two expressions that involve two sets are equal. For instance, to determine whether $(A \cup B)'$ and $A' \cap B'$ are equal for all sets A and B, we find what region or regions each of the expressions represents in Figure 2.1.

- If both expressions are represented by the same region(s), then the expressions are equal for all sets A and B.

- If both expressions are *not* represented by the same region(s), then the expressions are *not* equal for all sets A and B.

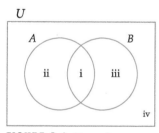

FIGURE 2.1 Venn diagram for two intersecting sets, in a universal set U.

▼ **example 4**   Equality of Sets

Determine whether $(A \cup B)' = A' \cap B'$ for all sets A and B.

Solution

To determine the region(s) in Figure 2.1, represented by $(A \cup B)'$, first determine the region(s) that are represented by $A \cup B$.

Set	Region or regions	Venn diagram
$A \cup B$	i, ii, iii The regions obtained by joining the regions represented by A (i, ii) and the regions represented by B (i, iii)	
$(A \cup B)'$	iv The region in U that is not in $(A \cup B)$	

Now determine the region(s) in Figure 2.1 that are represented by $A' \cap B'$.

Set	Region or regions	Venn diagram
A'	iii, iv The regions outside of A	
B'	ii, iv The regions outside of B	
$A' \cap B'$	iv The region common to A' and B'	

The expressions $(A \cup B)'$ and $A' \cap B'$ are both represented by region iv in Figure 2.1.

Thus $(A \cup B)' = A' \cap B'$ for all sets A and B.

▼ **check your progress** **4**   Determine whether $(A \cap B)' = A' \cup B'$ for all sets A and B.

Solution   *See page S5.*   ◄

The properties that were verified in Example 4 and Check Your Progress 4 are known as **De Morgan's laws**.

▼ **De Morgan's Laws**

For all sets A and B,

$$(A \cup B)' = A' \cap B' \text{ and } (A \cap B)' = A' \cup B'$$

De Morgan's law $(A \cup B)' = A' \cap B'$ can be stated as "the complement of the union of two sets is the intersection of the complements of the sets." De Morgan's law $(A \cap B)' = A' \cup B'$ can be stated as "the complement of the intersection of two sets is the union of the complements of the sets."

Venn Diagrams Involving Three Sets

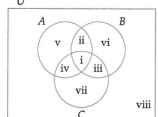

FIGURE 2.2 Venn diagram for three intersecting sets, in a universal set U.

The Venn diagram in Figure 2.2 shows the eight regions formed by three intersecting sets in a universal set U. It shows the eight possible relationships that can exist between an element of a universal set U and three sets A, B, and C.

An element of U:

- may be an element of A, B, and C.   Region i
- may be an element of A and B, but not C.   Region ii
- may be an element of B and C, but not A.   Region iii
- may be an element of A and C, but not B.   Region iv
- may be an element of A, but not B or C.   Region v
- may be an element of B, but not A or C.   Region vi
- may be an element of C, but not A or B.   Region vii
- may not be an element of A, B, or C.   Region viii

TAKE NOTE

The sets A, B, and C in Figure 2.2 separate the universal set, U, into eight regions. These eight regions can be numbered in any manner—that is, we can number any region as i, any region as ii, and so on. However, in this chapter we will use the numbering scheme shown in Figure 2.2 whenever we number the eight regions formed by three intersecting sets in a universal set.

▼ example 5 Determine Regions that Represent Sets

Use Figure 2.2 to answer each of the following.

a. Which regions represent $A \cap C$?   **b.** Which regions represent $A \cup C$?

c. Which regions represent $A \cap B'$?

Solution

a. $A \cap C$ is represented by all the regions common to circles A and C. Thus $A \cap C$ is represented by regions i and iv.

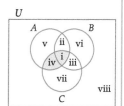

b. $A \cup C$ is represented by all the regions obtained by joining the regions in circle A (i, ii, iv, v) and the regions in circle C (i, iii, iv, vii). Thus $A \cup C$ is represented by regions i, ii, iii, iv, v, and vii.

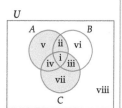

c. $A \cap B'$ is represented by all the regions common to circle A and the regions that are not in circle B. Thus $A \cap B'$ is represented by regions iv and v.

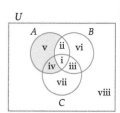

▼ check your progress 5 Use Figure 2.2 to answer each of the following.

a. Which region represents $(A \cap B) \cap C$?   **b.** Which regions represent $A \cup B'$?

c. Which regions represent $C' \cap B$?

Solution *See page S5.*

In Example 6, we use Figure 2.2 to determine whether two expressions that involve three sets are equal.

▼ **example 6** **Equality of Sets**

Determine whether $A \cup (B \cap C) = (A \cup B) \cap C$ for all sets A, B, and C.

Solution

To determine the region(s) in Figure 2.2 represented by $A \cup (B \cap C)$, we join the regions in A and the regions in $B \cap C$.

Set	Region or regions	Venn diagram
A	i, ii, iv, v The regions in A	
$B \cap C$	i, iii The regions common to B and C	
$A \cup (B \cap C)$	i, ii, iv, v, iii The regions in A joined with the regions in $B \cap C$	

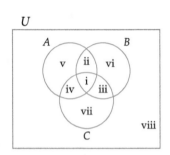

FIGURE 2.2 Displayed a second time for convenience

Now determine the region(s) in Figure 2.2 that are represented by $(A \cup B) \cap C$.

Set	Region or regions	Venn diagram
$A \cup B$	i, ii, iv, v, vi, iii The regions in A joined with the regions in B	
C	i, iii, iv, vii The regions in C	
$(A \cup B) \cap C$	i, iii, iv The regions common to $A \cup B$ and C	

The expressions $A \cup (B \cap C)$ and $(A \cup B) \cap C$ are not represented by the same regions.

Thus $A \cup (B \cap C) \neq (A \cup B) \cap C$ for all sets A, B, and C.

▼ **check your progress 6** Determine whether
$A \cap (B \cup C) = (A \cap B) \cup (A \cap C)$ for all sets A, B, and C.

Solution *See page S6.*

Venn diagrams can be used to verify each of the following properties.

▼ **Properties of Sets**

For all sets A and B:
Commutative Properties
$A \cap B = B \cap A$ Commutative property of intersection
$A \cup B = B \cup A$ Commutative property of union

For all sets A, B, and C:
Associative Properties
$(A \cap B) \cap C = A \cap (B \cap C)$ Associative property of intersection
$(A \cup B) \cup C = A \cup (B \cup C)$ Associative property of union
Distributive Properties
$A \cap (B \cup C) = (A \cap B) \cup (A \cap C)$ Distributive property of intersection over union

$A \cup (B \cap C) = (A \cup B) \cap (A \cup C)$ Distributive property of union over intersection

---

question Does $(B \cup C) \cap A = (A \cap B) \cup (A \cap C)$?

Application: Blood Groups and Blood Types

HISTORICAL NOTE

The Nobel Prize is an award granted to people who have made significant contributions to society. Nobel Prizes are awarded annually for achievements in physics, chemistry, physiology or medicine, literature, peace, and economics. The prizes were first established in 1901 by the Swedish industrialist Alfred Nobel, who invented dynamite.

Karl Landsteiner won a Nobel Prize in 1930 for his discovery of the four different human blood groups. He discovered that the blood of each individual contains exactly one of the following combinations of antigens.

- Only A antigens (blood group A)
- Only B antigens (blood group B)
- Both A and B antigens (blood group AB)
- No A antigens and no B antigens (blood group O)

These four blood groups are represented by the Venn diagram in the left margin of page 78.

In 1941, Landsteiner and Alexander Wiener discovered that human blood may or may not contain an Rh, or rhesus, factor. Blood with this factor is called

---

answer Yes. The commutative property of intersection allows us to write $(B \cup C) \cap A$ as $A \cap (B \cup C)$ and $A \cap (B \cup C) = (A \cap B) \cup (A \cap C)$ by the distributive property of intersection over union.

Rh-positive and is denoted by Rh+. Blood without this factor is called Rh-negative and is denoted by Rh−.

The Venn diagram in Figure 2.3 illustrates the eight blood types (A+, B+, AB+, O+, A−, B−, AB−, O−) that are possible if we consider antigens and the Rh factor.

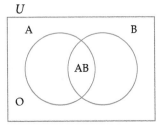

The four blood groups

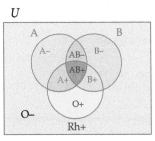

FIGURE 2.3   The eight blood types

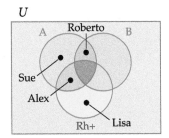

FIGURE 2.4

▼ example 7   Venn Diagrams and Blood Type

Use the Venn diagrams in Figures 2.3 and 2.4 to determine the blood type of each of the following people.

a. Sue   **b.** Lisa

Solution

a. Because Sue is in blood group A, not in blood group B, and not Rh+, her blood type is A−.

b. Lisa is in blood group O and she is Rh+, so her blood type is O+.

▼ check your progress 7   Use the Venn diagrams in Figures 2.3 and 2.4 to determine the blood type of each of the following people.

a. Alex   **b.** Roberto

Solution   *See page S6.*

The following table shows the blood types that can safely be given during a blood transfusion to persons of each of the eight blood types.

Blood Transfusion Table

Recipient blood type	Donor blood type
A+	A+, A−, O+, O−
B+	B+, B−, O+, O−
AB+	A+, A−, B+, B−, AB+, AB−, O+, O−
O+	O+, O−
A−	A−, O−
B−	B−, O−
AB−	A−, B−, AB−, O−
O−	O−

Source: American Red Cross

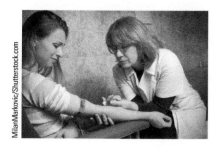

▼ **example 8**   **Applications of the Blood Transfusion Table**

Use the blood transfusion table and Figures 2.3 and 2.4 to answer the following questions.

a. Can Sue safely be given a type O+ blood transfusion?

b. Why is a person with type O− blood called a *universal donor?*

Solution

a. Sue's blood type is A−. The blood transfusion table shows that she can safely receive blood only if it is type A− or type O−. Thus it is not safe for Sue to receive type O+ blood in a blood transfusion.

b. The blood transfusion table shows that all eight blood types can safely receive type O− blood. Thus a person with type O− blood is said to be a universal donor.

▼ **check your progress 8**   Use the blood transfusion table and Figures 2.3 and 2.4 to answer the following questions.

a. Is it safe for Alex to receive type A− blood in a blood transfusion?

b. What blood type do you have if you are classified as a *universal recipient?*

Solution   *See page S6.*

MATHMATTERS   The Cantor Set

Consider the set of points formed by a line segment with a length of 1 unit. Remove the middle third of the line segment. Remove the middle third of each of the remaining 2 line segments. Remove the middle third of each of the remaining 4 line segments. Remove the middle third of each of the remaining 8 line segments. Remove the middle third of each of the remaining 16 line segments.

The first five steps in the formation of the Cantor set

The **Cantor set**, also known as **Cantor's Dust**, is the set of points that *remain* after the above process is repeated infinitely many times. You might conjecture that there are no points in the Cantor set, but it can be shown that there are just as many points in the Cantor set as in the original line segment! This is remarkable because it can also be shown that the sum of the lengths of the *removed* line segments equals 1 unit, which is the length of the original line segment. You can find additional information about the remarkable properties of the Cantor set on the Internet.

EXCURSION

Union and Intersection of Fuzzy Sets

This Excursion extends the concepts of fuzzy sets that were developed in Sections 2.1 and 2.2.

There are a number of ways in which the *union of two fuzzy sets* and the *intersection of two fuzzy sets* can be defined. The definitions we will use are called the **standard union operator** and the **standard intersection operator**. These standard operators preserve many of the set relations that exist in standard set theory.

▼ Union and Intersection of Two Fuzzy Sets

Let $A = \{(x_1, a_1), (x_2, a_2), (x_3, a_3), ...\}$ and $B = \{(x_1, b_1), (x_2, b_2), (x_3, b_3), ...\}$. Then

$$A \cup B = \{(x_1, c_1), (x_2, c_2), (x_3, c_3), ...\}$$

where c_i is the *maximum* of the two numbers a_i and b_i and

$$A \cap B = \{(x_1, c_1), (x_2, c_2), (x_3, c_3), ...\}$$

where c_i is the *minimum* of the two numbers a_i and b_i.

Each element of the fuzzy set $A \cup B$ has a membership value that is the *maximum* of its membership value in the fuzzy set A and its membership value in the fuzzy set B. Each element of the fuzzy set $A \cap B$ has a membership value that is the *minimum* of its membership value in fuzzy set A and its membership value in the fuzzy set B. In the following example, we form the union and intersection of two fuzzy sets. Let P and S be defined as follows.

Paul: $P = \{(\text{math}, 0.2), (\text{history}, 0.5), (\text{biology}, 0.7), (\text{art}, 0.8), (\text{music}, 0.9)\}$

Sally: $S = \{(\text{math}, 0.8), (\text{history}, 0.4), (\text{biology}, 0.3), (\text{art}, 0.1), (\text{music}, 0.7)\}$

Then

The maximum membership values for each of the given elements math, history, biology, art, and music

$P \cup S = \{(\text{math}, 0.8), (\text{history}, 0.5), (\text{biology}, 0.7), (\text{art}, 0.8), (\text{music}, 0.9)\}$
$P \cap S = \{(\text{math}, 0.2), (\text{history}, 0.4), (\text{biology}, 0.3), (\text{art}, 0.1), (\text{music}, 0.7)\}$

The minimum membership values for each of the given elements math, history, biology, art, and music

EXCURSION EXERCISES

In Excursion Exercise 1 of Section 2.1, we defined the following fuzzy sets.

Mark: $M = \{(A, 1), (B, 0.75), (C, 0.5), (D, 0.5), (F, 0)\}$
Erica: $E = \{(A, 1), (B, 0), (C, 0), (D, 0), (F, 0)\}$
Larry: $L = \{(A, 1), (B, 1), (C, 1), (D, 1), (F, 0)\}$
Jennifer: $J = \{(A, 1), (B, 0.8), (C, 0.6), (D, 0.1), (F, 0)\}$

Use these fuzzy sets to find each of the following.

1. $M \cup J$ **2.** $M \cap J$ **3.** $E \cup J'$ **4.** $J \cap L'$ **5.** $J \cap (M' \cup L')$

Photolibrary

Consider the following membership graphs.

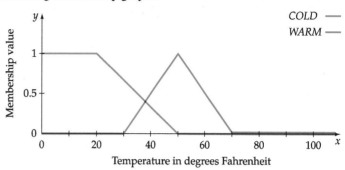

The membership graph of *COLD* ∪ *WARM* is shown in purple in the following figure. The membership graph of *COLD* ∪ *WARM* lies on either the membership graph of *COLD* or the membership graph of *WARM*, depending on which of these graphs is *higher* at any given temperature *x*.

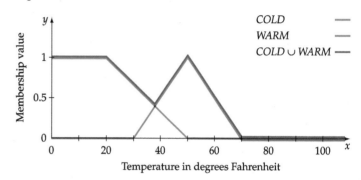

The membership graph of *COLD* ∩ *WARM* is shown in green in the following figure. The membership graph of *COLD* ∩ *WARM* lies on either the membership graph of *COLD* or the membership graph of *WARM*, depending on which of these graphs is *lower* at any given temperature *x*.

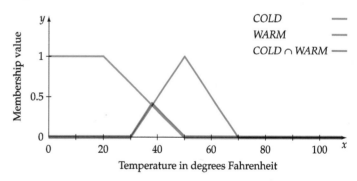

TAKE NOTE

The following lyrics from the old Scottish song *Loch Lomond* provide an easy way to remember how to draw the graph of the union or intersection of two membership graphs.

> Oh! ye'll take the high road and I'll take the low road, and I'll be in Scotland afore ye.

The graph of the union of two membership graphs takes the "high road" provided by the graphs, and the graph of the intersection of two membership graphs takes the "low road" provided by the graphs.

6. Use the following graphs to draw the membership graph of *WARM* ∪ *HOT*.

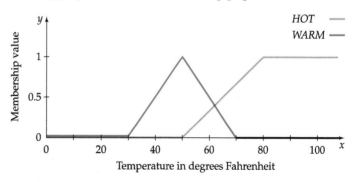

7. Let $X = \{a, b, c, d, e\}$ be the universal set. Determine whether De Morgan's law $(A \cap B)' = A' \cup B'$ holds true for the fuzzy sets $A = \{(a, 0.3), (b, 0.8), (c, 1), (d, 0.2), (e, 0.75)\}$ and $B = \{(a, 0.5), (b, 0.4), (c, 0.9), (d, 0.7), (e, 0.45)\}$.

EXERCISE SET 2.3

(Suggested Assignment: The Enhanced WebAssign Exercises and Exercises 21, 25, 35, 44, 53, 54, and 66)

■ In Exercises 1 to 20, let $U = \{1, 2, 3, 4, 5, 6, 7, 8\}$, $A = \{2, 4, 6\}$, $B = \{1, 2, 5, 8\}$, and $C = \{1, 3, 7\}$. Find each of the following.

1. $A \cup B$
2. $A \cap B$
3. $A \cap B'$
4. $B \cap C'$
5. $(A \cup B)'$
6. $(A' \cap B)'$
7. $A \cup (B \cup C)$
8. $A \cap (B \cup C)$
9. $A \cap (B \cap C)$
10. $A' \cup (B \cap C)$
11. $B \cap (B \cup C)$
12. $A \cap A'$
13. $B \cup B'$
14. $(A \cap (B \cup C))'$
15. $(A \cup C') \cap (B \cup A')$
16. $(A \cup C') \cup (B \cup A')$
17. $(C \cup B') \cup \varnothing$
18. $(A' \cup B) \cap \varnothing$
19. $(A \cup B) \cap (B \cap C')$
20. $(B \cap A') \cup (B' \cup C)$

In Exercises 21 to 28, write a sentence that describes the given mathematical expression.

21. $L' \cup T$
22. $J' \cap K$
23. $A \cup (B' \cap C)$
24. $(A \cup B) \cap C'$
25. $T \cap (J \cup K')$
26. $(A \cap B) \cup C$
27. $(W \cap V) \cup (W \cap Z)$
28. $D \cap (E \cup F)'$

■ In Exercises 29 to 36, draw a Venn diagram to show each of the following sets.

29. $A \cap B'$
30. $(A \cap B)'$
31. $(A \cup B)'$
32. $(A' \cap B) \cup B'$
33. $A \cap (B \cup C')$
34. $A \cap (B' \cap C)$
35. $(A \cup C) \cap (B \cup C')$
36. $(A' \cap B) \cup (A \cap C')$

■ In Exercises 37 to 40, draw two Venn diagrams to determine whether the following expressions are equal for all sets A and B.

37. $A \cap B'$; $A' \cup B$
38. $A' \cap B$; $A \cup B'$
39. $A \cup (A' \cap B)$; $A \cup B$
40. $A' \cap (B \cup B')$; $A' \cup (B \cap B')$

■ In Exercises 41 to 46, draw two Venn diagrams to determine whether the following expressions are equal for all sets A, B, and C.

41. $(A \cup C) \cap B'$; $A' \cup (B \cup C)$
42. $A' \cap (B \cap C)$, $(A \cup B') \cap C$
43. $(A' \cap B) \cup C$, $(A' \cap C) \cap (A' \cap B)$
44. $A' \cup (B' \cap C)$, $(A' \cup B') \cap (A' \cup C)$
45. $((A \cup B) \cap C)'$, $(A' \cap B') \cup C'$

46. $(A \cap B) \cap C$, $(A \cup B \cup C)'$

IN THE NEWS

Unemployment, Foreclosures, and the Economy

High unemployment drove up foreclosures in 72% of 206 leading metropolitan areas in 2010.

SOURCE: Julie Schmit, USA TODAY, 1/27/2011

Three topics (terms) in the news in August 2010 were unemployment, the economy, and foreclosures. Of these three topics, which one was being discussed the most by members of the general public? In the past, this question has been difficult to answer. However, Jeff Clark, a professional programmer, has created an Internet application called "Twitter Venn" that can answer this question. Twitter Venn works by searching for your topics in messages, called tweets, sent on the social networking service Twitter®. After a search is completed, a Venn diagram similar to the one below is displayed.

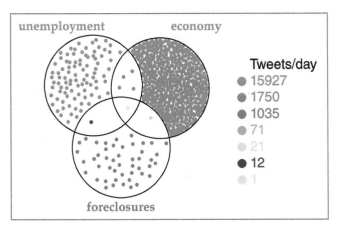

The results of a Twitter Venn search performed in August 2010

The number of dots indicate the relative number of tweets. Therefore, it is easy to see that at the time the above Twitter Venn search was performed, the economy was being discussed the most, followed by unemployment, and then foreclosures.

- The blue dots represent tweets that contain economy but neither of the other topics. (15,927 tweets per day)

- The red dots represent tweets that contain unemployment but neither of the other topics. (1750 tweets per day)

- The green dots represent tweets that contain foreclosures but neither of the other topics. (1035 tweets per day)

The Venn diagram, on the previous page, also shows the tweets that contain combinations of our topics. For instance, the intersection of the unemployment circle and the economy circle, but not the foreclosures circle, represents tweets that include both unemployment and economy, but not foreclosures. The legend to the right of the Venn diagram indicates 71 tweets per day for this particular combination of topics.

The Internet address (URL) for Twitter Venn is:

http://www.neoformix.com/Projects/TwitterVenn/view.php

The directions for using Twitter Venn are as follows. Enter two or three search terms, separated by commas, into the Search Terms box and press the Search button.

In Exercises 47 and 48, use Twitter Venn to perform a search for the given topics (terms) and then answer each question.

47. Topics: economy, Obama, China

  a. Which topic received the most tweets per day?

  b. Consider the combination of topics that includes both economy and Obama, but not China, and the combination of topics that includes both economy and China, but not Obama. Which one of these combinations received the most tweets per day?

48. Topics: Lady Gaga, Beyonce, Shakira

  a. Which topic received the most tweets per day?

  b. Consider the combination of topics that includes both Beyonce and Shakira, but not Lady Gaga, and the combination of topics that includes both Lady Gaga and Shakira, but not Beyonce. Which one of these combinations received the most tweets per day?

Additive Color Mixing Computers and televisions make use of *additive color mixing*. The following figure shows that when the *primary colors* red R, green G, and blue B are mixed together using additive color mixing, they produce white, W. Using set notation, we state this as $R \cap B \cap G = W$. The colors yellow Y, cyan C, and magenta M are called *secondary colors*. A secondary color is produced by mixing exactly two of the primary colors.

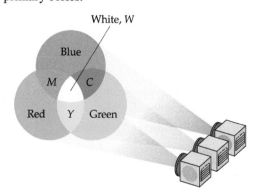

In Exercises 49 to 51, determine which color is represented by each of the following. Assume that the colors are being mixed using additive color mixing. (Use R for red, G for green, and B for blue.)

49. $R \cap G \cap B'$ **50.** $R \cap G' \cap B$
51. $R' \cap G \cap B$

Subtractive Color Mixing Artists who paint with pigments use *subtractive color mixing* to produce different colors. In a subtractive color mixing system, the primary colors are cyan C, magenta M, and yellow Y. The following figure shows that when the three primary colors are mixed in equal amounts, using subtractive color mixing, they form black, K. Using set notation, we state this as $C \cap M \cap Y = K$. In subtractive color mixing, the colors red R, blue B, and green G are the secondary colors. As mentioned previously, a secondary color is produced by mixing equal amounts of exactly two of the primary colors.

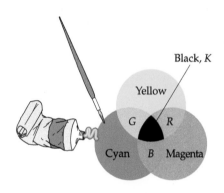

In Exercises 52 to 54, determine which color is represented by each of the following. Assume the colors are being mixed using subtractive color mixing. (Use C for cyan, M for magenta, and Y for yellow.)

52. $C \cap M \cap Y'$ **53.** $C' \cap M \cap Y$
54. $C \cap M' \cap Y$

■ In Exercises 55 to 64, use set notation to describe the shaded region. You may use any of the following symbols: A, B, C, $\cap$, $\cup$, and $'$. Keep in mind that each shaded region has more than one set description.

55.
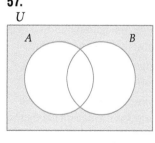

56.
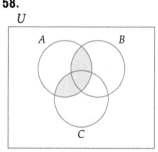

57.

58.
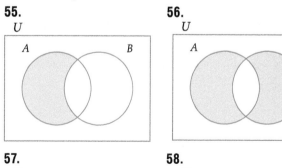

59.

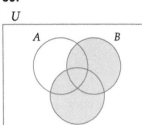

60.

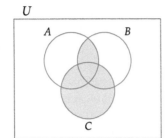

61.

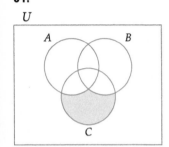

62.

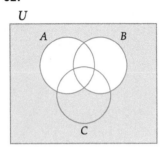

63.

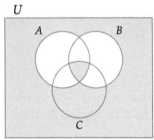

64.

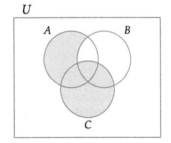

65. A Survey A special interest group plans to conduct a survey of households concerning a ban on handguns. The special interest group has decided to use the following Venn diagram to help illustrate the results of the survey. *Note:* A rifle is a gun, but it is not a hand gun.

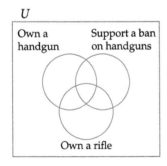

a. Shade in the regions that represent households that own a handgun and do not support the ban on handguns.

b. Shade in the region that represents households that own only a rifle and support the ban on handguns.

c. Shade in the region that represents households that do not own a gun and do not support the ban on handguns.

66. A Music Survey The administrators of an Internet music site plan to conduct a survey of college students to determine how the students acquire music. The administrators have decided to use the following Venn diagram to help tabulate the results of the survey.

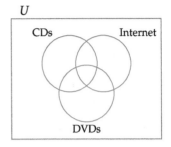

a. Shade in the region that represents students who acquire music from CDs and the Internet, but not from DVDs.

b. Shade in the regions that represent students who acquire music from CDs or the Internet.

c. Shade in the regions that represent students who acquire music from both CDs and DVDs.

■ In Exercises 67 to 70, draw a Venn diagram with each of the given elements placed in the correct region.

67. $U = \{-1, 0, 1, 2, 3, 4, 5, 6, 7, 8, 9\}$
$A = \{1, 3, 5\}$
$B = \{3, 5, 7, 8\}$
$C = \{-1, 8, 9\}$

68. $U = \{2, 4, 6, 8, 10, 12, 14\}$
$A = \{2, 10, 12\}$
$B = \{4, 8\}$
$C = \{6, 8, 10\}$

69. $U = \{$Sue, Bob, Al, Jo, Ann, Herb, Eric, Mike, Sal$\}$
$A = \{$Sue, Herb$\}$
$B = \{$Sue, Eric, Jo, Ann$\}$
$Rh+ = \{$Eric, Sal, Al, Herb$\}$

70. $U = \{$Hal, Marie, Rob, Armando, Joel, Juan, Melody$\}$
$A = \{$Marie, Armando, Melody$\}$
$B = \{$Rob, Juan, Hal$\}$
$Rh+ = \{$Hal, Marie, Rob, Joel, Juan, Melody$\}$

■ In Exercises 71 and 72, use two Venn diagrams to verify the following properties for all sets A, B, and C.

71. The associative property of intersection

$$(A \cap B) \cap C = A \cap (B \cap C)$$

72. The distributive property of intersection over union

$$A \cap (B \cup C) = (A \cap B) \cup (A \cap C)$$

EXTENSIONS

Critical thinking

Difference of Sets   Another operation that can be defined on sets A and B is the **difference of the sets**, denoted by $A - B$. Here is a formal definition of the difference of sets A and B.

$$A - B = \{x \mid x \in A \quad \text{and} \quad x \notin B\}$$

Thus $A - B$ is the set of elements that belong to A but not to B. For instance, let $A = \{1, 2, 3, 7, 8\}$ and $B = \{2, 7, 11\}$. Then $A - B = \{1, 3, 8\}$.

■ In Exercises 73 to 78, determine each difference, given that $U = \{1, 2, 3, 4, 5, 6, 7, 8, 9\}$, $A = \{2, 4, 6, 8\}$, and $B = \{2, 3, 8, 9\}$.

73. $B - A$ **74.** $A - B$

75. $A - B'$ **76.** $B' - A$

77. $A' - B'$ **78.** $A' - B$

79. 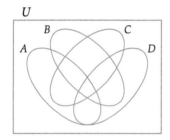 **John Venn**   Write a few paragraphs about the life of John Venn and his work in the area of mathematics.

The following Venn diagram illustrates that four sets can partition the universal set into 16 different regions.

■ In Exercises 80 and 81, use a Venn diagram similar to the one at the left below to shade in the region represented by the given expression.

80. $(A \cap B) \cup (C' \cap D)$

81. $(A \cup B)' \cap (C \cap D)$

Explorations

82. In an article in *New Scientist* magazine, Anthony W. F. Edwards illustrated how to construct Venn diagrams that involve many sets.[1] Search the Internet to find Edwards's method of constructing a Venn diagram for five sets and a Venn diagram for six sets. Use drawings to illustrate Edwards's method of constructing a Venn diagram for five sets and a Venn diagram for six sets. (*Source:* http://www.combinatorics.org/Surveys/ds5/VennWhatEJC.html)

---

section **2.4**   Applications of Sets

Surveys: An Application of Sets

Counting problems occur in many areas of applied mathematics. To solve these counting problems, we often make use of a Venn diagram and the inclusion-exclusion principle, which will be presented in this section.

▼ **example 1**   A Survey of Preferences

A movie company is making plans for future movies it wishes to produce. The company has done a random survey of 1000 people. The results of the survey are shown below.

    695 people like action adventures.

    340 people like comedies.

    180 people like both action adventures and comedies.

---
[1] Anthony W. F. Edwards, "Venn diagrams for many sets," *New Scientist,* 7 January 1989, pp. 51–56.

Of the people surveyed, how many people

a. like action adventures but not comedies?

b. like comedies but not action adventures?

c. do not like either of these types of movies?

Solution

A Venn diagram can be used to illustrate the results of the survey. We use two overlapping circles (see Figure 2.5). One circle represents the set of people who like action adventures and the other represents the set of people who like comedies. The region i where the circles intersect represents the set of people who like both types of movies.

We start with the information that 180 people like both types of movies and write 180 in region i. See Figure 2.6.

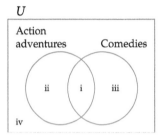

FIGURE 2.5

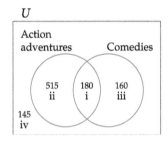

FIGURE 2.6

a. Regions i and ii have a total of 695 people. So far we have accounted for 180 of these people in region i. Thus the number of people in region ii, which is the set of people who like action adventures but do not like comedies, is $695 - 180 = 515$.

b. Regions i and iii have a total of 340 people. Thus the number of people in region iii, which is the set of people who like comedies but do not like action adventures, is $340 - 180 = 160$.

c. The number of people who do not like action adventure movies or comedies is represented by region iv. The number of people in region iv must be the total number of people, which is 1000, less the number of people accounted for in regions i, ii, and iii, which is 855. Thus the number of people who do not like either type of movie is $1000 - 855 = 145$.

▼ **check your progress** **1** The athletic director of a school has surveyed 200 students. The survey results are shown below.

    140 students like volleyball.

    120 students like basketball.

    85 students like both volleyball and basketball.

Of the students surveyed, how many students

a. like volleyball but not basketball?

b. like basketball but not volleyball?

c. do not like either of these sports?

Solution *See page S6.*

In the next example we consider a more complicated survey that involves three types of music.

▼ **example 2**   **A Music Survey**

A music teacher has surveyed 495 students. The results of the survey are listed below.

320 students like rap music.

395 students like rock music.

295 students like heavy metal music.

280 students like both rap music and rock music.

190 students like both rap music and heavy metal music.

245 students like both rock music and heavy metal music.

160 students like all three.

How many students

a. like exactly two of the three types of music?

b. like only rock music?

c. like only one of the three types of music?

Solution

The Venn diagram at the left shows three overlapping circles. Region i represents the set of students who like all three types of music. Each of the regions v, vi, and vii represent the students who like only one type of music.

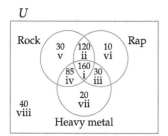

a. The survey shows that 245 students like rock and heavy metal music, so the numbers we place in regions i and iv must have a sum of 245. Since region i has 160 students, we see that region iv must have 245 − 160 = 85 students. In a similar manner, we can determine that region ii has 120 students and region iii has 30 students. Thus 85 + 120 + 30 = 235 students like exactly two of the three types of music.

b. The sum of the students represented by regions i, ii, iv, and v must be 395. The number of students in region v must be the difference between this total and the sum of the numbers of students in region i, ii, and iv. Thus the number of students who like only rock music is 395 − (160 + 120 + 85) = 30. See the Venn diagram at the left.

c. Using the same reasoning as in part b, we find that region vi has 10 students and region vii has 20 students. To find the number of students who like only one type of music, find the sum of the numbers of students in regions v, vi, and vii, which is 30 + 10 + 20 = 60. See the Venn diagram at the left.

▼ **check your progress 2**   An activities director for a cruise ship has surveyed 240 passengers. Of the 240 passengers,

135 like swimming.	80 like swimming and dancing.
150 like dancing.	40 like swimming and games.
65 like games.	25 like dancing and games.
	15 like all three activities.

How many passengers

a. like exactly two of the three types of activities?

b. like only swimming?

c. like none of these activities?

Solution   *See page S6.*

MATHMATTERS   Grace Chisholm Young (1868–1944)

Courtesy of Sylvia Wiegand

Grace Chisholm Young

Grace Chisholm Young studied mathematics at Girton College, which is part of Cambridge University. In England at that time, women were not allowed to earn a university degree, so she decided to continue her mathematical studies at the University of Göttingen in Germany, where her advisor was the renowned mathematician Felix Klein. She excelled while at Göttingen and at the age of 27 earned her doctorate in mathematics, magna cum laude. She was the first woman officially to earn a doctorate degree from a German university. Shortly after her graduation she married the mathematician William Young. Together they published several mathematical papers and books, one of which was the first textbook on set theory.

The Inclusion-Exclusion Principle

A music director wishes to take the band and the choir on a field trip. There are 65 students in the band and 30 students in the choir. The number of students in both the band and the choir is 16. How many students should the music director plan on taking on the field trip?

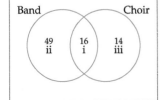

Using the process developed in the previous examples, we find that the number of students that are in only the band is $65 - 16 = 49$. The number of students that are in only the choir is $30 - 16 = 14$. See the Venn diagram at the left. Adding the numbers of students in regions i, ii, and iii gives us a total of $49 + 16 + 14 = 79$ students that might go on the field trip.

Although we can use Venn diagrams to solve counting problems, it is more convenient to make use of the following technique. First add the number of students in the band to the number of students in the choir. Then subtract the number of students who are in both the band and the choir. This technique gives us a total of $(65 + 30) - 16 = 79$ students, the same result as above. The reason we subtract the 16 students is that we have counted each of them twice. Note that first we include the students that are in both the band and the choir twice, and then we exclude them once. This procedure leads us to the following result.

TAKE NOTE

Recall that $n(A)$ represents the number of elements in set A.

▼ **The Inclusion-Exclusion Principle**

For all finite sets A and B,

$$n(A \cup B) = n(A) + n(B) - n(A \cap B)$$

   What must be true of the finite sets A and B if
$$n(A \cup B) = n(A) + n(B)?$$

▼ **example 3**   An Application of the Inclusion-Exclusion Principle

A school finds that 430 of its students are registered in chemistry, 560 are registered in mathematics, and 225 are registered in both chemistry and mathematics. How many students are registered in chemistry or mathematics?

answer A and B must be disjoint sets.

Solution

Let $C = \{$students registered in chemistry$\}$ and let
$M = \{$students registered in mathematics$\}$.

$$n(C \cup M) = n(C) + n(M) - n(C \cap M)$$
$$= 430 + 560 - 225$$
$$= 765$$

Using the inclusion-exclusion principle, we see that 765 students are registered in chemistry or mathematics.

▼ **check your progress 3**   A high school has 80 athletes who play basketball, 60 athletes who play soccer, and 24 athletes who play both basketball and soccer. How many athletes play either basketball or soccer?

Solution   *See page S7.*

The inclusion-exclusion principle can be used provided we know the number of elements in any three of the four sets in the formula.

▼ **example 4**   **An Application of the Inclusion-Exclusion Principle**

Given $n(A) = 15$, $n(B) = 32$, and $n(A \cup B) = 41$, find $n(A \cap B)$.

Solution

Substitute the given information in the inclusion-exclusion formula and solve for the unknown.

$$n(A \cup B) = n(A) + n(B) - n(A \cap B)$$
$$41 = 15 + 32 - n(A \cap B)$$
$$41 = 47 - n(A \cap B)$$

Thus

$$n(A \cap B) = 47 - 41$$
$$n(A \cap B) = 6$$

▼ **check your progress 4**   Given $n(A) = 785$, $n(B) = 162$, and $n(A \cup B) = 852$, find $n(A \cap B)$.

Solution   *See page S7.*

The inclusion-exclusion formula can be adjusted and applied to problems that involve percents. In the following formula we denote "the percent in set A" by the notation $p(A)$.

▼ **The Percent Inclusion-Exclusion Formula**

For all finite sets A and B,

$$p(A \cup B) = p(A) + p(B) - p(A \cap B)$$

▼ **example 5**   **An Application of the Percent Inclusion-Exclusion Formula**

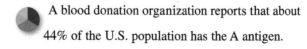 A blood donation organization reports that about
44% of the U.S. population has the A antigen.

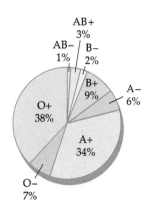

Approximate Percentage of
U.S. Population with Each
Blood Type

15% of the U.S. population has the B antigen.

4% of the U.S. population has both the A and the B antigen.

Use the percent inclusion-exclusion formula to estimate the percent of the U.S. population that has the A antigen or the B antigen.

Solution

We are given $p(A) = 44\%$, $p(B) = 15\%$, and $p(A \cap B) = 4\%$. Substituting in the percent inclusion-exclusion formula gives

$$p(A \cup B) = p(A) + p(B) - p(A \cap B)$$
$$= 44\% + 15\% - 4\%$$
$$= 55\%$$

Thus about 55% of the U.S. population has the A antigen or the B antigen.

▼ **check your progress 5** A blood donation organization reports that about

 44% of the U.S. population has the A antigen.

84% of the U.S. population is Rh+.

91% of the U.S. population either has the A antigen or is Rh+.

Use the percent inclusion-exclusion formula to estimate the percent of the U.S. population that has the A antigen *and* is Rh+.

Solution   *See page S7.*

In the next example, the data are provided in a table. The number in column G and row M represents the number of elements in $G \cap M$. The sum of all the numbers in column G and column B represents the number of elements in $G \cup B$.

▼ **example 6**   **A Survey Presented in Tabular Form**

A survey of men M, women W, and children C concerning the use of the Internet search engines Google G, Yahoo! Y, and Bing B yielded the following results.

	Google (*G*)	Yahoo! (*Y*)	Bing (*B*)
Men (*M*)	440	310	275
Women (*W*)	390	280	325
Children (*C*)	140	410	40

Use the data in the table to find each of the following.

a. $n(W \cap Y)$ **b.** $n(G \cap C')$ **c.** $n(M \cap (G \cup B))$

Solution

a. The table shows that 280 of the women surveyed use Yahoo! as a search engine. Thus $n(W \cap Y) = 280$.

b. The set $G \cap C'$ is the set of surveyed Google users who are men or women. The number in this set is $440 + 390 = 830$.

c. The number of men in the survey that use either Google or Bing is $440 + 275 = 715$.

▼ **check your progress** **6** Use the table in Example 6 to find each of the following.

a. $n(Y \cap C)$ **b.** $n(B \cap M')$ **c.** $n((G \cap M) \cup (G \cap W))$

Solution *See page S7.*

EXCURSION

© Jim West/Alamy

Voting Systems

There are many types of voting systems. When people are asked to vote for or against a resolution, a one-person, one-vote *majority system* is often used to decide the outcome. In this type of voting, each voter receives one vote, and the resolution passes only if it receives *most* of the votes.

In any voting system, the number of votes that is required to pass a resolution is called the **quota**. A **coalition** is a set of voters each of whom votes the same way, either for or against a resolution. A **winning coalition** is a set of voters the sum of whose votes is greater than or equal to the quota. A **losing coalition** is a set of voters the sum of whose votes is less than the quota.

Sometimes you can find all the winning coalitions in a voting process by making an organized list. For instance, consider the committee consisting of Alice, Barry, Cheryl, and Dylan. To decide on any issues, they use a one-person, one-vote majority voting system. Since each of the four voters has a single vote, the quota for this majority voting system is 3. The winning coalitions consist of all subsets of the voters that have three or more people. We list these winning coalitions in the table at the left below, where A represents Alice, B represents Barry, C represents Cheryl, and D represents Dylan.

A **weighted voting system** is one in which some voters' votes carry more weight regarding the outcome of an election. As an example, consider a selection committee that consists of four people designated by A, B, C, and D. Voter A's vote has a weight of 2, and the vote of each other member of the committee has a weight of 1. The quota for this weighted voting system is 3. A winning coalition must have a weighted voting sum of at least 3. The winning coalitions are listed in the table at the right below.

HISTORICAL NOTE

Reproduced with permission.

An ostrakon

In ancient Greece, the citizens of Athens adopted a procedure that allowed them to vote for the expulsion of any prominent person. The purpose of this procedure, known as an *ostracism*, was to limit the political power that any one person could attain.

In an ostracism, each voter turned in a *potsherd*, a piece of pottery fragment, on which was inscribed the name of the person the voter wished to ostracize. The pottery fragments used in the voting process became known as *ostrakon*.

The person who received the majority of the votes, above some set minimum, was exiled from Athens for a period of 10 years.

Winning coalition	Sum of the votes
{A, B, C}	3
{A, B, D}	3
{A, C, D}	3
{B, C, D}	3
{A, B, C, D}	4

Winning coalition	Sum of the weighted votes
{A, B}	3
{A, C}	3
{A, D}	3
{B, C, D}	3
{A, B, C}	4
{A, B, D}	4
{A, C, D}	4
{A, B, C, D}	5

A **minimal winning coalition** is a winning coalition that has no proper subset that is a winning coalition. In a minimal winning coalition each voter is said to be a **critical voter**, because if any of the voters leaves the coalition, the coalition will then become a losing coalition. In the table at the right on the previous page, the minimal winning coalitions are {A, B}, {A, C}, {A, D}, and {B, C, D}. If any single voter leaves one of these coalitions, then the coalition will become a losing coalition. The coalition {A, B, C, D} is not a minimal winning coalition, because it contains at least one proper subset, for instance {A, B, C}, that is a winning coalition.

EXCURSION EXERCISES

1. A selection committee consists of Ryan, Susan, and Trevor. To decide on issues, they use a one-person, one-vote majority voting system.

   a. Find all winning coalitions.

   b. Find all losing coalitions.

2. A selection committee consists of three people designated by M, N, and P. M's vote has a weight of 3, N's vote has a weight of 2, and P's vote has a weight of 1. The quota for this weighted voting system is 4. Find all winning coalitions.

3. Determine the minimal winning coalitions for the voting system in Excursion Exercise 2.

Additional information on the applications of mathematics to voting systems is given in Chapter 4.

EXERCISE SET 2.4

(Suggested Assignment: The Enhanced WebAssign Exercises and Exercises 9, 14, and 29)

■ In Exercises 1 to 8, let U = {English, French, History, Math, Physics, Chemistry, Psychology, Drama},
A = {English, History, Psychology, Drama},
B = {Math, Physics, Chemistry, Psychology, Drama}, and
C = {French, History, Chemistry}.
Find each of the following.

1. $n(B \cup C)$

2. $n(A \cup B)$

3. $n(B) + n(C)$

4. $n(A) + n(B)$

5. $n(A \cup B \cup C)$

6. $n(A \cap B)$

7. $n(A) + n(B) + n(C)$

8. $n(A \cap B \cap C)$

9. Verify that for A and B as defined in Exercises 1 to 8, $n(A \cup B) = n(A) + n(B) - n(A \cap B)$.

10. Verify that for A and C as defined in Exercises 1 to 8, $n(A \cup C) = n(A) + n(C) - n(A \cap C)$.

11. Given $n(J) = 245$, $n(K) = 178$, and $n(J \cup K) = 310$, find $n(J \cap K)$.

12. Given $n(L) = 780$, $n(M) = 240$, and $n(L \cap M) = 50$, find $n(L \cup M)$.

13. Given $n(A) = 1500$, $n(A \cup B) = 2250$, and $n(A \cap B) = 310$, find $n(B)$.

14. Given $n(A) = 640$, $n(B) = 280$, and $n(A \cup B) = 765$, find $n(A \cap B)$.

■ In Exercises 15 and 16, use the given information to find the number of elements in each of the regions labeled with a question mark.

15. $n(A) = 28$, $n(B) = 31$, $n(C) = 40$, $n(A \cap B) = 15$, $n(U) = 75$

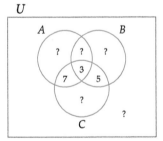

16. $n(A) = 610$, $n(B) = 440$, $n(C) = 1000$, $n(U) = 2900$

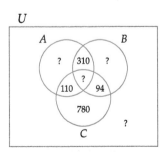

17. Investing In a survey of 600 investors, it was reported that 380 had invested in stocks, 325 had invested in bonds, and 75 had not invested in either stocks or bonds.

a. How many investors had invested in both stocks and bonds?

b. How many investors had invested only in stocks?

18. Commuting A survey of 1500 commuters in New York City showed that 1140 take the subway, 680 take the bus, and 120 do not take either the bus or the subway.

a. How many commuters take both the bus and the subway?

b. How many commuters take only the subway?

19. First Aid Treatments A team physician has determined that of all the athletes who were treated for minor back pain, 72% responded to an analgesic, 59% responded to a muscle relaxant, and 44% responded to both forms of treatment.

a. What percent of the athletes who were treated responded to the muscle relaxant but not to the analgesic?

b. What percent of the athletes who were treated did not respond to either form of treatment?

20. Gratuities The management of a hotel conducted a survey. It found that of the 2560 guests who were surveyed,

1785 tip the wait staff.
1219 tip the luggage handlers.
831 tip the maids.
275 tip the maids and the luggage handlers.
700 tip the wait staff and the maids.
755 tip the wait staff and the luggage handlers.
245 tip all three services.
210 do not tip these services.

How many of the surveyed guests tip

a. exactly two of the three services?

b. only the wait staff?

c. only one of the three services?

21. Advertising A computer company advertises its computers in *PC World*, in *PC Magazine*, and on television. A survey of 770 customers finds that the numbers of custom-

ers who are familiar with the company's computers because of the different forms of advertising are as follows.

305, *PC World*
290, *PC Magazine*
390, television
110, *PC World* and *PC Magazine*
135, *PC Magazine* and television
150, *PC World* and television
85, all three sources

How many of the surveyed customers know about the computers because of

a. exactly one of these forms of advertising?

b. exactly two of these forms of advertising?

c. *PC World* magazine and neither of the other two forms of advertising?

22. Blood Types During one month, a blood donation center found that

45.8% of the donors had the A antigen.
14.2% of the donors had the B antigen.
4.1% of the donors had the A antigen and the B antigen.
84.7% of the donors were Rh+.
87.2% of the donors had the B antigen or were Rh+.

Find the percent of the donors that

a. have the A antigen or the B antigen.

b. have the B antigen and are Rh+.

23. Gun Ownership A special interest group has conducted a survey concerning a ban on handguns. *Note:* A rifle is a gun, but it is not a handgun. The survey yielded the following results for the 1000 households that responded.

271 own a handgun.
437 own a rifle.
497 supported the ban on handguns.
140 own both a handgun and a rifle.
202 own a rifle but no handgun and do not support the ban on handguns.
74 own a handgun and support the ban on handguns.
52 own both a handgun and a rifle and also support the ban on handguns.

How many of the surveyed households

a. only own a handgun and do not support the ban on handguns?

b. do not own a gun and support the ban on handguns?

c. do not own a gun and do not support the ban on handguns?

24. **Acquisition of Music**   A survey of college students was taken to determine how the students acquired music. The survey showed the following results.

365 students acquired music from CDs.
298 students acquired music from the Internet.
268 students acquired music from DVDs.
212 students acquired music from both CDs and DVDs.
155 students acquired music from both CDs and the Internet.
36 students acquired music from DVDs, but not from CDs or the Internet.
98 students acquired music from CDs, DVDs, and the Internet.

Of those surveyed,

a. how many acquired music from CDs, but not from the Internet or DVDs?

b. how many acquired music from the Internet, but not from CDs or DVDs?

c. how many acquired music from CDs or the Internet?

d. how many acquired music from the Internet and DVDs?

25. **Diets**   A survey was completed by individuals who were currently on the Zone diet (Z), the South Beach diet (S), or the Weight Watchers diet (W). All persons surveyed were also asked whether they were currently in an exercise program (E), taking diet pills (P), or under medical supervision (M). The following table shows the results of the survey.

| | | Supplements | | | |
		E	P	M	Totals
Diet	Z	124	82	65	271
	S	101	66	51	218
	W	133	41	48	222
	Totals	358	189	164	711

Find the number of surveyed people in each of the following sets.

a. $S \cap E$

b. $Z \cup M$

c. $S' \cap (E \cup P)$

d. $(Z \cup S) \cap (M')$

e. $W' \cap (P \cup M)'$

f. $W' \cup P$

26. **Financial Assistance**   A college study categorized its seniors (S), juniors (J), and sophomores (M) who are currently receiving financial assistance. The types of financial assistance consist of full scholarships (F), partial scholarships (P), and government loans (G). The following table shows the results of the survey.

| | | Financial assistance | | | |
		F	P	G	Totals
Year	S	210	175	190	575
	J	180	162	110	452
	M	114	126	86	326
	Totals	504	463	386	1353

Find the number of students who are currently receiving financial assistance in each of the following sets.

a. $S \cap P$

b. $J \cup G$

c. $M \cup F'$

d. $S \cap (F \cup P)$

e. $J \cap (F \cup P)'$

f. $(S \cup J) \cap (F \cup P)$

EXTENSIONS

Critical Thinking

27. Given that set A has 47 elements and set B has 25 elements, determine each of the following.

  a. The maximum possible number of elements in $A \cup B$

  b. The minimum possible number of elements in $A \cup B$

  c. The maximum possible number of elements in $A \cap B$

  d. The minimum possible number of elements in $A \cap B$

28. Given that set A has 16 elements, set B has 12 elements, and set C has 7 elements, determine each of the following.

  a. The maximum possible number of elements in $A \cup B \cup C$

  b. The minimum possible number of elements in $A \cup B \cup C$

  c. The maximum possible number of elements in $A \cap (B \cup C)$

  d. The minimum possible number of elements in $A \cap (B \cup C)$

Cooperative Learning

29. Search Engines   The following Venn diagram displays U parceled into 16 distinct regions by four sets.

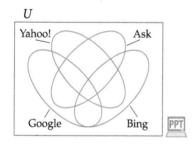

Use the preceding Venn diagram and the information below to answer the questions that follow.

A survey of 1250 Internet users shows the following results concerning the use of the search engines Google, Bing, Yahoo!, and Ask.

  585 use Google.
  620 use Yahoo!.
  560 use Ask.
  450 use Bing.
  100 use only Google, Yahoo!, and Ask.
  41 use only Google, Yahoo!, and Bing.
  50 use only Google, Ask, and Bing.
  80 use only Yahoo!, Ask, and Bing.
  55 use only Google and Yahoo!.
  34 use only Google and Ask.
  45 use only Google and Bing.
  50 use only Yahoo! and Ask.
  30 use only Yahoo! and Bing.
  45 use only Ask and Bing.
  60 use all four.

How many of the Internet users

  a. use only Google?

  b. use exactly three of the four search engines?

  c. do not use any of the four search engines?

Explorations

30. An Inclusion-Exclusion Formula for Three Sets   Exactly one of the following equations is a valid inclusion-exclusion formula for the union of three finite sets. Which equation do you think is the valid formula? *Hint:* Use the data in Example 2 on page 87 to check your choice.

  a. $n(A \cup B \cup C) = n(A) + n(B) + n(C)$

  b. $n(A \cup B \cup C) = n(A) + n(B) + n(C) - n(A \cap B \cap C)$

  c. $n(A \cup B \cup C) = n(A) + n(B) + n(C) - n(A \cap B) - n(A \cap C) - n(B \cap C)$

  d. $n(A \cup B \cup C) = n(A) + n(B) + n(C) - n(A \cap B) - n(A \cap C) - n(B \cap C) + n(A \cap B \cap C)$

section **2.5** Infinite Sets

One-to-One Correspondences

Much of Georg Cantor's work with sets concerned infinite sets. Some of Cantor's work with infinite sets was so revolutionary that it was not readily accepted by his contemporaries. Today, however, his work is generally accepted, and it provides unifying ideas in several diverse areas of mathematics.

Much of Cantor's set theory is based on the simple concept of a *one-to-one correspondence*.

▼ One-to-One Correspondence

A **one-to-one correspondence** (or 1–1 correspondence) between two sets A and B is a rule or procedure that pairs each element of A with exactly one element of B and each element of B with exactly one element of A.

Many practical problems can be solved by applying the concept of a one-to-one correspondence. For instance, consider a concert hall that has 890 seats. During a performance the manager of the concert hall observes that every person occupies exactly one seat and that every seat is occupied. Thus, without doing any counting, the manager knows that there are 890 people in attendance. During a different performance the manager notes that all but six seats are filled, and thus there are $890 - 6 = 884$ people in attendance.

Recall that two sets are equivalent if and only if they have the same number of elements. One method of showing that two sets are equivalent is to establish a one-to-one correspondence between the elements of the sets.

▼ One-to-One Correspondence and Equivalent Sets

Two sets A and B are equivalent, denoted by $A \sim B$, if and only if A and B can be placed in a one-to-one correspondence.

Set {a, b, c, d, e} is equivalent to set {1, 2, 3, 4, 5} because we can show that the elements of each set can be placed in a one-to-one correspondence. One method of establishing this one-to-one correspondence is shown in the following figure.

$$\{a, \quad b, \quad c, \quad d, \quad e\}$$
$$\updownarrow \quad \updownarrow \quad \updownarrow \quad \updownarrow \quad \updownarrow$$
$$\{1, \quad 2, \quad 3, \quad 4, \quad 5\}$$

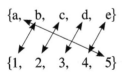

Each element of {a, b, c, d, e} has been paired with exactly one element of {1, 2, 3, 4, 5}, and each element of {1, 2, 3, 4, 5} has been paired with exactly one element of {a, b, c, d, e}. This is not the only one-to-one correspondence that we can establish. The figure at the left shows another one-to-one correspondence between the sets. In any case, we know that both sets have the same number of elements because we have established a one-to-one correspondence between the sets.

Sometimes a set is defined by including a general element. For instance, in the set {3, 6, 9, 12, 15, ... , 3n, ...}, the 3n (where n is a natural number) indicates that all the elements of the set are multiples of 3.

Some sets can be placed in a one-to-one correspondence with a proper subset of themselves. Example 1 illustrates this concept for the set of natural numbers.

▼ **example** **1** Establish a One-to-One Correspondence

Establish a one-to-one correspondence between the set of natural numbers $N = \{1, 2, 3, 4, 5, \ldots, n, \ldots\}$ and the set of even natural numbers $E = \{2, 4, 6, 8, 10, \ldots, 2n, \ldots\}$.

Solution

Write the sets so that one is aligned below the other. Draw arrows to show how you wish to pair the elements of each set. One possible method is shown in the following figure.

$$N = \{1, 2, 3, 4, \ldots, n, \ldots\}$$
$$\updownarrow \updownarrow \updownarrow \updownarrow \quad\quad \updownarrow$$
$$E = \{2, 4, 6, 8, \ldots, 2n, \ldots\}$$

In the above correspondence, each natural number $n \in N$ is paired with the even number $(2n) \in E$. The *general correspondence* $n \leftrightarrow (2n)$ enables us to determine exactly which element of E will be paired with any given element of N, and vice versa. For instance, under this correspondence, $19 \in N$ is paired with the even number $2 \cdot 19 = 38 \in E$, and $100 \in E$ is paired with the natural number $\frac{1}{2} \cdot 100 = 50 \in N$.

The general correspondence $n \leftrightarrow (2n)$ establishes a one-to-one correspondence between the sets.

▼ **check your progress** **1** Establish a one-to-one correspondence between the set of natural numbers $N = \{1, 2, 3, 4, 5, \ldots, n, \ldots\}$ and the set of odd natural numbers $D = \{1, 3, 5, 7, 9, \ldots, 2n - 1, \ldots\}$.

Solution *See page S7.* ◄

Infinite Sets

> ▼ **Infinite Set**
>
> A set is an **infinite set** if it can be placed in a one-to-one correspondence with a proper subset of itself.

We know that the set of natural numbers N is an infinite set because in Example 1 we were able to establish a one-to-one correspondence between the elements of N and the elements of one of its proper subsets, E.

question Can the set $\{1, 2, 3\}$ be placed in a one-to-one correspondence with one of its proper subsets?

▼ **example** **2** Verify That a Set Is an Infinite Set

Verify that $S = \{5, 10, 15, 20, \ldots, 5n, \ldots\}$ is an infinite set.

answer No. The set $\{1, 2, 3\}$ is a finite set with three elements. Every proper subset of $\{1, 2, 3\}$ has two or fewer elements.

The solution shown in Example 2 is not the only way to establish that S is an infinite set. For instance, $R = \{10, 15, 20, \ldots, 5n + 5, \ldots\}$ is also a proper set of S, and the sets S and R can be placed in a one-to-one correspondence as follows.

$$S = \{5, \quad 10, \quad 15, \quad 20, \ldots, \quad 5n, \quad \ldots\}$$

$$R = \{10, \quad 15, \quad 20, \quad 25, \ldots, \quad 5n + 5, \ldots\}$$

This one-to-one correspondence between S and one of its proper subsets R also establishes that S is an infinite set.

Solution

One proper subset of S is $T = \{10, 20, 30, 40, \ldots, 10n, \ldots\}$, which was produced by deleting the odd numbers in S. To establish a one-to-one correspondence between set S and set T, consider the following diagram.

$$S = \{5, \quad 10, \quad 15, \quad 20, \ldots, \quad 5n, \quad \ldots\}$$

$$T = \{10, \quad 20, \quad 30, \quad 40, \ldots, \quad 10n, \quad \ldots\}$$

In the above correspondence, each $(5n) \in S$ is paired with $(10n) \in T$. The *general correspondence* $(5n) \leftrightarrow (10n)$ establishes a one-to-one correspondence between S and one of its proper subsets, namely T. Thus S is an infinite set.

▼ check your progress 2   Verify that $V = \{40, 41, 42, 43, \ldots, 39 + n, \ldots\}$ is an infinite set.

Solution   *See page S7.*   ◄

The Cardinality of Infinite Sets

The symbol $\aleph_0$ is used to represent the cardinal number for the set N of natural numbers. $\aleph$ is the first letter of the Hebrew alphabet and is pronounced *aleph*. $\aleph_0$ is read as "aleph-null." Using mathematical notation, we write this concept as $n(N) = \aleph_0$. Since $\aleph_0$ represents a cardinality larger than any finite number, it is called a **transfinite number**. Many infinite sets have a cardinality of $\aleph_0$. In Example 3, for instance, we show that the cardinality of the set of integers is $\aleph_0$ by establishing a one-to-one correspondence between the elements of the set of integers and the elements of the set of natural numbers.

▼ example 3   Establish the Cardinality of the Set of Integers

Show that the set of integers $I = \{\ldots, -5, -4, -3, -2, -1, 0, 1, 2, 3, 4, 5, \ldots\}$ has a cardinality of $\aleph_0$.

Solution

First we try to establish a one-to-one correspondence between I and N, with the elements in each set arranged as shown below. No general method of pairing the elements of N with the elements of I seems to emerge from this figure.

$$N = \{1, 2, 3, 4, 5, 6, 7, 8, 9, 10, 11, \ldots\}$$
$$?$$
$$I = \{\ldots, -5, -4, -3, -2, -1, 0, 1, 2, 3, 4, 5, \ldots\}$$

If we arrange the elements of I as shown in the figure below, then *two* general correspondences, shown by the blue arrows and the red arrows, can be identified.

$$N = \{1, 2, \quad 3, \quad 4, \quad 5, \quad 6, \quad 7, \quad 8, \quad 9, \quad 10, 11, \ldots, \quad 2n - 1, 2n, \ldots\}$$

$$I = \{0, 1, \quad -1, 2, \quad -2, 3, \quad -3, 4, \quad -4, \quad 5, \quad -5, \ldots, \quad -n + 1, \quad n, \ldots\}$$

- Each even natural number $2n$ of N is paired with the integer n of I. This correspondence is shown by the blue arrows.

- Each odd natural number $2n - 1$ of N is paired with the integer $-n + 1$ of I. This correspondence is shown by the red arrows.

Together the two general correspondences $(2n) \leftrightarrow n$ and $(2n - 1) \leftrightarrow (-n + 1)$ establish a one-to-one correspondence between the elements of I and the elements of N. Thus the cardinality of the set of integers must be the same as the cardinality of the set of natural numbers, which is $\aleph_0$.

▼ check your progress 3 Show that $M = \{\frac{1}{2}, \frac{1}{3}, \frac{1}{4}, \frac{1}{5}, ..., \frac{1}{n + 1}, ...\}$ has a cardinality of $\aleph_0$.

Solution   *See page S7.*

Cantor was also able to show that the set of positive rational numbers is equivalent to the set of natural numbers. Recall that a rational number is a number that can be written as a fraction $\frac{p}{q}$, where p and q are integers and $q \neq 0$. Cantor's proof used an array of rational numbers similar to the array shown below.

Theorem   The set $Q+$ of positive rational numbers is equivalent to the set N of natural numbers.

Proof   Consider the following array of positive rational numbers.

TAKE NOTE

The rational number $\frac{2}{2}$ is not listed in the second row because $\frac{2}{2} = 1 = \frac{1}{1}$, which is already listed in the first row.

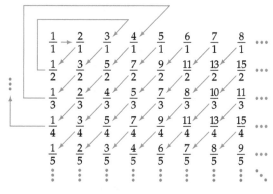

An array of all the positive rational numbers

The first row of the above array contains, in order from smallest to largest, all the positive rational numbers which, *when expressed in lowest terms,* have a denominator of 1. The second row contains the positive rational numbers which, *when expressed in lowest terms,* have a denominator of 2. The third row contains the positive rational numbers which, *when expressed in lowest terms,* have a denominator of 3. This process continues indefinitely.

Cantor reasoned that every positive rational number appears once and only once in this array. Note that $\frac{3}{5}$ appears in the fifth row. In general, if $\frac{p}{q}$ is in lowest terms, then it appears in row q.

At this point, Cantor used a numbering procedure that establishes a one-to-one correspondence between the natural numbers and the positive rational numbers in the array. The numbering procedure starts in the upper left corner with $\frac{1}{1}$. Cantor considered this to be the first number in the array, so he assigned the natural number 1 to this rational number. He then moved to the right and assigned the natural number 2 to the rational number $\frac{2}{1}$. From this point on, he followed the diagonal paths shown by the arrows and assigned each number he encountered to the next consecutive natural number. When he reached the bottom of a diagonal, he moved up to the top of the array and continued to number the rational numbers in the next diagonal. The following table shows the first 10 rational numbers Cantor numbered using this scheme.

Rational number in the array	$\frac{1}{1}$	$\frac{2}{1}$	$\frac{1}{2}$	$\frac{3}{1}$	$\frac{3}{2}$	$\frac{1}{3}$	$\frac{4}{1}$	$\frac{5}{2}$	$\frac{2}{3}$	$\frac{1}{4}$
Corresponding natural number	1	2	3	4	5	6	7	8	9	10

This numbering procedure shows that each element of $Q+$ can be paired with exactly one element of N, and each element of N can be paired with exactly one element of $Q+$. Thus $Q+$ and N are equivalent sets.

The negative rational numbers $Q-$ can also be placed in a one-to-one correspondence with the set of natural numbers in a similar manner.

question Using Cantor's numbering scheme, which rational numbers in the array shown on page 99 would be assigned the natural numbers 11, 12, 13, 14, and 15?

▼ Countable Set

A set is a **countable set** if and only if it is a finite set or an infinite set that is equivalent to the set of natural numbers.

Every infinite set that is countable has a cardinality of $\aleph_0$. Every infinite set that we have considered up to this point is countable. You might think that all infinite sets are countable; however, Cantor was able to show that this is not the case. Consider, for example, $A = \{x \mid x \in R \text{ and } 0 < x < 1\}$. To show that A is *not* a countable set, we use a *proof by contradiction*, where we assume that A is countable and then proceed until we arrive at a contradiction.

To better understand the concept of a proof by contradiction, consider the situation in which you are at a point where a road splits into two roads. See the figure at the left. Assume you know that only one of the two roads leads to your desired destination. If you can show that one of the roads cannot get you to your destination, then you know, without ever traveling down the other road, that it is the road that leads to your destination. In the following proof, we know that either set A is a countable set or set A is not a countable set. To establish that A is *not* countable, we show that the assumption that A is countable leads to a contradiction. In other words, our assumption that A is countable must be incorrect, and we are forced to conclude that A is not countable.

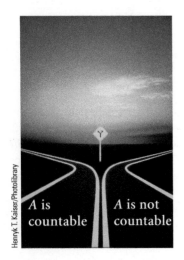

A is countable   A is not countable

Theorem The set $A = \{x \mid x \in R \text{ and } 0 < x < 1\}$ is not a countable set.

Proof by contradiction Either A is countable or A is not countable. Assume A is countable. Then we can place the elements of A, which we will represent by $a_1, a_2, a_3, a_4, \ldots$, in a one-to-one correspondence with the elements of the natural numbers as shown below.

$$N = \{1, 2, 3, 4, \ldots, n, \ldots\}$$

$$A = \{a_1, a_2, a_3, a_4, \ldots, a_n, \ldots\}$$

answer The rational numbers in the next diagonal, namely $\frac{5}{1}, \frac{7}{2}, \frac{4}{3}, \frac{3}{4}$, and $\frac{1}{5}$, would be assigned to the natural numbers 11, 12, 13, 14, and 15, respectively.

For example, the numbers $a_1, a_2, a_3, a_4, \ldots, a_n, \ldots$ could be as shown below.

$$1 \leftrightarrow a_1 = 0 . \boxed{3}\, 5\, 7\, 3\, 4\, 8\, 5 \ldots$$
$$2 \leftrightarrow a_2 = 0 . 0\, \boxed{6}\, 5\, 2\, 8\, 9\, 1 \ldots$$
$$3 \leftrightarrow a_3 = 0 . 6\, 8\, \boxed{2}\, 3\, 5\, 1\, 4 \ldots$$
$$4 \leftrightarrow a_4 = 0 . 0\, 5\, 0\, \boxed{0}\, 3\, 1\, 0 \ldots$$
$$\vdots$$
$$n \leftrightarrow a_n = 0 . 3\, 1\, 5\, 5\, 7\, 2\, 8 \ldots \boxed{5} \ldots$$
$$\vdots$$

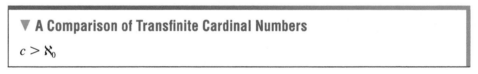
*n*th decimal digit of a_n

At this point we use a "diagonal technique" to construct a real number *d* that is greater than 0 and less than 1 and is not in the above list. We construct *d* by writing a decimal that *differs* from a_1 in the first decimal place, differs from a_2 in the second decimal place, differs from a_3 in the third decimal place, and, in general, differs from a_n in the *n*th decimal place. For instance, in the above list, a_1 has 3 as its first decimal digit. The first decimal digit of *d* can be any digit other than 3, say 4. The real number a_2 has 6 as its second decimal digit. The second decimal digit of *d* can be any digit other than 6, say 7. The real number a_3 has 2 as its third decimal digit. The third decimal digit of *d* can be any digit other than 2, say 3. Continue in this manner to determine the decimal digits of *d*. Now $d = 0.473\ldots$ must be in *A* because $0 < d < 1$. However, *d* is not in *A*, because *d* differs from each of the numbers in *A* in at least one decimal place.

We have reached a contradiction. Our assumption that the elements of *A* could be placed in a one-to-one correspondence with the elements of the natural numbers must be false. Thus *A* is not a countable set.

An infinite set that is not countable is said to be **uncountable**. Because the set $A = \{x \,|\, x \in R \text{ and } 0 < x < 1\}$ is uncountable, the cardinality of *A* is not $\aleph_0$. Cantor used the letter *c*, which is the first letter of the word *continuum*, to represent the cardinality of *A*. Cantor was also able to show that set *A* is equivalent to the set of all real numbers *R*. Thus the cardinality of *R* is also *c*. Cantor was able to prove that $c > \aleph_0$.

> **▼ A Comparison of Transfinite Cardinal Numbers**
>
> $c > \aleph_0$

Up to this point, all of the infinite sets we have considered have a cardinality of either $\aleph_0$ or *c*. The following table lists several infinite sets and the transfinite cardinal number that is associated with each set.

The Cardinality of Some Infinite Sets

Set	Cardinal number
Natural numbers, *N*	$\aleph_0$
Integers, *I*	$\aleph_0$
Rational numbers, *Q*	$\aleph_0$
Irrational numbers, $\mathcal{I}$	*c*
Any set of the form $\{x \,|\, a \le x \le b\}$, where *a* and *b* are real numbers such that $a < b$	*c*
Real numbers, *R*	*c*

Your intuition may suggest that $\aleph_0$ and c are the only two cardinal numbers associated with infinite sets; however, this is not the case. In fact, Cantor was able to show that no matter how large the cardinal number of a set, we can find a set that has a larger cardinal number. Thus there are infinitely many transfinite numbers. Cantor's proof of this concept is now known as *Cantor's theorem*.

▼ Cantor's Theorem

Let S be any set. The set of all subsets of S has a cardinal number that is larger than the cardinal number of S.

The set of all subsets of S is called the **power set** of S and is denoted by $P(S)$. We can see that Cantor's theorem is true for the finite set $S = \{a, b, c\}$ because the cardinality of S is 3 and S has $2^3 = 8$ subsets. The interesting part of Cantor's theorem is that it also applies to infinite sets.

Some of the following theorems can be established by using the techniques illustrated in the Excursion that follows.

▼ Transfinite Arithmetic Theorems

- For any whole number a, $\aleph_0 + a = \aleph_0$ and $\aleph_0 - a = \aleph_0$.

- $\aleph_0 + \aleph_0 = \aleph_0$ and, in general, $\underbrace{\aleph_0 + \aleph_0 + \aleph_0 + \cdots + \aleph_0}_{\text{a finite number of aleph nulls}} = \aleph_0$.

- $c + c = c$ and, in general, $\underbrace{c + c + c + \cdots + c}_{\text{a finite number of c's}} = c$.

- $\aleph_0 + c = c$.

- $\aleph_0 c = c$.

MATHMATTERS   Criticism and Praise of Cantor's Work

Georg Cantor's work in the area of infinite sets was not well received by some of his colleagues. For instance, the mathematician Leopold Kronecker tried to stop the publication of some of Cantor's work. He felt many of Cantor's theorems were ridiculous and asked, "How can one infinity be greater than another?" The following quote illustrates that Cantor was aware that his work would attract harsh criticism.

> ... I realize that in this undertaking I place myself in a certain opposition to views widely held concerning the mathematical infinite and to opinions frequently defended on the nature of numbers.[2]

A few mathematicians were willing to show support for Cantor's work. For instance, the famous mathematician David Hilbert stated that Cantor's work was

> ... the finest product of mathematical genius and one of the supreme achievements of purely intellectual human activity.[3]

---

[2] *Source:* http://www-groups.dcs.st-and.ac.uk/%7Ehistory/Biographies/Cantor.html
[3] See note 2.

Photolibrary

EXCURSION

Transfinite Arithmetic

Disjoint sets are often used to explain addition. The sum $4 + 3$, for example, can be deduced by selecting two disjoint sets, one with exactly four elements and one with exactly three elements. See the Venn diagram at the left. Now form the union of the two sets. The union of the two sets has exactly seven elements; thus $4 + 3 = 7$. In mathematical notation, we write

$$n(A) + n(B) = n(A \cup B)$$
$$4 \ + \ 3 \ = \ \ \ 7$$

Cantor extended this idea to infinite sets. He reasoned that the sum $\aleph_0 + 1$ could be determined by selecting two disjoint sets, one with cardinality of $\aleph_0$ and one with cardinality of 1. In this case the set N of natural numbers and the set $Z = \{0\}$ are appropriate choices. Thus

$$n(N) + n(Z) = n(N \cup Z)$$
$$= n(W) \qquad \bullet \ W \text{ represents the set of whole numbers}$$
$$\aleph_0 \ + \ 1 \ = \ \ \ \aleph_0$$

and, in general, for any whole number a, $\aleph_0 + a = \aleph_0$.

To find the sum $\aleph_0 + \aleph_0$, use two disjoint sets, each with cardinality of $\aleph_0$. The set E of even natural numbers and the set D of odd natural numbers satisfy the necessary conditions. Since E and D are disjoint sets, we know

$$n(E) + n(D) = n(E \cup D)$$
$$= n(W)$$
$$\aleph_0 \ + \ \aleph_0 \ = \ \aleph_0$$

Thus $\aleph_0 + \aleph_0 = \aleph_0$ and, in general,

$$\underbrace{\aleph_0 + \aleph_0 + \aleph_0 + \cdots + \aleph_0}_{\text{a finite number of aleph-nulls}} = \aleph_0$$

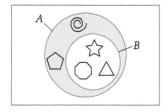

To determine a difference such as $5 - 3$ using sets, we first select a set A that has exactly five elements. We then find a subset B of this set that has exactly three elements. The difference $5 - 3$ is the cardinal number of the set $A \cap B'$, which is shown in blue in the figure at the left.

To determine $\aleph_0 - 3$, select a set with $\aleph_0$ elements, such as N, and then select a subset of this set that has exactly three elements. One such subset is $C = \{1, 2, 3\}$. The difference $\aleph_0 - 3$ is the cardinal number of the set $N \cap C' = \{4, 5, 6, 7, 8, ...\}$. Since $N \cap C'$ is a countably infinite set, we can conclude that $\aleph_0 - 3 = \aleph_0$. This procedure can be generalized to show that for any whole number a, $\aleph_0 - a = \aleph_0$.

EXCURSION EXERCISES

1. Use two disjoint sets to show that $\aleph_0 + 2 = \aleph_0$.

2. Use two disjoint sets other than the set of even natural numbers and the set of odd natural numbers to show that $\aleph_0 + \aleph_0 = \aleph_0$.

3. Use sets to show that $\aleph_0 - 6 = \aleph_0$.

4. **a.** Find two sets that can be used to show that $\aleph_0 - \aleph_0 = \aleph_0$. Now find another two sets that can be used to show that $\aleph_0 - \aleph_0 = 1$.

   b. Use the results of Excursion Exercise 4a to explain why subtraction of two transfinite numbers is an undefined operation.

EXERCISE SET 2.5

(Suggested Assignment: The Enhanced WebAssign Exercises and Exercises 1, 11, 29, and 30)

1. **a.** Use arrows to establish a one-to-one correspondence between $V = \{a, e, i\}$ and $M = \{3, 6, 9\}$.

   b. How many different one-to-one correspondences between V and M can be established?

2. Establish a one-to-one correspondence between the set of natural numbers $N = \{1, 2, 3, 4, 5, ..., n, ...\}$ and $F = \{5, 10, 15, 20, ..., 5n, ...\}$ by stating a general rule that can be used to pair the elements of the sets.

3. Establish a one-to-one correspondence between $D = \{1, 3, 5, ..., 2n - 1, ...\}$ and $M = \{3, 6, 9, ..., 3n, ...\}$ by stating a general rule that can be used to pair the elements of the sets.

■ In Exercises 4 to 10, state the cardinality of each set.

4. $\{2, 11, 19, 31\}$

5. $\{2, 9, 16, ..., 7n - 5, ...\}$, where n is a natural number

6. The set Q of rational numbers

7. The set R of real numbers

8. The set $\mathscr{I}$ of irrational numbers

9. $\{x \mid 5 \leq x \leq 9\}$

10. The set of subsets of $\{1, 5, 9, 11\}$

■ In Exercises 11 to 14, determine whether the given sets are equivalent.

11. The set of natural numbers and the set of integers

12. The set of whole numbers and the set of real numbers

13. The set of rational numbers and the set of integers

14. The set of rational numbers and the set of real numbers

■ In Exercises 15 to 18, show that the given set is an infinite set by placing it in a one-to-one correspondence with a proper subset of itself.

15. $A = \{5, 10, 15, 20, 25, 30, ..., 5n, ...\}$

16. $B = \{11, 15, 19, 23, 27, 31, ..., 4n + 7, ...\}$

17. $C = \left\{\dfrac{1}{2}, \dfrac{3}{4}, \dfrac{5}{6}, \dfrac{7}{8}, \dfrac{9}{10}, ..., \dfrac{2n-1}{2n}, ...\right\}$

18. $D = \left\{\dfrac{1}{2}, \dfrac{1}{3}, \dfrac{1}{4}, \dfrac{1}{5}, \dfrac{1}{6}, ..., \dfrac{1}{n+1}, ...\right\}$

■ In Exercises 19 to 26, show that the given set has a cardinality of $\aleph_0$ by establishing a one-to-one correspondence between the elements of the given set and the elements of N.

19. $\{50, 51, 52, 53, ..., n + 49, ...\}$

20. $\{10, 5, 0, -5, -10, -15, ..., -5n + 15, ...\}$

21. $\left\{1, \dfrac{1}{3}, \dfrac{1}{9}, \dfrac{1}{27}, ..., \dfrac{1}{3^{n-1}}, ...\right\}$

22. $\{-12, -18, -24, -30, ..., -6n - 6, ...\}$

23. $\{10, 100, 1000, ..., 10^n, ...\}$

24. $\left\{1, \dfrac{1}{2}, \dfrac{1}{4}, \dfrac{1}{8}, ..., \dfrac{1}{2^{n-1}}, ...\right\}$

25. $\{1, 8, 27, 64, ..., n^3, ...\}$

26. $\{0.1, 0.01, 0.001, 0.0001, ..., 10^{-n}, ...\}$

EXTENSIONS

Critical Thinking

27. **a.** Place the set $M = \{3, 6, 9, 12, 15, ...\}$ of positive multiples of 3 in a one-to-one correspondence with the set K of all natural numbers that are not multiples of 3. Write a sentence or two that explains the general rule you used to establish the one-to-one correspondence.

   b. Use your rule to determine what number from K is paired with the number 606 from M.

   c. Use your rule to determine what number from M is paired with the number 899 from K.

In the figure below, every point on line segment AB corresponds to a real number from 0 to 1 and every real number from 0 to 1 corresponds to a point on line segment AB.

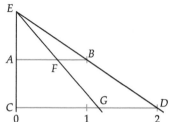

The line segment CD represents the real numbers from 0 to 2. Note that any point F on line segment AB can be paired with a unique point G on line segment CD by drawing a line from E through F. Also, any arbitrary point G on line segment CD can be paired with a unique point F on line segment AB by drawing the line EG. This geometric procedure establishes a one-to-one correspondence between the set $\{x \mid 0 \le x \le 1\}$ and the set $\{x \mid 0 \le x \le 2\}$. Thus $\{x \mid 0 \le x \le 1\} \sim \{x \mid 0 \le x \le 2\}$.

28. Draw a figure that can be used to verify each of the following.

a. $\{x \mid 0 \le x \le 1\} \sim \{x \mid 0 \le x \le 5\}$

b. $\{x \mid 2 \le x \le 5\} \sim \{x \mid 1 \le x \le 8\}$

29. Consider the semicircle with arc length π and center C and the line L_1 in the following figure. Each point on the semicircle, *other than the endpoints,* represents a unique real number between 0 and π. Each point on line L_1 represents a unique real number.

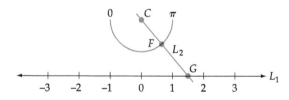

Any line through C that intersects the semicircle at a point other than one of its endpoints will intersect line L_1 at a unique point. Also, any line through C that intersects line L_1 will intersect the semicircle at a unique point that is not an endpoint of the semicircle. What can we conclude from this correspondence?

30. Explain how to use the figure below to verify that the set of all points on the circle is equivalent to the set of all points on the square.

Explorations

31. The Hilbert Hotel   The Hilbert Hotel is an imaginary hotel created by the mathematician David Hilbert (1862–1943). The hotel has an infinite number of rooms. Each room is numbered with a natural number—room 1, room 2, room 3, and so on. Search the Internet for information on Hilbert's Hotel. Write a few paragraphs that explain some of the interesting questions that arise when guests arrive to stay at the hotel.

Mary Pat Campbell has written a song about a hotel with an infinite number of rooms. Her song is titled *Hotel Aleph Null—yeah.* Here are the lyrics for the chorus of her song, which is to be sung to the tune of *Hotel California* by the Eagles (*Source:* http://www.marypat.org/mathcamp/doc2001/hellrelays.html#hotel).[4]

Hotel Aleph Null—yeah
Welcome to the Hotel Aleph Null—yeah
What a lovely place (what a lovely place)
Got a lot of space
Packin' em in at the Hotel Aleph Null—yeah
Any time of year
You can find space here

32. The Continuum Hypothesis   Cantor conjectured that no set can have a cardinality larger than $\aleph_0$ but smaller than c. This conjecture has become known as the *Continuum Hypothesis.* Search the Internet for information on the Continuum Hypothesis and write a short report that explains how the Continuum Hypothesis was resolved.

[4] Reprinted by permission of Mary Pat Campbell.

SUMMARY

The following table summarizes essential concepts in this chapter. The references given in the right-hand column list Examples and Exercises that can be used to test your understanding of a concept.

2.1 Basic Properties of Sets

The Roster Method The roster method is used to represent a set by listing each element of the set inside a pair of braces. Commas are used to separate the elements.	See **Example 1** on page 52, and then try Exercises 1 and 2 on page 108.
Basic Number Sets Natural Numbers or Counting Numbers $N = \{1, 2, 3, 4, 5, ...\}$ Whole Numbers $W = \{0, 1, 2, 3, 4, 5, ...\}$ Integers $I = \{..., -4, -3, -2, -1, 0, 1, 2, 3, 4, ...\}$ Rational Numbers $Q =$ the set of all terminating or repeating decimals Irrational Numbers $\mathcal{I} =$ the set of all nonterminating, nonrepeating decimals Real Numbers $R =$ the set of all rational or irrational numbers	See **Example 3** and **Check Your Progress 3** on pages 53 and 54, and then try Exercises 3 to 6 on page 108.
Set-Builder Notation Set-builder notation is used to represent a set, by describing its elements.	See **Example 5** on page 55, and then try Exercises 7 to 10 on page 108.
Cardinal Number of a Finite Set The cardinal number of a finite set is the number of elements in the set. The cardinal number of a finite set A is denoted by the notation $n(A)$.	See **Example 6** on page 55, and then try Exercises 63 to 67 on page 109.
Equal Sets and Equivalent Sets Two sets are equal if and only if they have exactly the same elements. Two sets are equivalent if and only if they have the same number of elements.	See **Example 7** on page 56, and then try Exercises 11 and 12 on page 108.

2.2 Complements, Subsets, and Venn Diagrams

The Universal Set and the Complement of a Set The universal set, denoted by U, is the set of all elements that are under consideration. The complement of set A, denoted by A', is the set of all elements of the universal set that are not elements of A.	See **Example 1** on page 62, and then try Exercises 22 to 24 on page 108.
Subset of a Set Set A is a subset of set B, denoted by $A \subseteq B$ if and only if every element of A is also an element of B.	See **Example 2** on page 63, and then try Exercises 25 and 26 on page 108.
Proper Subset of a Set Set A is a proper subset of set B, denoted by $A \subset B$, if every element of A is an element of B and $A \neq B$.	See **Example 3** on page 65, and then try Exercises 27 to 30 on page 108.
The Number of Subsets of a Set A set with n elements has 2^n subsets.	See **Example 5** on page 66, and then try Exercise 38 on page 108.

2.3 Set Operations

Intersection of Sets The intersection of sets A and B, denoted by $A \cap B$, is the set of elements common to both A and B.

$$A \cap B = \{x \mid x \in A \quad \text{and} \quad x \in B\}$$

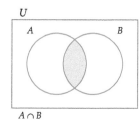

See **Example 1** on page 72, and then try Exercises 17 and 19 on page 108.

Union of Sets The union of sets A and B, denoted by $A \cup B$, is the set that contains all the elements that belong to A or to B or to both.

$$A \cup B = \{x \mid x \in A \quad \text{or} \quad x \in B\}$$

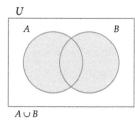

See **Example 2** on page 72, and then try Exercises 18 and 20 on page 108.

De Morgan's Laws For all sets A and B,

$$(A \cup B)' = A' \cap B' \quad \text{and} \quad (A \cap B)' = A' \cup B'$$

See **Example 4** and **Check Your Progress 4** on pages 73 and 74, and then try Exercises 39 and 40 on page 108.

Venn Diagrams and the Equality of Set Expressions Two sets are equal if and only if they each represent the same region(s) on a Venn diagram. Venn diagrams can be used to verify each of the following properties.

For all sets A, B, and C:

Commutative Properties
$A \cap B = B \cap A$
$A \cup B = B \cup A$

Associative Properties
$(A \cap B) \cap C = A \cap (B \cap C)$
$(A \cup B) \cup C = A \cup (B \cup C)$

Distributive Properties
$A \cap (B \cup C) = (A \cap B) \cup (A \cap C)$
$A \cup (B \cap C) = (A \cup B) \cap (A \cup C)$

See **Example 6** and **Check Your Progress 6** on pages 76 and 77, and then try Exercises 45 to 48 on pages 108 and 109.

2.4 Applications of Sets

Applications Many counting problems that arise in applications involving surveys can be solved by using sets and Venn diagrams.

See **Examples 1 and 2** on pages 85 and 87, and then try Exercises 53 and 54 on page 109.

The Inclusion-Exculsion Formula For all finite sets A and B,

$$n(A \cup B) = n(A) + n(B) - n(A \cap B)$$

See **Examples 3 and 4** on pages 88 and 89, and then try Exercises 55 and 56 on page 109.

2.5 Infinite Sets

One-to-One Correspondence and Equivalent Sets Two sets A and B are equivalent, denoted by $A \sim B$, if and only if A and B can be placed in a one-to-one correspondence.	See **Examples 1 and 3** on pages 97 and 98, and then try Exercises 57 to 60 on page 109.
Infinite Set A set is an infinite set if it can be placed in a one-to-one correspondence with a proper subset of itself.	See **Example 2** on page 97, and then try Exercises 61 and 62 on page 109.

CHAPTER 2 REVIEW EXERCISES

■ In Exercises 1 to 6, use the roster method to represent each set.

1. The set of months of the year with a name that starts with the letter J

2. The set of states in the United States that do not share a common border with another state

3. The set of whole numbers less than 8

4. The set of integers that satisfy $x^2 = 64$

5. The set of natural numbers that satisfy $x + 3 \le 7$

6. The set of counting numbers larger than -3 and less than or equal to 6

■ In Exercises 7 to 10, use set-builder notation to write each set.

7. The set of integers greater than -6

8. {April, June, September, November}

9. {Kansas, Kentucky}

10. {1, 8, 27, 64, 125}

■ In Exercises 11 and 12, state whether each of the following sets are equal, equivalent, both, or neither.

11. {2, 4, 6, 8}, {$x \mid x \in N$ and $x < 5$}

12. {8, 9}, the set of single digit whole numbers greater than 7

■ In Exercises 13 to 16, determine whether the statement is true or false.

13. {3} $\in$ {1, 2, 3, 4}

14. $-11 \in I$

15. {a, b, c} $\sim$ {1, 5, 9}

16. The set of small numbers is a well-defined set

■ In Exercises 17 to 24, let U = {2, 6, 8, 10, 12, 14, 16, 18}, A = {2, 6, 10}, B = {6, 10, 16, 18}, and C = {14, 16}. Find each of the following.

17. $A \cap B$

18. $A \cup B$

19. $A' \cap C$

20. $B \cup C'$

21. $A \cup (B \cap C)$

22. $(A \cup C)' \cap B'$

23. $(A \cap B')'$

24. $(A \cup B \cup C)'$

■ In Exercises 25 and 26, determine whether the first set is a subset of the second set.

25. {0, 1, 5, 9}, the set of natural numbers

26. {1, 2, 4, 8, 9.5}, the set of integers

■ In Exercises 27 to 30, determine whether the first set is a proper subset of the second set.

27. The set of natural numbers; the set of whole numbers

28. The set of integers; the set of real numbers

29. The set of counting numbers; the set of natural numbers

30. The set of real numbers; the set of rational numbers

■ In Exercises 31 to 34, list all the subsets of the given set.

31. {I, II}

32. {s, u, n}

33. {penny, nickel, dime, quarter}

34. {A, B, C, D, E}

■ In Exercises 35 to 38, find the number of subsets of the given set.

35. The set of the four musketeers

36. The set of the letters of the English alphabet

37. The set of the letters of "uncopyrightable," which is the longest English word with no repeated letters

38. The set of the seven dwarfs

■ In Exercises 39 and 40, determine whether each statement is true or false for all sets A and B.

39. $(A \cup B')' = A' \cap B$

40. $(A' \cap B')' = A \cup B$

■ In Exercises 41 to 44, draw a Venn diagram to represent the given set.

41. $A \cap B'$

42. $A' \cup B'$

43. $(A \cup B) \cup C'$

44. $A \cap (B' \cup C)$

■ In Exercises 45 to 48, draw Venn diagrams to determine whether the expressions are equal for all sets A, B, and C.

45. $A' \cup (B \cup C)$; $(A' \cup B) \cup (A' \cup C)$

46. $(A \cap B) \cap C'$; $(A' \cup B') \cup C$

47. $A \cap (B' \cap C); (A \cup B') \cap (A \cup C)$

48. $A \cap (B \cup C); A' \cap (B \cup C)$

■ In Exercises 49 and 50, use set notation to describe the shaded region.

49. *U*

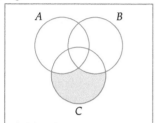

50. *U*

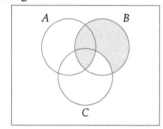

■ In Exercises 51 and 52, draw a Venn diagram with each of the given elements placed in the correct region.

51. $U = \{e, h, r, d, w, s, t\}$
$A = \{t, r, e\}$
$B = \{w, s, r, e\}$
$C' = \{s, r, d, h\}$

52. $U = \{\alpha, \beta, \Gamma, \gamma, \Delta, \delta, \varepsilon, \theta\}$
$A' = \{\beta, \Delta, \theta, \gamma\}$
$B = \{\delta, \varepsilon\}$
$C = \{\beta, \varepsilon, \Gamma\}$

53. An Exercise Survey   In a survey at a health club, 208 members indicated that they enjoy aerobic exercises, 145 indicated that they enjoy weight training, 97 indicated that they enjoy both aerobics and weight training, and 135 indicated that they do not enjoy either of these types of exercise. How many members were surveyed?

Tomasz Trojanowski/Shutterstock.com

54. A Coffee Survey   A gourmet coffee bar conducted a survey to determine the preferences of its customers. Of the customers surveyed,

221 like espresso.
127 like cappuccino and chocolate-flavored coffee.
182 like cappuccino.
136 like espresso and chocolate-flavored coffee.
209 like chocolate-flavored coffee.
96 like all three types of coffee.
116 like espresso and cappuccino.
82 like none of these types of coffee.

How many of the customers in the survey

a. like only chocolate-flavored coffee?

b. like cappuccino and chocolate-flavored coffee but not espresso?

c. like espresso and cappuccino but not chocolate-flavored coffee?

d. like exactly one of the three types of coffee?

■ In Exercises 55 and 56, use the inclusion-exclusion formula to answer each application.

55. On a football team, 27 of its athletes play on offense, 22 play on defense, and 43 play on offense or defense. How many of the athletes play both on offense and on defense?

56. A college finds that 625 of its students are registered in biology, 433 are registered in psychology, and 184 are registered in both biology and psychology. How many students are registered in biology or psychology?

■ In Exercises 57 to 60, establish a one-to-one correspondence between the sets.

57. $\{1, 3, 6, 10\}, \{1, 2, 3, 4\}$

58. $\{x \mid x > 10 \text{ and } x \in N\}, \{2, 4, 6, 8, \dots, 2n, \dots\}$

59. $\{3, 6, 9, 12, \dots, 3n, \dots\}, \{10, 100, 1000, \dots, 10^n, \dots\}$

60. $\{x \mid 0 \leq x \leq 1\}, \{x \mid 0 \leq x \leq 4\}$ (*Hint:* Use a drawing.)

■ In Exercises 61 and 62, show that the given set is an infinite set.

61. $A = \{6, 10, 14, 18, \dots, 4n + 2, \dots\}$

62. $B = \left\{1, \dfrac{1}{2}, \dfrac{1}{4}, \dfrac{1}{8}, \dots, \dfrac{1}{2^{n-1}}, \dots\right\}$

■ In Exercises 63 to 70, state the cardinality of each set.

63. $\{5, 6, 7, 8, 6\}$

64. $\{4, 6, 8, 10, 12, \dots, 22\}$

65. $\{0, \varnothing\}$

66. The set of all states in the United States that border the Gulf of Mexico

67. The set of integers less than 1,000,000

68. The set of rational numbers between 0 and 1

69. The set of irrational numbers

70. The set of real numbers between 0 and 1

■ In Exercises 71 and 72, find each of the following, where $\aleph_0$ and c are transfinite cardinal numbers.

71. $\aleph_0 + (\aleph_0 + \aleph_0)$

72. $c + (c + c)$

CHAPTER 2   TEST

■ In Exercises 1 to 4, let $U = \{1, 2, 3, 4, 5, 6, 7, 8, 9, 10\}$, $A = \{3, 5, 7, 8\}$, $B = \{2, 3, 8, 9, 10\}$, and $C = \{1, 4, 7, 8\}$. Use the roster method to write each of the following sets.

1. $(A \cap B)'$

2. $A' \cap B$

3. $A' \cup (B \cap C')$

4. $A \cap (B' \cup C)$

■ In Exercises 5 and 6, use set-builder notation to write each of the given sets.

5. $\{0, 1, 2, 3, 4, 5, 6\}$

6. $\{-3, -2, -1, 0, 1, 2\}$

7. State the cardinality of each set.

a. The set of whole numbers less than 4

b. The set of integers

■ In Exercises 8 and 9, state whether the given sets are equal, equivalent, both, or neither.

8. a. $\{j, k, l\}$, $\{a, e, i, o, u\}$

b. $\{x \mid x \in W \text{ and } x < 4\}$, $\{9, 10, 11, 12\}$

9. a. the set of natural numbers; the set of integers

b. the set of whole numbers; the set of positive integers

10. List all of the subsets of $\{a, b, c, d\}$.

11. Determine the number of subsets of a set with 21 elements.

■ In Exercises 12 and 13, draw a Venn diagram to represent the given set.

12. $(A \cup B') \cap C$

13. $(A' \cap B) \cup (A \cap C')$

14. Write $(A \cup B)'$ as the intersection of two sets.

15. Upgrade Options An automobile company makes a sedan with nine upgrade options.

a. How many different versions of this sedan can the company produce?

b. What is the minimum number of upgrade options the company must provide if it wishes to offer at least 2500 versions of this sedan?

16. Student Demographics A college finds that 841 of its students are receiving financial aid, 525 students are business majors, and 202 students are receiving financial aid and are business majors. How many students are receiving financial aid or are business majors?

17. A Survey In the town of LeMars, 385 families have a DVD player, 142 families have a Blu-ray player, 41 families have both a DVD player and a Blu-ray player, and 55 families do not have a DVD player or a Blu-ray player. How many families live in LeMars?

18. A Survey A survey of 1000 households was taken to determine how they obtained news about current events. The survey considered only television, newspapers, and the Internet as sources for news. Of the households surveyed,

724 obtained news from television.

545 obtained news from newspapers.

280 obtained news from the Internet.

412 obtained news from both television and newspapers.

185 obtained news from both television and the Internet.

105 obtained news from television, newspapers, and the Internet.

64 obtained news from the Internet but not from television or newspapers.

Of those households that were surveyed,

a. how many obtained news from television but not from newspapers or the Internet?

b. how many obtained news from newspapers but not from television or the Internet?

c. how many obtained news from television or newspapers?

d. how many did not acquire news from television, newspapers, or the Internet?

19. Show a method that can be used to establish a one-to-one correspondence between the elements of the following sets.

$$\{5, 10, 15, 20, 25, \ldots, 5n, \ldots\}, W$$

20. Prove that the following set is an infinite set by illustrating a one-to-one correspondence between the elements of the set and the elements of one of the set's proper subsets.

$$\{3, 6, 9, 12, \ldots, 3n, \ldots\}$$

3

Logic

In today's complex world, it is not easy to summarize in a few paragraphs the subject matter known as logic. For lawyers and judges, logic is the science of correct reasoning. They often use logic to communicate more effectively, construct valid arguments, analyze legal contracts, and make decisions. Law schools consider a knowledge of logic to be one of the most important predictors of future success for their new students. A sizeable portion of the LSAT (Law School Admission Test), which is required by law school applicants as part of their admission process, concerns logical reasoning. A typical LSAT logic problem is presented in Exercise 29, page 165.

LSAT
**The Law School
Admission Test**

Many other professions also make extensive use of logic. For instance, programmers use logic to design computer software, electrical engineers use logic to design circuits for smart phones, and mathematicians use logic to solve problems and construct mathematical proofs.

In this chapter, you will encounter several facets of logic. Specifically, you will use logic to

- analyze information and the relationship between statements,
- determine the validity of arguments,
- determine valid conclusions based on given assumptions, and
- analyze electronic circuits.

HISTORICAL NOTE

George Boole

(bool) was born in 1815 in Lincoln, England. He was raised in poverty, but he was very industrious and had learned Latin and Greek by the age of 12. Later he mastered German, French, and Italian. His first profession, at the young age of 16, was that of an assistant school teacher. At the age of 20 he started his own school.

In 1849 Boole was appointed the chairperson of mathematics at Queens College in Cork, Ireland.

Many of Boole's mathematical ideas, such as Boolean algebra, have applications in the areas of computer programming and the design of electronic circuits.

Logic Statements and Quantifiers

One of the first mathematicians to make a serious study of symbolic logic was Gottfried Wilhelm Leibniz (1646–1716). Leibniz tried to advance the study of logic from a merely philosophical subject to a formal mathematical subject. Leibniz never completely achieved this goal; however, several mathematicians, such as Augustus De Morgan (1806–1871) and George Boole (1815–1864), contributed to the advancement of symbolic logic as a mathematical discipline.

Boole published *The Mathematical Analysis of Logic* in 1848. In 1854 he published the more extensive work, *An Investigation of the Laws of Thought*. Concerning this document, the mathematician Bertrand Russell stated, "Pure mathematics was discovered by Boole in a work which is called *The Laws of Thought*."

Logic Statements

Every language contains different types of sentences, such as statements, questions, and commands. For instance,

"Is the test today?" is a question.

"Go get the newspaper" is a command.

"This is a nice car" is an opinion.

"Denver is the capital of Colorado" is a statement of fact.

The symbolic logic that Boole was instrumental in creating applies only to sentences that are *statements* as defined below.

▼ A Statement

A **statement** is a declarative sentence that is either true or false, but not both true and false.

It may not be necessary to determine whether a sentence is true to determine whether it is a statement. For instance, consider the following sentence.

American Shaun White won an Olympic gold medal in speed skating.

You may not know if the sentence is true, but you do know that the sentence is either true or it is false, and that it is not both true and false. Thus, you know that the sentence is a statement.

Shaun White

▼ example 1 Identify Statements

Determine whether each sentence is a statement.

a. Florida is a state in the United States.

b. How are you?

c. $9^9 + 2$ is a prime number.

d. $x + 1 = 5$.

Solution

a. Florida is one of the 50 states in the United States, so this sentence is true and it is a statement.

b. The sentence "How are you?" is a question; it is not a declarative sentence. Thus it is not a statement.

c. You may not know whether $9^9 + 2$ is a prime number; however, you do know that it is a whole number larger than 1, so it is either a prime number or it is not a prime number. The sentence is either true or it is false, and it is not both true and false, so it is a statement.

d. $x + 1 = 5$ is a statement. It is known as an *open statement*. It is true for $x = 4$, and it is false for any other values of x. For any given value of x, it is true or false but not both.

▼ **check your progress** **1** Determine whether each sentence is a statement.

a. Open the door.

b. 7055 is a large number.

c. In the year 2020, the president of the United States will be a woman.

d. $x > 3$.

Solution *See page S7.*

MATHMATTERS   Charles Dodgson

Charles Dodgson (Lewis Carroll)

One of the best-known logicians is Charles Dodgson (1832–1898). His mathematical works include *A Syllabus of Plane Algebraical Geometry, The Fifth Book of Euclid Treated Algebraically,* and *Symbolic Logic.* Although Dodgson was a distinguished mathematician in his time, he is best known by his pen name Lewis Carroll, which he used when he published *Alice's Adventures in Wonderland* and *Through the Looking-Glass.*

Queen Victoria of the United Kingdom enjoyed *Alice's Adventures in Wonderland* to the extent that she told Dodgson she was looking forward to reading another of his books. He promptly sent her his *Syllabus of Plane Algebraical Geometry,* and it was reported that she was less than enthusiastic about the latter book.

Simple Statements and Compound Statements

▼ **Simple Statements and Compound Statements**
A **simple statement** is a statement that conveys a single idea. A **compound statement** is a statement that conveys two or more ideas.

Connecting simple statements with words and phrases such as *and, or, if … then,* and *if and only if* creates a compound statement. For instance, "I will attend the meeting or I will go to school." is a compound statement. It is composed of the two simple statements, "I will attend the meeting." and "I will go to school." The word *or* is a connective for the two simple statements.

George Boole used symbols such as *p, q, r,* and *s* to represent simple statements and the symbols $\land, \lor, \sim, \rightarrow,$ and $\leftrightarrow$ to represent connectives. See Table 3.1.

TABLE 3.1 Logic Connectives and Symbols

Statement	Connective	Symbolic form	Type of statement
not p	not	$\sim p$	negation
p and q	and	$p \wedge q$	conjunction
p or q	or	$p \vee q$	disjunction
If p, then q	If … then	$p \rightarrow q$	conditional
p if and only if q	if and only if	$p \leftrightarrow q$	biconditional

question What connective is used in a conjunction?

▼ Truth Value and Truth Tables

The **truth value** of a simple statement is either true (T) or false (F).

The **truth value** of a compound statement depends on the truth values of its simple statements and its connectives.

A **truth table** is a table that shows the truth value of a compound statement for all possible truth values of its simple statements.

Truth Table for $\sim p$

p	$\sim p$
T	F
F	T

The *negation* of the statement "Today is Friday." is the statement "Today is not Friday." In symbolic logic, the tilde symbol $\sim$ is used to denote the negation of a statement. If a statement p is true, its negation $\sim p$ is false, and if a statement p is false, its negation $\sim p$ is true. See the table at the left. The negation of the negation of a statement is the original statement. Thus $\sim(\sim p)$ can be replaced by p in any statement.

▼ example 2 Write the Negation of a Statement

Write the negation of each statement.

a. Bill Gates has a yacht.

b. *Avatar* was not selected as best picture at the 82nd Academy Awards ceremony.

Solution

a. Bill Gates does not have a yacht.

b. *Avatar* was selected as best picture at the 82nd Academy Awards ceremony.

▼ check your progress 2 Write the negation of each statement.

a. The *Queen Mary 2* is the world's largest cruise ship.

b. The fire engine is not red.

Solution *See page S7.*

The *Queen Mary 2*

Luboslav Tiles/Shutterstock.com

We will often find it useful to write compound statements in symbolic form.

---

answer The connective *and*.

▼ example **3**   **Write Compound Statements in Symbolic Form**

Consider the following simple statements.

> p: Today is Friday.
>
> q: It is raining.
>
> r: I am going to a movie.
>
> s: I am not going to the basketball game.

Write the following compound statements in symbolic form.

a. Today is Friday and it is raining.

b. It is not raining and I am going to a movie.

c. I am going to the basketball game or I am going to a movie.

d. If it is raining, then I am not going to the basketball game.

Solution

a. $p \land q$ **b.** $\sim q \land r$ **c.** $\sim s \lor r$ **d.** $q \rightarrow s$

▼ check your progress **3**   Use p, q, r, and s as defined in Example 3 to write the following compound statements in symbolic form.

a. Today is not Friday and I am going to a movie.

b. I am going to the basketball game and I am not going to a movie.

c. I am going to the movie if and only if it is raining.

d. If today is Friday, then I am not going to a movie.

Solution   *See page S7.*

In the next example, we translate symbolic statements into English sentences.

▼ example **4**   **Translate Symbolic Statements**

Consider the following statements.

> p: The game will be played in Atlanta.
>
> q: The game will be shown on CBS.
>
> r: The game will not be shown on ESPN.
>
> s: The Dodgers are favored to win.

Write each of the following symbolic statements in words.

a. $q \land p$ **b.** $\sim r \land s$ **c.** $s \leftrightarrow \sim p$

Solution

a. The game will be shown on CBS and the game will be played in Atlanta.

b. The game will be shown on ESPN and the Dodgers are favored to win.

c. The Dodgers are favored to win if and only if the game will not be played in Atlanta.

▼ check your progress **4**   Consider the following statements.

> e: All men are created equal.
>
> t: I am trading places.
>
> a: I get Abe's place.
>
> g: I get George's place.

Use the above information to translate the dialogue in the speech bubbles at the left.

Solution   *See page S8.*

Compound Statements and Grouping Symbols

If a compound statement is written in symbolic form, then parentheses are used to indicate which simple statements are grouped together. Table 3.2 illustrates the use of parentheses to indicate groupings for some statements in symbolic form.

TABLE 3.2

Symbolic form	The parentheses indicate that:
$p \wedge (q \vee \sim r)$	q and $\sim r$ are grouped together.
$(p \wedge q) \vee r$	p and q are grouped together.
$(p \wedge \sim q) \to (r \vee s)$	p and $\sim q$ are grouped together. r and s are also grouped together.

If a compound statement is written as an English sentence, then a comma is used to indicate which simple statements are grouped together. **Statements on the same side of a comma are grouped together.** See Table 3.3.

TABLE 3.3

English sentence	The comma indicates that:
p, and q or not r.	q and $\sim r$ are grouped together because they are both on the same side of the comma.
p and q, or r.	p and q are grouped together because they are both on the same side of the comma.
If p and not q, then r or s.	p and $\sim q$ are grouped together because they are both to the left of the comma. r and s are grouped together because they are both to the right of the comma.

If a statement in symbolic form is written as an English sentence, then the simple statements that appear together in parentheses in the symbolic form will all be on the same side of the comma that appears in the English sentence.

▼ example 5   Translate Compound Statements

Let p, q, and r represent the following.

  p: You get a promotion.
  q: You complete the training.
  r: You will receive a bonus.

a. Write $(p \wedge q) \to r$ as an English sentence.

b. Write "If you do not complete the training, then you will not get a promotion and you will not receive a bonus." in symbolic form.

Solution

a. Because the p and the q statements both appear in parentheses in the symbolic form, they are placed to the left of the comma in the English sentence.

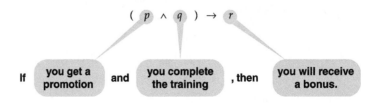

Thus the translation is: If you get a promotion and complete the training, then you will receive a bonus.

b. Because the not *p* and the not *r* statements are both to the right of the comma in the English sentence, they are grouped together in parentheses in the symbolic form.

Thus the translation is: $\sim q \rightarrow (\sim p \wedge \sim r)$

▼ **check your progress 5**   Let *p*, *q*, and *r* represent the following.

p: Kesha's singing style is similar to Uffie's.
q: Kesha has messy hair.
r: Kesha is a rapper.

a. Write $(p \wedge q) \rightarrow r$ as an English sentence.

b. Write "If Kesha is not a rapper, then Kesha does not have messy hair and Kesha's singing style is not similar to Uffie's." in symbolic form.

Solution   *See page S8.*

Kesha

The use of parentheses in a symbolic statement may affect the meaning of the statement. For instance, $\sim(p \vee q)$ indicates the negation of the compound statement $p \vee q$. However, $\sim p \vee q$ indicates that only the *p* statement is negated.

The statement $\sim(p \vee q)$ is read as, "It is not true that, *p* or *q*." The statement $\sim p \vee q$ is read as, "Not *p* or *q*."

If you order cake *and* ice cream in a restaurant, the waiter will bring *both* cake and ice cream. In general, the **conjunction** $p \wedge q$ is true if both *p* and *q* are true, and the conjunction is false if either *p* or *q* is false. The truth table at the left shows the four possible cases that arise when we form a conjunction of two statements.

Truth Table for $p \wedge q$

p	*q*	$p \wedge q$
T	T	T
T	F	F
F	T	F
F	F	F

▼ **Truth Value of a Conjunction**

The conjunction $p \wedge q$ is true if and only if both *p* and *q* are true.

Sometimes the word *but* is used in place of the connective *and*. For instance, "I ride my bike to school, but I ride the bus to work," is equivalent to the conjunction, "I ride my bike to school and I ride the bus to work."

Any **disjunction** $p \vee q$ is true if *p* is true or *q* is true or both *p* and *q* are true. The truth table at the left shows that the disjunction *p* or *q* is false if both *p* and *q* are false; however, it is true in all other cases.

Truth Table for $p \vee q$

p	*q*	$p \vee q$
T	T	T
T	F	T
F	T	T
F	F	F

▼ **Truth Value of a Disjunction**

The disjunction $p \vee q$ is true if and only if *p* is true, *q* is true, or both *p* and *q* are true.

▼ **example 6** Determine the Truth Value of a Statement

Determine whether each statement is true or false.

a. $7 \geq 5$.

b. 5 is a whole number and 5 is an even number.

c. 2 is a prime number and 2 is an even number.

Solution

a. $7 \geq 5$ means $7 > 5$ or $7 = 5$. Because $7 > 5$ is true, the statement $7 \geq 5$ is a true statement.

b. This is a false statement because 5 is not an even number.

c. This is a true statement because each simple statement is true.

▼ **check your progress 6** Determine whether each statement is true or false.

a. 21 is a rational number and 21 is a natural number.

b. $4 \leq 9$.

c. $-7 \geq -3$.

Solution *See page S8.* ◀

Truth tables for the conditional and biconditional are given in Section 3.3.

Quantifiers and Negation

In a statement, the word *some* and the phrases *there exists* and *at least one* are called **existential quantifiers**. Existential quantifiers are used as prefixes to assert the existence of something.

In a statement, the words *none*, *no*, *all*, and *every* are called **universal quantifiers**. The universal quantifiers *none* and *no* deny the existence of something, whereas the universal quantifiers *all* and *every* are used to assert that every element of a given set satisfies some condition.

Recall that the negation of a false statement is a true statement and the negation of a true statement is a false statement. It is important to remember this fact when forming the negation of a quantified statement. For instance, what is the negation of the false statement, "All dogs are mean"? You may think that the negation is "No dogs are mean," but this is also a false statement. Thus the statement "No dogs are mean" is not the negation of "All dogs are mean." The negation of "All dogs are mean," which is a false statement, is in fact "Some dogs are not mean," which is a true statement. The statement "Some dogs are not mean" can also be stated as "At least one dog is not mean" or "There exists a dog that is not mean."

What is the negation of the false statement, "No doctors write in a legible manner"? Whatever the negation is, we know it must be a true statement. The negation cannot be "All doctors write in a legible manner," because this is also a false statement. The negation is "Some doctors write in a legible manner." This can also be stated as, "There exists at least one doctor who writes in a legible manner."

Table 3.4A illustrates how to write the negation of some quantified statements.

TABLE 3.4A Quantified Statements and Their Negations

Statement	Negation
All X are Y.	Some X are not Y.
No X are Y.	Some X are Y.
Some X are not Y.	All X are Y.
Some X are Y.	No X are Y.

In Table 3.4A, the negations of the statements in the first column are shown in the second column. Also, the negation of the statements in the second column are the statements in the first column. Thus the information in Table 3.4A can be shown more compactly as in Table 3.4B.

TABLE 3.4B Quantified Statements and Their Negations Displayed in a Compact Format

All *X* are *Y*.	negation	Some *X* are not *Y*.
No *X* are *Y*.	negation	Some *X* are *Y*.

▼ **example 7** **Write the Negation of a Quantified Statement**

Write the negation of each of the following statements.

a. Some airports are open.

b. All movies are worth the price of admission.

c. No odd numbers are divisible by 2.

Solution

a. No airports are open.

b. Some movies are not worth the price of admission.

c. Some odd numbers are divisible by 2.

▼ **check your progress 7** Write the negation of the following statements.

a. All bears are brown.

b. No smartphones are expensive.

c. Some vegetables are not green.

Solution   *See page S8.*

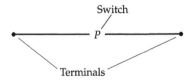

EXCURSION

Claude E. Shannon

Switching Networks

In 1939, Claude E. Shannon (1916–2001) wrote a thesis on an application of symbolic logic to *switching networks*. A switching network consists of wires and switches that can open and close. Switching networks are used in many electrical appliances, as well as in telephone equipment and computers. Figure 3.1 shows a switching network that consists of a single switch *P* that connects two terminals. An electric current can flow from one terminal to the other terminal provided that the switch *P* is in the closed position. If *P* is in the open position, then the current cannot flow from one terminal to the other. If a current can flow between the terminals, we say that a network is closed, and if a current cannot flow between the terminals, we say that the network is open. We designate this network by the letter *P*. There exists an analogy between a network *P* and a statement *p* in that a network is either open or it is closed, and a statement is either true or it is false.

Figure 3.2 shows two switches *P* and *Q* connected in **series**. This series network is closed if and only if both switches are closed. We will use $P \wedge Q$ to denote this series network because it is analogous to the logic statement $p \wedge q$, which is true if and only if both *p* and *q* are true.

Switch

P

Terminals

FIGURE 3.1

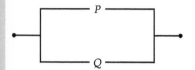

FIGURE 3.2 A series network

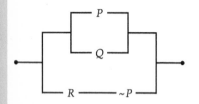

FIGURE 3.3 A parallel network

Figure 3.3 shows two switches P and Q connected in **parallel**. This parallel network is closed if either P or Q is closed. We will designate this parallel network by $P \vee Q$ because it is analogous to the logic statement $p \vee q$, which is true if p is true or if q is true.

Series and parallel networks can be combined to produce more complicated networks, as shown in Figure 3.4.

The network shown in Figure 3.4 is closed provided P or Q is closed or provided that both R and $\sim P$ are closed. Note that the switch $\sim P$ is closed if P is open, and $\sim P$ is open if P is closed. We use the symbolic statement $(P \vee Q) \vee (R \wedge \sim P)$ to represent this network.

If two switches are always open at the same time and always closed at the same time, then we will use the same letter to designate both switches.

FIGURE 3.4

EXCURSION EXERCISES

Write a symbolic statement to represent each of the networks in Excursion Exercises 1 to 6.

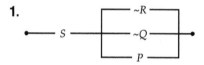

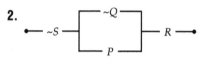

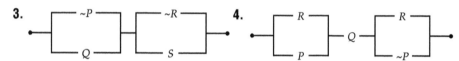

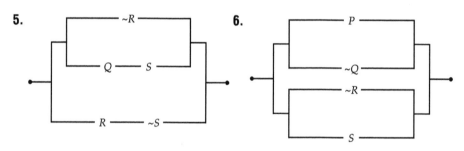

7. Which of the networks in Excursion Exercises 1 to 6 are closed networks, given that P is closed, Q is open, R is closed, and S is open?

8. Which of the networks in Excursion Exercises 1 to 6 are closed networks, given that P is open, Q is closed, R is closed, and S is closed?

In Excursion Exercises 9 to 14, draw a network to represent each statement.

9. $(\sim P \vee Q) \wedge (R \wedge P)$ **10.** $P \wedge [(Q \wedge \sim R) \vee R]$

11. $[\sim P \wedge Q \wedge R] \vee (P \wedge R)$ **12.** $(Q \vee R) \vee (S \vee \sim P)$

13. $[(\sim P \wedge R) \vee Q] \vee (\sim R)$ **14.** $(P \vee Q \vee R) \wedge S \wedge (\sim Q \vee R)$

Warning Circuits   The circuits shown in Excursion Exercises 15 and 16 include a switching network, a warning light, and a battery. In each circuit the warning light will turn on only when the switching network is closed.

15. Consider the following circuit.

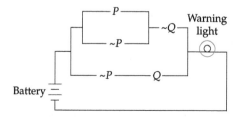

For each of the following conditions, determine whether the warning light will be on or off.

a. *P* is closed and *Q* is open.   **b.** *P* is closed and *Q* is closed.

c. *P* is open and *Q* is closed.   **d.** *P* is open and *Q* is open.

16. An engineer thinks that the following circuit can be used in place of the circuit shown in Excursion Exercise 15. Do you agree? Explain.

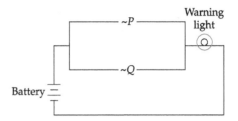

EXERCISE SET 3.1   (Suggested Assignment: The Enhanced WebAssign Exercises and Exercises 4, 7, 10, 14, 15, 33, 51, 56, 57, and 69)

■ In Exercises 1 to 8, determine whether each sentence is a statement.

1. *The Dark Knight* is the greatest movie of all time.

2. Harvey Mudd college is in Oregon.

3. The area code for Storm Lake, Iowa, is 512.

4. January 1, 2021, will be a Sunday.

5. Have a fun trip.

6. Do you like to read?

7. Mickey Mouse was the first animated character to receive a star on the Hollywood Walk of Fame.

8. Drew Brees is the starting quarterback of the San Diego Chargers.

■ In Exercises 9 to 12, determine the simple statements in each compound statement.

9. The principal will attend the class on Tuesday or Wednesday.

10. 5 is an odd number and 6 is an even number.

11. A triangle is an acute triangle if and only if it has three acute angles.

12. If this is Saturday, then tomorrow is Sunday.

■ In Exercises 13 to 16, write the negation of each statement.

13. The Giants lost the game.

14. The lunch was served at noon.

15. The game did not go into overtime.

16. The game was not shown on ABC.

■ In Exercises 17 to 24, write each sentence in symbolic form. Represent each simple statement in the sentence with the letter indicated in the parentheses. Also state whether the sentence is a conjunction, a disjunction, a negation, a conditional, or a biconditional.

17. If today is Wednesday (*w*), then tomorrow is Thursday (*t*).

18. I went to the post office (*p*) and the bookstore (*s*).

19. A triangle is an equilateral triangle (*l*) if and only if it is an equiangular triangle (*a*).

20. A number is an even number (*e*) if and only if it has a factor of 2 (*t*).

21. If it is a dog (*d*), it has fleas (*f*).

22. Polynomials that have exactly three terms (*p*) are called trinomials (*t*).

23. I will major in mathematics (*m*) or computer science (*c*).

24. All pentagons (*p*) have exactly five sides (*s*).

■ In Exercises 25 to 30, write each symbolic statement in words. Use *p*, *q*, *r*, *s*, *t*, and *u* as defined below.

p:	The tour goes to Italy.
q:	The tour goes to Spain.
r:	We go to Venice.
s:	We go to Florence.
t:	The hotel fees are included.
u:	The meals are not included.

25. $p \wedge \sim q$ **26.** $r \vee s$

27. $r \rightarrow \sim s$ **28.** $p \rightarrow r$

29. $s \leftrightarrow \sim r$ **30.** $\sim t \wedge u$

■ In Exercises 31 to 36, write each symbolic statement as an English sentence. Use *p*, *q*, *r*, *s*, and *t* as defined below.

p:	Taylor Swift is a singer.
q:	Taylor Swift is not a songwriter.
r:	Taylor Swift is an actress.
s:	Taylor Swift plays the piano.
t:	Taylor Swift does not play the guitar.

31. $(p \vee r) \wedge q$ **32.** $\sim s \rightarrow (p \wedge \sim q)$

33. $p \to (q \land \sim r)$ **34.** $(s \land \sim q) \to t$

35. $(r \land p) \leftrightarrow q$ **36.** $t \leftrightarrow (\sim r \land \sim p)$

■ In Exercises 37 to 42, write each sentence in symbolic form. Use *p, q, r* and *s* as defined below.

> *p:* Dwyane Wade is a football player.
>
> *q:* Dwyane Wade is a basketball player.
>
> *r:* Dwyane Wade is a rock star.
>
> *s:* Dwyane Wade plays for the Miami Heat.

37. Dwyane Wade is a football player or a basketball player, and he is not a rock star.

38. Dwyane Wade is a rock star, and he is not a basketball player or a football player.

39. If Dwyane Wade is a basketball player and a rock star, then he is not a football player.

40. Dwyane Wade is a basketball player, if and only if he is not a football player and he is not a rock star.

41. If Dwyane Wade plays for the Miami Heat, then he is a basketball player and he is not a football player.

42. It is not true that, Dwyane Wade is a football player or a rock star.

■ In Exercises 43 to 50, determine whether each statement is true or false.

43. $7 < 5$ or $3 > 1$.

44. $3 \leq 9$.

45. $(-1)^{50} = 1$ and $(-1)^{99} = -1$.

46. $7 \neq 3$ or 9 is a prime number.

47. $-5 \geq -11$.

48. $4.5 \leq 5.4$.

49. 2 is an odd number or 2 is an even number.

50. The square of any real number is a positive number.

■ In Exercises 51 to 58, write the negation of each quantified statement. Start each negation with "Some," "No," or "All."

51. Some lions are playful.

52. Some dogs are not friendly.

53. All classic movies were first produced in black and white.

54. Everybody enjoyed the dinner.

55. No even numbers are odd numbers.

56. Some actors are not rich.

57. All cars run on gasoline.

58. None of the students took my advice.

Critical Thinking

Write Quotations in Symbolic Form   In Exercises 59 to 62, translate each quotation into symbolic form. For each simple statement in the quotation, indicate what letter you used to represent the simple statement.

59. If you can count your money, you don't have a billion dollars. *J. Paul Getty*

60. If you aren't fired with enthusiasm, then you will be fired with enthusiasm. *Vince Lombardi*

61. If people concentrated on the really important things in life, there'd be a shortage of fishing poles. *Doug Larson*

62. If you're killed, you've lost a very important part of your life. *Brooke Shields*

Write Statements in Symbolic Form   In Exercises 63 to 68, translate each mathematical statement into symbolic form. For each simple statement in the given statement, indicate what letter you used to represent the simple statement.

63. An angle is a right angle if and only if its measure is 90°.

64. Any angle inscribed in a semicircle is a right angle.

65. If two sides of a triangle are equal in length, then the angles opposite those sides are congruent.

66. The sum of the measures of the three angles of any triangle is 180°.

67. All squares are rectangles.

68. If the corresponding sides of two triangles are proportional, then the triangles are similar.

Cooperative Learning

69. **Recreational Logic**   The following diagram shows two cylindrical teapots. The yellow teapot has the same diameter as the green teapot, but it is one and one-half times as tall as the green teapot.

If the green teapot can hold a maximum of 6 cups of tea, then estimate the maximum number of cups of tea that the yellow teapot can hold. Explain your reasoning.

section 3.2 Truth Tables, Equivalent Statements, and Tautologies

Truth Tables

In Section 3.1, we defined truth tables for the negation of a statement, the conjunction of two statements, and the disjunction of two statements. Each of these truth tables is shown below for review purposes.

Negation

p	$\sim p$
T	F
F	T

Conjunction

p	q	$p \wedge q$
T	T	T
T	F	F
F	T	F
F	F	F

Disjunction

p	q	$p \vee q$
T	T	T
T	F	T
F	T	T
F	F	F

p	q	Given statement
T	T	
T	F	
F	T	
F	F	

Standard truth table form for a given statement that involves only the two simple statements p and q

In this section, we consider methods of constructing truth tables for a statement that involves a combination of conjunctions, disjunctions, and/or negations. If the given statement involves only two simple statements, then start with a table with four rows (see the table at the left), called the **standard truth table form**, and proceed as shown in Example 1.

▼ example 1 Truth Tables

a. Construct a table for $\sim(\sim p \vee q) \vee q$.

b. Use the truth table from part a to determine the truth value of $\sim(\sim p \vee q) \vee q$, given that p is true and q is false.

Solution

a. Start with the standard truth table form and then include a $\sim p$ column.

p	q	$\sim p$
T	T	F
T	F	F
F	T	T
F	F	T

Now use the truth values from the $\sim p$ and q columns to produce the truth values for $\sim p \vee q$, as shown in the rightmost column of the following table.

p	q	$\sim p$	$\sim p \vee q$
T	T	F	T
T	F	F	F
F	T	T	T
F	F	T	T

Negate the truth values in the $\sim p \vee q$ column to produce the following.

p	q	$\sim p$	$\sim p \vee q$	$\sim(\sim p \vee q)$
T	T	F	T	F
T	F	F	F	T
F	T	T	T	F
F	F	T	T	F

As our last step, we form the disjunction of $\sim(\sim p \vee q)$ with q and place the results in the rightmost column of the table. See the following table. The shaded column is the truth table for $\sim(\sim p \vee q) \vee q$.

p	q	$\sim p$	$\sim p \vee q$	$\sim(\sim p \vee q)$	$\sim(\sim p \vee q) \vee q$	
T	T	F	T	F	T	row 1
T	F	F	F	T	T	row 2
F	T	T	T	F	T	row 3
F	F	T	T	F	F	row 4

b. In row 2 of the above truth table, we see that when p is true, and q is false, the statement $\sim(\sim p \vee q) \vee q$ in the rightmost column is true.

▼ check your progress 1

a. Construct a truth table for $(p \wedge \sim q) \vee (\sim p \vee q)$.

b. Use the truth table that you constructed in part a to determine the truth value of $(p \wedge \sim q) \vee (\sim p \vee q)$, given that p is true and q is false.

Solution   *See page S8.*

p	q	r	Given statement
T	T	T	
T	T	F	
T	F	T	
T	F	F	
F	T	T	
F	T	F	
F	F	T	
F	F	F	

Standard truth table form for a statement that involves the three simple statements p, q, and r

Compound statements that involve exactly three simple statements require a standard truth table form with $2^3 = 8$ rows, as shown at the left.

▼ example 2   Truth Tables

a. Construct a truth table for $(p \wedge q) \wedge (\sim r \vee q)$.

b. Use the truth table from part a to determine the truth value of $(p \wedge q) \wedge (\sim r \vee q)$, given that p is true, q is true, and r is false.

Solution

a. Using the procedures developed in Example 1, we can produce the following table. The shaded column is the truth table for $(p \wedge q) \wedge (\sim r \vee q)$. The numbers in the squares below the columns denote the order in which the columns were constructed. Each truth value in the column numbered 4 is the conjunction of the truth values to its left in the columns numbered 1 and 3.

p	q	r	$p \wedge q$	$\sim r$	$\sim r \vee q$	$(p \wedge q) \wedge (\sim r \vee q)$	
T	T	T	T	F	T	T	row 1
T	T	F	T	T	T	T	row 2
T	F	T	F	F	F	F	row 3
T	F	F	F	T	T	F	row 4
F	T	T	F	F	T	F	row 5
F	T	F	F	T	T	F	row 6
F	F	T	F	F	F	F	row 7
F	F	F	F	T	T	F	row 8
			[1]	[2]	[3]	[4]	

b. In row 2 of the above truth table, we see that $(p \wedge q) \wedge (\sim r \vee q)$ is true when p is true, q is true, and r is false.

a. Construct a truth table for $(\sim p \wedge r) \vee (q \wedge \sim r)$.

b. Use the truth table that you constructed in part a to determine the truth value of $(\sim p \wedge r) \vee (q \wedge \sim r)$, given that p is false, q is true, and r is false.

Solution   *See page S8.*

Alternative Method for the Construction of a Truth Table

In Example 3 we use an *alternative procedure* to construct a truth table.

TAKE NOTE

The alternative procedure for constructing a truth table, as described to the right, generally requires less writing, less time, and less effort than the truth table procedure that was used in Examples 1 and 2.

▼ Alternative Procedure for Constructing a Truth Table

1. If the given statement has n simple statements, then start with a standard form that has 2^n rows. Enter the truth values for each simple statement and their negations.

2. Use the truth values for each simple statement and their negations to enter the truth values under each connective within a pair of grouping symbols, including parentheses ( ), brackets [ ], and braces { }. If some grouping symbols are nested inside other grouping symbols, then work from the inside out. In any situation in which grouping symbols have not been used, then we use the following **order of precedence agreement**.

First assign truth values to negations from left to right, followed by conjunctions from left to right, followed by disjunctions from left to right, followed by conditionals from left to right, and finally by biconditionals from left to right.

3. The truth values that are entered into the column under the connective for which truth values are assigned *last*, form the truth table for the given statement.

▼ example 3   Use the Alternative Procedure to Construct a Truth Table

Construct a truth table for $p \vee [\sim(p \wedge \sim q)]$.

Solution

Step 1: The given statement $p \vee [\sim(p \wedge \sim q)]$ has the two simple statements p and q. Thus we start with a standard form that has $2^2 = 4$ rows. In each column, enter the truth values for the statements p and $\sim q$, as shown in the columns numbered 1, 2, and 3 of the following table.

p	q	p	$\vee$	$[\sim$	$(p$	$\wedge$	$\sim q)]$
T	T	T				T	F
T	F	T				T	T
F	T	F				F	F
F	F	F				F	T
		⌐1⌐				⌐2⌐	⌐3⌐

Step 2: Use the truth values in columns 2 and 3 to determine the truth values to enter under the "and" connective. See column 4 in the following truth table. Now negate the truth values in the column numbered 4 to produce the truth values in the column numbered 5.

p	*q*	*p*	∨	[~	(*p*	∧	~*q*)]
T	T	T		T	T	F	F
T	F	T		F	T	T	T
F	T	F		T	F	F	F
F	F	F		T	F	F	T
		1		5	2	4	3

Step 3: Use the truth values in the columns numbered 1 and 5 to determine the truth values to enter under the "or" connective. See the column numbered 6, which is the truth table for $p \vee [\sim(p \wedge \sim q)]$.

p	*q*	*p*	∨	[~	(*p*	∧	~*q*)]
T	T	T	T	T	T	F	F
T	F	T	T	F	T	T	T
F	T	F	T	T	F	F	F
F	F	F	T	T	F	F	T
		1	6	5	2	4	3

▼**check your progress 3** Construct a truth table for $\sim p \vee (p \wedge q)$.

Solution *See page S9.*

MATHMATTERS A Three-Valued Logic

Jan Lukasiewicz (loo-kä-shä-vēch) (1878–1956) was the Polish minister of education in 1919 and served as a professor of mathematics at Warsaw University from 1920 to 1939. Most of Lukasiewicz's work was in the area of logic. He is well known for developing *polish notation,* which was first used in logic to eliminate the need for parentheses in symbolic statements. Today *reverse polish notation* is used by many computers and calculators to perform computations without the need to enter parentheses.

Jan Lukasiewicz was one of the first mathematicians to consider a three-valued logic in which a statement is true, false, or "somewhere between true and false." In his three-valued logic, Lukasiewicz classified the truth value of a statement as true (T), false (F), or maybe (M). The following table shows truth values for negation, conjunction, and disjunction in this three-valued logic.

p	*q*	Negation ~*p*	Conjunction *p* ∧ *q*	Disjunction *p* ∨ *q*
T	T	F	T	T
T	M	F	M	T
T	F	F	F	T
M	T	M	M	T
M	M	M	M	M
M	F	M	F	M
F	T	T	F	T
F	M	T	F	M
F	F	T	F	F

Equivalent Statements

Two statements are **equivalent** if they both have the same truth value for all possible truth values of their simple statements. Equivalent statements have identical truth values in the final columns of their truth tables. The notation $p \equiv q$ is used to indicate that the statements p and q are equivalent.

 example 4   Verify That Two Statements Are Equivalent

Show that $\sim(p \vee \sim q)$ and $\sim p \wedge q$ are equivalent statements.

Solution
Construct two truth tables and compare the results. The truth tables below show that $\sim(p \vee \sim q)$ and $\sim p \wedge q$ have the same truth values for all possible truth values of their simple statements. Thus the statements are equivalent.

p	q	$\sim(p \vee \sim q)$
T	T	F
T	F	F
F	T	T
F	F	F

p	q	$\sim p \wedge q$
T	T	F
T	F	F
F	T	T
F	F	F

———— identical truth values ————
Thus $\sim(p \vee \sim q) \equiv \sim p \wedge q$.

▼ check your progress 4   Show that $p \vee (p \wedge \sim q)$ and p are equivalent.

Solution   *See page S9.*

The truth tables in Table 3.5 show that $\sim(p \vee q)$ and $\sim p \wedge \sim q$ are equivalent statements. The truth tables in Table 3.6 show that $\sim(p \wedge q)$ and $\sim p \vee \sim q$ are equivalent statements.

TABLE 3.5

p	q	$\sim(p \vee q)$	$\sim p \wedge \sim q$
T	T	F	F
T	F	F	F
F	T	F	F
F	F	T	T

TABLE 3.6

p	q	$\sim(p \wedge q)$	$\sim p \vee \sim q$
T	T	F	F
T	F	T	T
F	T	T	T
F	F	T	T

These equivalences are known as **De Morgan's laws for statements**.

▼ De Morgan's Laws for Statements

For any statements p and q,

$$\sim(p \vee q) \equiv \sim p \wedge \sim q$$

$$\sim(p \wedge q) \equiv \sim p \vee \sim q$$

De Morgan's laws can be used to restate certain English sentences in an equivalent form.

▼ **example 5** State an Equivalent Form

Use one of De Morgan's laws to restate the following sentence in an equivalent form.

It is not true that, I graduated or I got a job.

Solution
Let p represent the statement "I graduated." Let q represent the statement "I got a job." In symbolic form, the original sentence is $\sim(p \vee q)$. One of De Morgan's laws states that this is equivalent to $\sim p \wedge \sim q$. Thus a sentence that is equivalent to the original sentence is "I did not graduate and I did not get a job."

▼ **check your progress 5** Use one of De Morgan's laws to restate the following sentence in an equivalent form.

It is not true that, I am going to the dance and I am going to the game.

Solution *See page S9.* ◀

Tautologies and Self-Contradictions

A **tautology** is a statement that is always true. A **self-contradiction** is a statement that is always false.

▼ **example 6** Verify Tautologies and Self-Contradictions

Show that $p \vee (\sim p \vee q)$ is a tautology.

Solution

Construct a truth table as shown below.

p	q	p	$\vee$	$(\sim p$	$\vee$	$q)$
T	T	T	T	F	T	T
T	F	T	T	F	F	F
F	T	F	T	T	T	T
F	F	F	T	T	T	F
		1	5	2	4	3

The table shows that $p \vee (\sim p \vee q)$ is always true. Thus $p \vee (\sim p \vee q)$ is a tautology.

▼ **check your progress 6** Show that $p \wedge (\sim p \wedge q)$ is a self-contradiction.

Solution *See page S9.* ◀

question Is the statement $x + 2 = 5$ a tautology or a self-contradiction?

---

answer Neither. The statement is not true for all values of x, and it is not false for all values of x.

EXCURSION

Switching Networks—Part II

The Excursion in Section 3.1 introduced the application of symbolic logic to switching networks. This Excursion makes use of *closure tables* to determine under what conditions a switching network is open or closed. **In a closure table, we use a 1 to designate that a switch or switching network is closed and a 0 to indicate that it is open.**

Figure 3.5 shows a switching network that consists of the single switch P and a second network that consists of the single switch $\sim P$. The table below shows that the switching network $\sim P$ is open when P is closed and is closed when P is open.

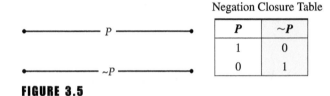

Negation Closure Table

P	$\sim P$
1	0
0	1

FIGURE 3.5

Figure 3.6 shows switches P and Q connected to form a series network. The table below shows that this series network is closed if and only if both P and Q are closed.

Series Network Closure Table

P	Q	$P \wedge Q$
1	1	1
1	0	0
0	1	0
0	0	0

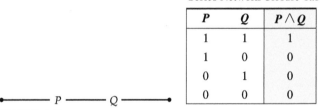

FIGURE 3.6   A series network

Figure 3.7 shows switches P and Q connected to form a parallel network. The table below shows that this parallel network is closed if P is closed or if Q is closed.

Parallel Network Closure Table

P	Q	$P \vee Q$
1	1	1
1	0	1
0	1	1
0	0	0

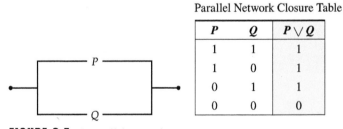

FIGURE 3.7   A parallel network

Now consider the network shown in Figure 3.8. To determine the required conditions under which the network is closed, we first write a symbolic statement that represents the network, and then we construct a closure table.

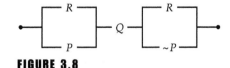

FIGURE 3.8

A symbolic statement that represents the network in Figure 3.8 is

$$[(R \lor P) \land Q] \land (R \lor \sim P)$$

The closure table for this network is shown below.

P	Q	R	[(R	∨	P)	∧	Q]	∧	(R	∨	~P)	
1	1	1	1	1	1	1	1	1	1	1	0	row 1
1	1	0	0	1	1	1	1	0	0	0	0	row 2
1	0	1	1	1	1	0	0	0	1	1	0	row 3
1	0	0	0	1	1	0	0	0	0	0	0	row 4
0	1	1	1	1	0	1	1	1	1	1	1	row 5
0	1	0	0	0	0	0	1	0	0	1	1	row 6
0	0	1	1	1	0	0	0	0	1	1	1	row 7
0	0	0	0	0	0	0	0	0	0	1	1	row 8

1	6	2	7	3	9	4	8	5

The rows numbered 1 and 5 of the above table show that the network is closed whenever

- P is closed, Q is closed, and R is closed, or
- P is open, Q is closed, and R is closed.

Thus the switching network in Figure 3.8 is closed provided Q is closed and R is closed. The switching network is open under all other conditions.

EXCURSION EXERCISES

Construct a closure table for each of the following switching networks. Use the closure table to determine the required conditions for the network to be closed.

1. 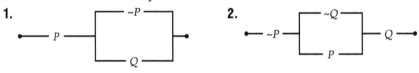   **2.**

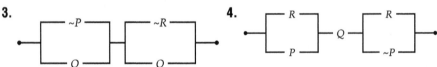

3.   **4.**

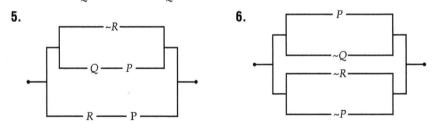

5.   **6.**

7. *Warning Circuits*

  a. The following circuit shows a switching network used in an automobile. The warning buzzer will buzz only when the switching network is closed. Construct a closure table for the switching network.

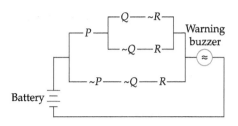

$$\{P \wedge [(Q \wedge \sim R) \vee (\sim Q \wedge R)]\} \vee [(\sim P \wedge \sim Q) \wedge R]$$

b. An engineer thinks that the following circuit can be used in place of the circuit in part a. Do you agree? *Hint:* Construct a closure table for the switching network and compare your closure table with the closure table in part a.

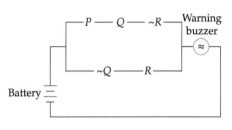

$$[(P \wedge Q) \wedge \sim R] \vee (\sim Q \wedge R)$$

EXERCISE SET 3.2 (Suggested Assignment: The Enhanced WebAssign Exercises and Exercises 11, 19, 23, 27, 31, 47, and 61)

■ In Exercises 1 to 10, determine the truth value of the compound statement given that p is a false statement, q is a true statement, and r is a true statement.

1. $p \vee (\sim q \vee r)$

2. $r \wedge \sim (p \vee r)$

3. $(p \wedge q) \vee (\sim p \wedge \sim q)$

4. $(p \wedge q) \vee [(\sim p \wedge \sim q) \vee q]$

5. $[\sim (p \wedge \sim q) \vee r] \wedge (p \wedge \sim r)$

6. $(p \wedge \sim q) \vee [(p \wedge \sim q) \vee r]$

7. $[(p \wedge \sim q) \vee \sim r] \wedge (q \wedge r)$

8. $(\sim p \wedge q) \wedge [(p \wedge \sim q) \vee r]$

9. $[(p \wedge q) \wedge r] \vee [p \vee (q \wedge \sim r)]$

10. $\{[(\sim p \wedge q) \wedge r] \vee [(p \wedge q) \wedge \sim r]\} \vee [p \wedge (q \wedge r)]$

11. a. Given that p is a false statement, what can be said about $p \wedge (q \vee r)$?

b. Explain why it is not necessary to know the truth values of q and r to determine the truth value of $p \wedge (q \vee r)$ in part a above.

12. a. Given that q is a true statement, what can be said about $q \vee \sim r$?

b. Explain why it is not necessary to know the truth value of r to determine the truth value of $q \vee \sim r$ in part a above.

■ In Exercises 13 to 28, construct a truth table for each compound statement.

13. $\sim p \vee q$

14. $(q \wedge \sim p) \vee \sim q$

15. $p \wedge \sim q$

16. $p \vee [\sim (p \wedge \sim q)]$

17. $(p \wedge \sim q) \vee [\sim (p \wedge q)]$

18. $(p \vee q) \wedge [\sim (p \vee \sim q)]$

19. $\sim (p \vee q) \wedge (\sim r \vee q)$

20. $[\sim (r \wedge \sim q)] \vee (\sim p \vee q)$

21. $(p \wedge \sim r) \vee [\sim q \vee (p \wedge r)]$

22. $[r \wedge (\sim p \vee q)] \wedge (r \vee \sim q)$

23. $[(p \wedge q) \vee (r \wedge \sim p)] \wedge (r \vee \sim q)$

24. $(p \wedge q) \wedge \{[\sim(\sim p \vee r)] \wedge q\}$

25. $q \vee [\sim r \vee (p \wedge r)]$

26. $\{[\sim(p \vee \sim r)] \wedge \sim q\} \vee r$

27. $(\sim q \wedge r) \vee [p \wedge (q \wedge \sim r)]$

28. $\sim [\sim p \wedge (q \wedge r)]$

■ In Exercises 29 to 36, use two truth tables to show that each of the statements are equivalent.

29. $p \vee (p \wedge r), p$

30. $q \wedge (q \vee r), q$

31. $p \wedge (q \vee r), (p \wedge q) \vee (p \wedge r)$

32. $p \vee (q \wedge r), (p \vee q) \wedge (p \vee r)$

33. $p \vee (q \wedge \sim p), p \vee q$

34. $\sim [p \vee (q \wedge r)], \sim p \wedge (\sim q \vee \sim r)$

35. $[(p \wedge q) \wedge r] \vee [p \wedge (q \wedge \sim r)], p \wedge q$

36. $[(\sim p \wedge \sim q) \wedge r] \vee [(p \wedge q) \wedge \sim r] \vee [p \wedge (q \wedge r)],$ $(p \wedge q) \vee [(\sim p \wedge \sim q) \wedge r]$

■ In Exercises 37 to 42, make use of one of De Morgan's laws to write the given statement in an equivalent form.

37. It is not true that, it rained or it snowed.

38. I did not pass the test and I did not complete the course.

39. She did not visit France and she did not visit Italy.

40. It is not true that, I bought a new car and I moved to Florida.

41. It is not true that, she received a promotion or that she received a raise.

42. It is not the case that, the students cut classes or took part in the demonstration.

■ In Exercises 43 to 48, use a truth table to determine whether the given statement is a tautology.

43. $p \vee \sim p$

44. $q \vee [\sim(q \wedge r) \wedge \sim q]$

45. $(p \vee q) \vee (\sim p \vee q)$

46. $(p \wedge q) \vee (\sim p \vee \sim q)$

47. $(\sim p \vee q) \vee (\sim q \vee r)$

48. $\sim [p \wedge (\sim p \vee q)] \vee q$

■ In Exercises 49 to 54, use a truth table to determine whether the given statement is a self-contradiction.

49. $\sim r \wedge r$

50. $\sim(p \vee \sim p)$

51. $p \wedge (\sim p \wedge q)$

52. $\sim [(p \vee q) \vee (\sim p \vee q)]$

53. $[p \wedge (\sim p \vee q)] \vee q$

54. $\sim [p \vee (\sim p \vee q)]$

55. Explain why the statement $7 \leq 8$ is a disjunction.

56. a. Why is the statement $5 \leq 7$ true?

   b. Why is the statement $7 \leq 7$ true?

EXTENSIONS

Critical Thinking

57. How many rows are needed to construct a truth table for the statement $[p \wedge (q \vee \sim r)] \vee (s \wedge \sim t)$?

58. Explain why no truth table can have exactly 100 rows.

Cooperative Learning

In Exercises 59 and 60, construct a truth table for the given compound statement. *Hint:* Use a table with 16 rows.

59. $[(p \wedge \sim q) \vee (q \wedge \sim r)] \wedge (r \vee \sim s)$

60. $s \wedge [\sim(\sim r \vee q) \vee \sim p]$

61. **Recreational Logic** A friend hands you the slip of paper shown below and challenges you to circle exactly four digits that have a sum of 19.

$$\boxed{1 \quad 3 \quad 3 \quad 5 \quad 5 \quad 7 \quad 7 \quad 9 \quad 9 \quad 9}$$

Explain how you can meet this challenge.

section **3.3**   The Conditional and the Biconditional

Conditional Statements

Humphrey Bogart and Ingrid Bergman star in *Casablanca* (1942).

If you don't get in that plane, you'll regret it. Maybe not today, maybe not tomorrow, but soon, and for the rest of your life.

The above quotation is from the movie *Casablanca*. Rick, played by Humphrey Bogart, is trying to convince Ilsa, played by Ingrid Bergman, to get on the plane with Laszlo. The sentence, "If you don't get in that plane, you'll regret it," is a *conditional statement*. **Conditional statements** can be written in *if p, then q* form or in *if p, q* form. For instance, all of the following are conditional statements.

If we order pizza, then we can have it delivered.

If you go to the movie, you will not be able to meet us for dinner.

If *n* is a prime number greater than 2, then *n* is an odd number.

In any conditional statement represented by "If *p*, then *q*" or by "If *p*, *q*," the *p* statement is called the **antecedent** and the *q* statement is called the **consequent**.

▼ **example 1**   Identify the Antecedent and Consequent of a Conditional

Identify the antecedent and consequent in the following statements.

a. If our school was this nice, I would go there more than once a week.
—*The Basketball Diaries*

b. If you don't stop and look around once in a while, you could miss it.
—Ferris in *Ferris Bueller's Day Off*

c. If you strike me down, I shall become more powerful than you can possibly imagine.—Obi-Wan Kenobi, Star Wars, Episode IV, *A New Hope*

Solution

a. *Antecedent:* our school was this nice

   Consequent: I would go there more than once a week

b. *Antecedent:* you don't stop and look around once in a while

   Consequent: you could miss it

c. *Antecedent:* you strike me down

   Consequent: I shall become more powerful than you can possibly imagine

▼ **check your progress 1**   Identify the antecedent and consequent in each of the following conditional statements.

a. If I study for at least 6 hours, then I will get an A on the test.

b. If I get the job, I will buy a new car.

c. If you can dream it, you can do it.

Solution   *See page S9.*

▼ **Arrow Notation**

The conditional statement, "If *p*, then *q*," can be written using the **arrow notation** $p \rightarrow q$. The arrow notation $p \rightarrow q$ is read as "if *p*, then *q*" or as "*p* implies *q*."

The Truth Table for the Conditional $p \rightarrow q$

To determine the truth table for $p \rightarrow q$, consider the advertising slogan for a web authoring software product that states, "If you can use a word processor, you can create a webpage." This slogan is a conditional statement. The antecedent is p, "you can use a word processor," and the consequent is q, "you can create a webpage." Now consider the truth value of $p \rightarrow q$ for each of the following four possibilities.

TABLE 3.7

Antecedent p: you can use a word processor	Consequent q: you can create a webpage	$p \rightarrow q$	
T	T	?	row 1
T	F	?	row 2
F	T	?	row 3
F	F	?	row 4

Row 1: Antecedent T, Consequent T   You can use a word processor, and you can create a webpage. In this case the truth value of the advertisement is true. To complete Table 3.7, we place a T in place of the question mark in row 1.

Row 2: Antecedent T, Consequent F   You can use a word processor, but you cannot create a webpage. In this case the advertisement is false. We put an F in place of the question mark in row 2 of Table 3.7.

Row 3: Antecedent F, Consequent T   You cannot use a word processor, but you can create a webpage. Because the advertisement does not make any statement about what you might or might not be able to do if you cannot use a word processor, we cannot state that the advertisement is false, and we are compelled to place a T in place of the question mark in row 3 of Table 3.7.

Row 4: Antecedent F, Consequent F   You cannot use a word processor, and you cannot create a webpage. Once again we must consider the truth value in this case to be true because the advertisement does not make any statement about what you might or might not be able to do if you cannot use a word processor. We place a T in place of the question mark in row 4 of Table 3.7.

TABLE 3.8 Truth Table for $p \rightarrow q$

p	q	$p \rightarrow q$
T	T	T
T	F	F
F	T	T
F	F	T

The truth table for the conditional $p \rightarrow q$ is given in Table 3.8.

> ▼ **Truth Value of the Conditional $p \rightarrow q$**
>
> The conditional $p \rightarrow q$ is false if p is true and q is false. It is true in all other cases.

▼ **example** **2**   Find the Truth Value of a Conditional

Determine the truth value of each of the following.

a. If 2 is an integer, then 2 is a rational number.

b. If 3 is a negative number, then $5 > 7$.

c. If $5 > 3$, then $2 + 7 = 4$.

Solution

a. Because the consequent is true, this is a true statement.

CALCULATOR NOTE

Program FACTOR

```
0→dim (L1)
Prompt N
1→S: 2→F:0→E
√(N)→M
While F≤ M
While fPart (N/F)=0
E+1→E:N/F→N
End
If E>0
Then
F→L1(S)
E→L1(S+1)
S+2→S:0→E
√(N)→M
End
If F=2
Then
3→F
Else
F+2→F
End: End
If N≠1
Then
N→L1(S)
1→L1(S+1)
End
If S=1
Then
Disp N, "IS PRIME"
Else
Disp L1
```

b. Because the antecedent is false, this is a true statement.

c. Because the antecedent is true and the consequent is false, this is a false statement.

▼ **check your progress 2** Determine the truth value of each of the following.

a. If $4 \geq 3$, then $2 + 5 = 6$.

b. If $5 > 9$, then $4 > 9$.

c. If Tuesday follows Monday, then April follows March.

Solution *See page S9.*

▼ **example 3** **Construct a Truth Table for a Statement Involving a Conditional**

Construct a truth table for $[p \wedge (q \vee \sim p)] \rightarrow \sim p$.

Solution
Using the generalized procedure for truth table construction, we produce the following table.

p	q	$[p$	$\wedge$	$(q$	$\vee$	$\sim p)]$	$\rightarrow$	$\sim p$
T	T	T	T	T	T	F	F	F
T	F	T	F	F	F	F	T	F
F	T	F	F	T	T	T	T	T
F	F	F	F	F	T	T	T	T

| 1 | | 6 | | 2 | | 5 | | 3 | | 7 | | 4 |

▼ **check your progress 3** Construct a truth table for $[p \wedge (p \rightarrow q)] \rightarrow q$.

Solution *See page S9.*

MATHMATTERS Use Conditional Statements to Control a Calculator Program

Reprinted with permission of Texas Instruments

TI-83

Computer and calculator programs use conditional statements to control the flow of a program. For instance, the "If...Then" instruction in a TI-83 or TI-84 calculator program directs the calculator to execute a group of commands if a condition is true and to skip to the End statement if the condition is false. See the program steps below.

:If *condition*

:Then (skip to End if *condition* is false)

:*command* if *condition* is true

:End

:*command*

The TI-83/84 program FACTOR shown above, in the left margin, factors a natural number N into its prime factors. Note the use of the "If...Then" instructions highlighted in red.

An Equivalent Form of the Conditional

TABLE 3.9 Truth Table
for $\sim p \vee q$

p	q	$\sim p \vee q$
T	T	T
T	F	F
F	T	T
F	F	T

The truth table for $\sim p \vee q$ is shown in Table 3.9. The truth values in this table are identical to the truth values in Table 3.8 on page 134. Hence, the conditional $p \rightarrow q$ is equivalent to the disjunction $\sim p \vee q$.

▼ **An Equivalent Form of the Conditional $p \rightarrow q$**

$$p \rightarrow q \equiv \sim p \vee q$$

▼ **example 4**   **Write a Conditional in Its Equivalent Disjunctive Form**

Write each of the following in its equivalent disjunctive form.

a. If I could play the guitar, I would join the band.

b. If David Beckham cannot play, then his team will lose.

Solution

In each case we write the disjunction of the negation of the antecedent and the consequent.

a. I cannot play the guitar or I would join the band.

b. David Beckham can play or his team will lose.

▼ **check your progress 4**   Write each of the following in its equivalent disjunctive form.

a. If I don't move to Georgia, I will live in Houston.

b. If the number is divisible by 2, then the number is even.

Solution   *See page S9.*

The Negation of the Conditional

Because $p \rightarrow q \equiv \sim p \vee q$, an equivalent form of $\sim(p \rightarrow q)$ is given by $\sim(\sim p \vee q)$, which, by one of De Morgan's laws, can be expressed as the conjunction $p \wedge \sim q$.

▼ **The Negation of $p \rightarrow q$**

$$\sim(p \rightarrow q) \equiv p \wedge \sim q$$

▼ **example 5**   **Write the Negation of a Conditional Statement**

Write the negation of each conditional statement.

a. If they pay me the money, I will sign the contract.

b. If the lines are parallel, then they do not intersect.

Solution

In each case, we write the conjunction of the antecedent and the negation of the consequent.

a. They paid me the money and I did not sign the contract.

b. The lines are parallel and they intersect.

▼ **check your progress 5**   Write the negation of each conditional statement.

a. If I finish the report, I will go to the concert.

b. If the square of n is 25, then n is 5 or -5.

Solution   *See page S9.*

The Biconditional

The statement $(p \rightarrow q) \wedge (q \rightarrow p)$ is called a **biconditional** and is denoted by $p \leftrightarrow q$, which is read as "*p* if and only if *q*."

> ▼ **The Biconditional $p \leftrightarrow q$**
>
> $$p \leftrightarrow q \equiv [(p \rightarrow q) \wedge (q \rightarrow p)]$$

TABLE 3.10 Truth Table for $p \leftrightarrow q$

p	q	$p \leftrightarrow q$
T	T	T
T	F	F
F	T	F
F	F	T

Table 3.10 shows that $p \leftrightarrow q$ is true only when p and q have the same truth value.

▼ **example 6**   Determine the Truth Value of a Biconditional

State whether each biconditional is true or false.

a. $x + 4 = 7$ if and only if $x = 3$.

b. $x^2 = 36$ if and only if $x = 6$.

Solution

a. Both equations are true when $x = 3$, and both are false when $x \neq 3$. Both equations have the same truth value for any value of x, so this is a true statement.

b. If $x = -6$, the first equation is true and the second equation is false. Thus this is a false statement.

▼ **check your progress 6**   State whether each biconditional is true or false.

a. $x > 7$ if and only if $x > 6$.

b. $x + 5 > 7$ if and only if $x > 2$.

Solution   *See page S9.*

EXCURSION

Logic Gates

Modern digital computers use *gates* to process information. These gates are designed to receive two types of electronic impulses, which are generally represented as a 1 or a 0. Figure 3.9, on page 138, shows a *NOT gate*. It is constructed so that a stream of impulses that enter the gate will exit the gate as a stream of impulses in which each 1 is converted to a 0 and each 0 is converted to a 1.

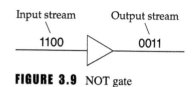

Input stream   Output stream

1100   0011

FIGURE 3.9 NOT gate

Note the similarity between the logical connective *not* and the logic gate NOT. The *not* connective converts the sequence of truth values T F to F T. The NOT gate converts the input stream 1 0 to 0 1. If the 1s are replaced with Ts and the 0s with Fs, then the NOT logic gate yields the same results as the *not* connective.

Many gates are designed so that two input streams are converted to one output stream. For instance, Figure 3.10 shows an *AND gate*. The AND gate is constructed so that a 1 is the output if and only if both input streams have a 1. In any other situation, a 0 is produced as the output.

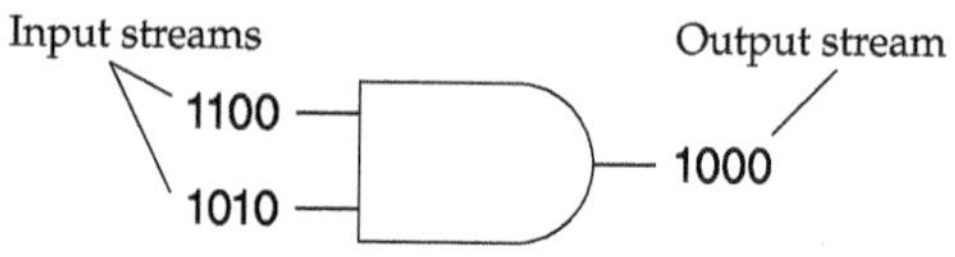

Input streams   Output stream

1100

1010   1000

FIGURE 3.10 AND gate

Note the similarity between the logical connective *and* and the logic gate AND. The *and* connective combines the sequence of truth values T T F F with the truth values T F T F to produce T F F F. The AND gate combines the input stream 1 1 0 0 with the input stream 1 0 1 0 to produce 1 0 0 0. If the 1s are replaced with Ts and the 0s with Fs, then the AND logic gate yields the same result as the *and* connective.

The *OR gate* is constructed so that its output is a 0 if and only if both input streams have a 0. All other situations yield a 1 as the output. See Figure 3.11.

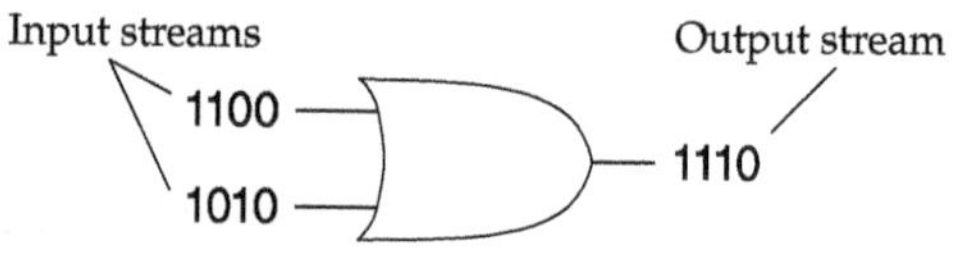

Input streams   Output stream

1100

1010   1110

FIGURE 3.11 OR gate

Figure 3.12 shows a network that consists of a NOT gate and an AND gate.

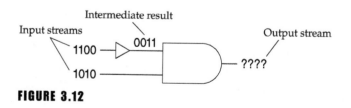

Intermediate result

Input streams   Output stream

1100   0011

1010   ????

FIGURE 3.12

question What is the output stream for the network in Figure 3.12?

answer 0 0 1 0

EXCURSION EXERCISES

1. For each of the following, determine the output stream for the given input streams.

    a. Input streams   Output stream

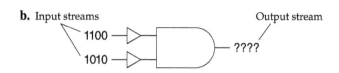

    b. Input streams   Output stream

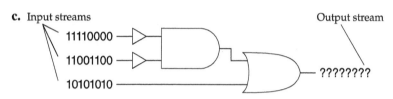

    c. Input streams   Output stream

2. Construct a network using NOT, AND, and OR gates as needed that accepts the two input streams 1 1 0 0 and 1 0 1 0 and produces the output stream 0 1 1 1.

EXERCISE SET 3.3   (Suggested Assignment: The Enhanced WebAssign Exercises and Exercises 5, 17, 35, 51, and 63)

■ In Exercises 1 to 6, identify the antecedent and the consequent of each conditional statement.

1. If I had the money, I would buy the painting.

2. If Shelly goes on the trip, she will not be able to take part in the graduation ceremony.

3. If they had a guard dog, then no one would trespass on their property.

4. If I don't get to school before 7:30, I won't be able to find a parking place.

5. If I change my major, I must reapply for admission.

6. If your blood type is type O−, then you are classified as a universal blood donor.

■ In Exercises 7 to 14, determine the truth value of the given statement.

7. If x is an even integer, then x^2 is an even integer.

8. If x is a prime number, then $x + 2$ is a prime number.

9. If all frogs can dance, then today is Monday.

10. If all cats are black, then I am a millionaire.

11. If $4 < 3$, then $7 = 8$.

12. If $x < 2$, then $x + 5 < 7$.

13. If $|x| = 6$, then $x = 6$.

14. If $\pi = 3$, then $2\pi = 6$.

■ In Exercises 15 to 24, construct a truth table for the given statement.

15. $(p \wedge \sim q) \rightarrow [\sim(p \wedge q)]$

16. $[(p \rightarrow q) \wedge p] \rightarrow p$

17. $[(p \rightarrow q) \wedge p] \rightarrow q$

18. $(\sim p \vee \sim q) \rightarrow \sim(p \wedge q)$

19. $[r \wedge (\sim p \vee q)] \rightarrow (r \vee \sim q)$

20. $[(p \rightarrow \sim r) \wedge q] \rightarrow \sim r$

21. $[(p \rightarrow q) \vee (r \wedge \sim p)] \rightarrow (r \vee \sim q)$

22. $\{p \wedge [(p \rightarrow q) \wedge (q \rightarrow r)]\} \rightarrow r$

23. $[\sim(p \rightarrow \sim r) \wedge \sim q] \rightarrow r$

24. $[p \wedge (r \rightarrow \sim q)] \rightarrow (r \vee q)$

■ In Exercises 25 to 30, write each conditional statement in its equivalent disjunctive form.

25. If she could sing, she would be perfect for the part.

26. If he does not get frustrated, he will be able to complete the job.

27. If x is an irrational number, then x is not a terminating decimal.

28. If Mr. Hyde had a brain, he would be dangerous.

29. If the fog does not lift, our flight will be cancelled.

30. If the Yankees win the pennant, Carol will be happy.

■ In Exercises 31 to 36, write the negation of each conditional statement in its equivalent conjunctive form.

31. If they offer me the contract, I will accept.

32. If I paint the house, I will get the money.

33. If pigs had wings, pigs could fly.

34. If we had a telescope, we could see that comet.

35. If she travels to Italy, she will visit her relatives.

36. If Paul could play better defense, he could be a professional basketball player.

■ In Exercises 37 to 46, state whether the given biconditional is true or false. Assume that x and y are real numbers.

37. $x^2 = 9$ if and only if $x = 3$.

38. x is a positive number if and only if $x > 0$.

39. $|x|$ is a positive number if and only if $x \neq 0$.

40. $|x + y| = x + y$ if and only if $x + y > 0$.

41. A number is a rational number if and only if the number can be written as a terminating decimal.

42. $0.\overline{3}$ is a rational number if and only if $\frac{1}{3}$ is a rational number.

43. $4 = 7$ if and only if $2 = 3$.

44. x is an even number if and only if x is not an odd number.

45. Triangle ABC is an equilateral triangle if and only if triangle ABC is an equiangular triangle.

46. Today is March 1 if and only if yesterday was February 28.

■ In Exercises 47 to 52, let v represent "I will take a vacation," let p represent "I get the promotion," and let t represent "I am transferred." Write each of the following statements in symbolic form.

47. If I get the promotion, I will take a vacation.

48. If I am not transferred, I will take a vacation.

49. If I am transferred, then I will not take a vacation.

50. If I will not take a vacation, then I will not be transferred and I get the promotion.

51. If I am not transferred and I get the promotion, then I will take a vacation.

52. If I get the promotion, then I am transferred and I will take a vacation.

■ In Exercises 53 to 58, construct a truth table for each statement to determine if the statements are equivalent.

53. $p \to \sim r, r \lor \sim p$

54. $p \to q, q \to p$

55. $\sim p \to (p \lor r), r$

56. $p \to q, \sim q \to \sim p$

57. $p \to (q \lor r), (p \to q) \lor (p \to r)$

58. $\sim q \to p, p \lor q$

EXTENSIONS

Critical Thinking

The statement, "All squares are rectangles," can be written as "If a figure is a square, then it is a rectangle." In Exercises 59 to 62, write each statement given in "All Xs are Ys" form in the form "If it is an X, then it is a Y."

59. All rational numbers are real numbers.

60. All whole numbers are integers.

61. All Sauropods are herbivorous.

62. All paintings by Vincent van Gogh are valuable.

Coperative Learning

63. **Recreational Logic** The field of a new soccer stadium is watered by three individual sprinkler systems, as shown by the A, B, and C regions in the figure at the right. Each sprinkler system is controlled by exactly one of three on-off valves in an underground maintenance room, and each sprinkler system can be turned on without turning on the other two systems. Each of the valves is presently in the off position, and the field is dry. The valves have not been labeled, so you do not know which valve controls which sprinkler system. You want to correctly label the valves as A, B, and C. You also want to do it by making only one trip up to

the field. You cannot see the field from the maintenance room, and no one is available to help you. What procedure can you use to determine how to correctly label the valves? Assume that all of the valves and all of the sprinkler systems are operating properly. Also assume that the sprinklers are either completely off or completely on. Explain your reasoning.

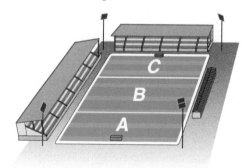

Explorations

64. A Factor Program If you have access to a TI-83 or a TI-84 calculator, enter the program FACTOR on page 135 into the calculator and demonstrate the program to your classmates.

section 3.4   The Conditional and Related Statements

Equivalent Forms of the Conditional

Every conditional statement can be stated in many equivalent forms. It is not even necessary to state the antecedent before the consequent. For instance, the conditional "If I live in Boston, then I must live in Massachusetts" can also be stated as

I must live in Massachusetts, if I live in Boston.

Table 3.11 lists some of the various forms that may be used to write a conditional statement.

TABLE 3.11 Common Forms of $p \rightarrow q$

Every conditional statement $p \rightarrow q$ can be written in the following equivalent forms.	
If p, then q.	Every p is a q.
If p, q.	q, if p.
p only if q.	q provided that p.
p implies q.	q is a necessary condition for p.
Not p or q.	p is a sufficient condition for q.

▼ example 1   Write a Statement in an Equivalent Form

Write each of the following in "If p, then q" form.

a. The number is an even number provided that it is divisible by 2.

b. Today is Friday, only if yesterday was Thursday.

Solution

a. The statement, "The number is an even number provided that it is divisible by 2," is in "q provided that p" form. The antecedent is "it is divisible by 2," and the consequent is "the number is an even number." Thus its "If p, then q" form is

   If it is divisible by 2, then the number is an even number.

b. The statement, "Today is Friday, only if yesterday was Thursday," is in "p only if q" form. The antecedent is "today is Friday." The consequent is "yesterday was Thursday." Its "If p, then q" form is

   If today is Friday, then yesterday was Thursday.

▼ check your progress 1   Write each of the following in "If p, then q" form.

a. Every square is a rectangle.

b. Being older than 30 is sufficient to show that I am at least 21.

Solution   *See page S9.*   ◄

The Converse, the Inverse, and the Contrapositive

Every conditional statement has three related statements. They are called the *converse,* the *inverse,* and the *contrapositive.*

▼ Statements Related to the Conditional Statement

The **converse** of $p \rightarrow q$ is $q \rightarrow p$.

The **inverse** of $p \rightarrow q$ is $\sim p \rightarrow \sim q$.

The **contrapositive** of $p \rightarrow q$ is $\sim q \rightarrow \sim p$.

The above definitions show the following:

- The converse of $p \rightarrow q$ is formed by interchanging the antecedent p with the consequent q.
- The inverse of $p \rightarrow q$ is formed by negating the antecedent p and negating the consequent q.
- The contrapositive of $p \rightarrow q$ is formed by negating both the antecedent p and the consequent q and interchanging these negated statements.

▼ example 2   Write the Converse, Inverse, and Contrapositive of a Conditional

Write the converse, inverse, and contrapositive of

If I get the job, then I will rent the apartment.

Solution
Converse: If I rent the apartment, then I get the job.
Inverse: If I do not get the job, then I will not rent the apartment.
Contrapositive: If I do not rent the apartment, then I did not get the job.

▼ check your progress 2   Write the converse, inverse, and contrapositive of

If we have a quiz today, then we will not have a quiz tomorrow.

Solution   *See page S10.*   ◀

Table 3.12 shows that any conditional statement is equivalent to its contrapositive and that the converse of a conditional statement is equivalent to the inverse of the conditional statement.

TABLE 3.12   Truth Tables for Conditional and Related Statements

p	q	Conditional $p \rightarrow q$	Converse $q \rightarrow p$	Inverse $\sim p \rightarrow \sim q$	Contrapositive $\sim q \rightarrow \sim p$
T	T	T	T	T	T
T	F	F	T	T	F
F	T	T	F	F	T
F	F	T	T	T	T

$$q \rightarrow p \equiv \sim p \rightarrow \sim q$$
$$p \rightarrow q \equiv \sim q \rightarrow \sim p$$

▼ example 3   Determine Whether Related Statements Are Equivalent

Determine whether the given statements are equivalent.

a. If a number ends with a 5, then the number is divisible by 5.
   If a number is divisible by 5, then the number ends with a 5.

b. If two lines in a plane do not intersect, then the lines are parallel.

If two lines in a plane are not parallel, then the lines intersect.

Solution

a. The second statement is the converse of the first. The statements are not equivalent.

b. The second statement is the contrapositive of the first. The statements are equivalent.

▼ check your progress 3   Determine whether the given statements are equivalent.

a. If $a = b$, then $a \cdot c = b \cdot c$.

If $a \neq b$, then $a \cdot c \neq b \cdot c$.

b. If I live in Nashville, then I live in Tennessee.

If I do not live in Tennessee, then I do not live in Nashville.

Solution   *See page S10.*   ◄

In mathematics, it is often necessary to prove statements that are in "If p, then q" form. If a proof cannot be readily produced, mathematicians often try to prove the contrapositive "If $\sim q$, then $\sim p$." Because a conditional and its contrapositive are equivalent statements, a proof of either statement also establishes the proof of the other statement.

question   A mathematician wishes to prove the following statement about the integer x.

Statement (I):   If x^2 is an odd integer, then x is an odd integer.

If the mathematician is able to prove the statement, "If x is an even integer, then x^2 is an even integer," does this also prove statement (I)?

▼ example 4   **Use the Contrapositive to Determine a Truth Value**

Write the contrapositive of each statement and use the contrapositive to determine whether the original statement is true or false.

a. If $a + b$ is not divisible by 5, then a and b are not both divisible by 5.

b. If x^3 is an odd integer, then x is an odd integer. (Assume x is an integer.)

c. If a geometric figure is not a rectangle, then it is not a square.

Solution

a. If a and b are both divisible by 5, then $a + b$ is divisible by 5. This is a true statement, so the original statement is also true.

b. If x is an even integer, then x^3 is an even integer. This is a true statement, so the original statement is also true.

c. If a geometric figure is a square, then it is a rectangle. This is a true statement, so the original statement is also true.

▼ check your progress 4   Write the contrapositive of each statement and use the contrapositive to determine whether the original statement is true or false.

a. If $3 + x$ is an odd integer, then x is an even integer. (Assume x is an integer.)

b. If two triangles are not similar triangles, then they are not congruent triangles. *Note:* Similar triangles have the same shape. Congruent triangles have the same size and shape.

c. If today is not Wednesday, then tomorrow is not Thursday.

Solution   *See page S10.*   ◄

answer   Yes, because the second statement is the contrapositive of statement (I).

MATHMATTERS Grace Hopper

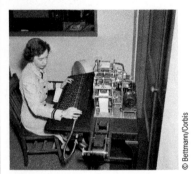

Rear Admiral Grace Hopper

Grace Hopper (1906–1992) was a visionary in the field of computer programming. She was a mathematics professor at Vassar from 1931 to 1943, but she retired from teaching to start a career in the U.S. Navy at the age of 37.

The Navy assigned Hopper to the Bureau of Ordnance Computation at Harvard University. It was here that she was given the opportunity to program computers. It has often been reported that she was the third person to program the world's first large-scale digital computer. Grace Hopper had a passion for computers and computer programming. She wanted to develop a computer language that would be user-friendly and enable people to use computers in a more productive manner.

Grace Hopper had a long list of accomplishments. She designed some of the first computer compilers, she was one of the first to introduce English commands into computer languages, and she wrote the precursor to the computer language COBOL.

Grace Hopper retired from the Navy (for the first time) in 1966. In 1967 she was recalled to active duty and continued to serve in the Navy until 1986, at which time she was the nation's oldest active duty officer.

In 1951, the UNIVAC I computer that Grace Hopper was programming started to malfunction. The malfunction was caused by a moth that had become lodged in one of the computer's relays. Grace Hopper pasted the moth into the UNIVAC I logbook with a label that read, "computer bug." Since then computer programmers have used the word *bug* to indicate any problem associated with a computer program. Modern computers use logic gates instead of relays to process information, so actual bugs are not a problem; however, bugs such as the "Year 2000 bug" can cause serious problems.

EXCURSION

Sheffer's Stroke and the NAND Gate

In 1913, the logician Henry M. Sheffer created a connective that we now refer to as *Sheffer's stroke* (or *NAND*). This connective is often denoted by the symbol $|$. Table 3.13 shows that $p\,|\,q$ is equivalent to $\sim(p \wedge q)$. Sheffer's stroke $p\,|\,q$ is false when both p and q are true, and it is true in all other cases.

Any logic statement can be written using only Sheffer's stroke connectives. For instance, Table 3.14 shows that $p\,|\,p \equiv \sim p$ and $(p\,|\,p)\,|\,(q\,|\,q) \equiv p \vee q$.

Figure 3.13 shows a logic gate called a *NAND gate*. This gate models the Sheffer's stroke connective in that its output is 0 when both input streams are 1 and its output is 1 in all other cases.

TABLE 3.13
Sheffer's Stroke

p	q	$p\,|\,q$
T	T	F
T	F	T
F	T	T
F	F	T

TABLE 3.14

p	q	$p\,|\,p$	$(p\,|\,p)\,|\,(q\,|\,q)$
T	T	F	T
T	F	F	T
F	T	T	T
F	F	T	F

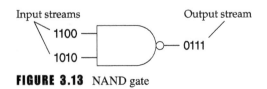

Input streams → 1100, 1010 — NAND gate — Output stream → 0111

FIGURE 3.13 NAND gate

EXCURSION EXERCISES

1. a. Complete a truth table for $p|(q|q)$.

   b. Use the results of Excursion Exercise 1a to determine an equivalent statement for $p|(q|q)$.

2. a. Complete a truth table for $(p|q)|(p|q)$.

   b. Use the results of Excursion Exercise 2a to determine an equivalent statement for $(p|q)|(p|q)$.

3. a. Determine the output stream for the following network of NAND gates. *Note:* In a network of logic gates, a solid circle • is used to indicate a connection. A symbol such as ⇇ is used to indicate "no connection."

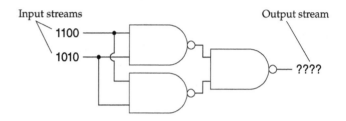

FIGURE 3.14

   b. What logic gate is modeled by the network in Figure 3.14?

4. NAND gates are functionally complete in that any logic gate can be constructed using only NAND gates. Construct a network of NAND gates that would produce the same output stream as an OR gate.

EXERCISE SET 3.4   (Suggested Assignment: The Enhanced WebAssign Exercises and Exercises 37, 41, and 43)

■ In Exercises 1 to 10, write each statement in "If p, then q" form.

1. We will be in good shape for the ski trip provided that we take the aerobics class.

2. We can get a dog only if we install a fence around the backyard.

3. Every odd prime number is greater than 2.

4. The triangle is a 30°-60°-90° triangle, if the length of the hypotenuse is twice the length of the shorter leg.

5. He can join the band, if he has the talent to play a keyboard.

6. Every theropod is carnivorous.

7. I will be able to prepare for the test only if I have the textbook.

8. I will be able to receive my credential provided that Education 147 is offered in the spring semester.

9. Being in excellent shape is a necessary condition for running the Boston marathon.

10. If it is an ankylosaur, it is quadrupedal.

■ In Exercises 11 to 24, write the **a.** converse, **b.** inverse, and **c.** contrapositive of the given statement.

11. If I were rich, I would quit this job.

12. If we had a car, then we would be able to take the class.

13. If she does not return soon, we will not be able to attend the party.

14. I will be in the talent show only if I can do the same comedy routine I did for the banquet.

15. Every parallelogram is a quadrilateral.

16. If you get the promotion, you will need to move to Denver.

17. I would be able to get current information about astronomy provided I had access to the Internet.

18. You need four-wheel drive to make the trip to Death Valley.

19. We will not have enough money for dinner, if we take a taxi.

20. If you are the president of the United States, then your age is at least 35.

21. She will visit Kauai only if she can extend her vacation for at least two days.

22. In a right triangle, the acute angles are complementary.

23. Two lines perpendicular to a given line are parallel.

24. If $x + 5 = 12$, then $x = 7$.

■ In Exercises 25 to 30, determine whether the given statements are equivalent.

25. If Kevin wins, we will celebrate.
If we celebrate, then Kevin will win.

26. If I save $1000, I will go on the field trip.
If I go on the field trip, then I saved $1000.

27. If she attends the meeting, she will make the sale.
If she does not make the sale, then she did not attend the meeting.

28. If you understand algebra, you can remember algebra.
If you do not understand algebra, you cannot remember algebra.

29. If $a > b$, then $ac > bc$.
If $a \le b$, then $ac \le bc$.

30. If $a < b$, then $\dfrac{1}{a} > \dfrac{1}{b}$.
If $\dfrac{1}{a} \le \dfrac{1}{b}$, then $a \ge b$.
(Assume $a \ne 0$ and $b \ne 0$.)

■ In Exercises 31 to 36, write the contrapositive of the statement and use the contrapositive to determine whether the given statement is true or false.

31. If $3x - 7 = 11$, then $x \ne 7$.

32. If $x \ne 3$, then $5x + 7 \ne 22$.

33. If $a \ne 3$, then $|a| \ne 3$.

34. If $a + b$ is divisible by 3, then a is divisible by 3 and b is divisible by 3.

35. If $\sqrt{a + b} \ne 5$, then $a + b \ne 25$.

36. If x^2 is an even integer, then x is an even integer. (Assume x is an integer.)

37. What is the converse of the inverse of the contrapositive of $p \to q$?

38. What is the inverse of the converse of the contrapositive of $p \to q$?

EXTENSIONS

Critical Thinking

39. Give an example of a true conditional statement whose
a. converse is true.　　**b.** converse is false.

40. Give an example of a true conditional statement whose
a. inverse is true.　　**b.** inverse is false.

■ In Exercises 41 to 44, determine the original statement if the given statement is related to the original in the manner indicated.

41. *Converse:* If you can do it, you can dream it.

42. *Inverse:* If I did not have a dime, I would not spend it.

43. *Contrapositive:* If I were a singer, I would not be a dancer.

44. *Negation:* Pigs have wings and pigs cannot fly.

45. Explain why it is not possible to find an example of a true conditional statement whose contrapositive is false.

46. If a conditional statement is false, must its converse be true? Explain.

47. **A Puzzle** Lewis Carroll (Charles Dodgson) wrote many puzzles, many of which he recorded in his diaries. Solve the following puzzle, which appears in one of his diaries.

The Granger Collection

The Dodo says that the Hatter tells lies.
The Hatter says that the March Hare tells lies.
The March Hare says that both the Dodo and the Hatter tell lies.
Who is telling the truth?[1]

Hint: Consider the three different cases in which only one of the characters is telling the truth. In only one of these cases can all three of the statements be true.

---

[1] This puzzle is from *Lewis Carroll's Games and Puzzles*, compiled and edited by Edward Wakeling. New York: Dover Publications, Inc., copyright 1992, p. 11, puzzle 9, "Who's Telling the Truth?"

Cooperative Learning

48. Recreational Logic Consider a checkerboard with two red squares on opposite corners removed, as shown in the figure at the right. Determine whether it is possible to completely cover the checker board with 31 dominoes if each domino is placed horizontally or vertically and each domino covers exactly two squares. If it is possible, show how to do it. If it is not possible, explain why it cannot be done.

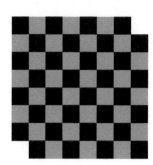

<hr>

section 3.5 Symbolic Arguments

Arguments

HISTORICAL NOTE

Aristotle
(ăr'ĭ-stŏt'l)
(384–322 B.C.) was an ancient Greek philosopher who studied under Plato. He wrote about many subjects, including logic, biology, politics, astronomy, metaphysics, and ethics. His ideas about logic and the reasoning process have had a major impact on mathematics and philosophy.

In this section we consider methods of analyzing arguments to determine whether they are *valid* or *invalid*. For instance, consider the following argument.

> If Aristotle was human, then Aristotle was mortal. Aristotle was human. Therefore, Aristotle was mortal.

To determine whether the above argument is a valid argument, we must first define the terms *argument* and *valid argument*.

> ### ▼ An Argument and a Valid Argument
>
> An **argument** consists of a set of statements called **premises** and another statement called the **conclusion**. An argument is **valid** if the conclusion is true whenever all the premises are assumed to be true. An argument is **invalid** if it is not a valid argument.

In the argument about Aristotle, the two premises and the conclusion are shown below. It is customary to place a horizontal line between the premises and the conclusion.

First Premise:	If Aristotle was human, then Aristotle was mortal.
Second Premise:	Aristotle was human.
Conclusion:	Therefore, Aristotle was mortal.

Arguments can be written in **symbolic form**. For instance, if we let h represent the statement "Aristotle was human" and m represent the statement "Aristotle was mortal," then the argument can be expressed as

$$h \rightarrow m$$
$$h$$
$$\therefore m$$

The three dots $\therefore$ are a symbol for "therefore."

▼ example 1 Write an Argument in Symbolic Form

Write the following argument in symbolic form.

> The fish is fresh or I will not order it. The fish is fresh. Therefore I will order it.

Solution

Let *f* represent the statement "The fish is fresh." Let *o* represent the statement "I will order it." The symbolic form of the argument is

$$f \lor \sim o$$
$$\underline{f}$$
$$\therefore o$$

▼ **check your progress 1** Write the following argument in symbolic form.

If she doesn't get on the plane, she will regret it. She does not regret it.
Therefore, she got on the plane.

Solution *See page S10.*

Arguments and Truth Tables

The following truth table procedure can be used to determine whether an argument is valid or invalid.

▼ **Truth Table Procedure to Determine the Validity of an Argument**

1. Write the argument in symbolic form.

2. Construct a truth table that shows the truth value of each premise and the truth value of the conclusion for all combinations of truth values of the simple statements.

3. If the conclusion is true in every row of the truth table in which all the premises are true, the argument is valid. If the conclusion is false in any row in which all of the premises are true, the argument is invalid.

We will now use the above truth table procedure to determine the validity of the argument about Aristotle.

1. Once again we let *h* represent the statement "Aristotle was human" and *m* represent the statement "Aristotle was mortal." In symbolic form the argument is

$h \to m$	First premise
$\underline{h}$	Second premise
$\therefore m$	Conclusion

2. Construct a truth table as shown below.

h	*m*	First premise $h \to m$	Second premise h	Conclusion m	
T	T	T	T	T	row 1
T	F	F	T	F	row 2
F	T	T	F	T	row 3
F	F	T	F	F	row 4

3. Row 1 is the only row in which all the premises are true, so it is the only row that we examine. Because the conclusion is true in row 1, the argument is valid.

In Example 2, we use the truth table method to determine the validity of a more complicated argument.

▼ example 2   Determine the Validity of an Argument

Determine whether the following argument is valid or invalid.

> If it rains, then the game will not be played. It is not raining. Therefore, the game will be played.

Solution

If we let r represent "it rains" and g represent "the game will be played," then the symbolic form is

$$r \rightarrow \sim g$$
$$\underline{\sim r}$$
$$\therefore g$$

The truth table for this argument follows.

r	g	First premise $r \rightarrow \sim g$	Second premise $\sim r$	Conclusion g	
T	T	F	F	T	row 1
T	F	T	F	F	row 2
F	T	T	T	T	row 3
F	F	T	T	F	row 4

question   Why do we need to examine only rows 3 and 4?

Because the conclusion in row 4 is false and the premises are both true, the argument is invalid.

▼ check your progress 2   Determine the validity of the following argument.

> If the stock market rises, then the bond market will fall.
> The bond market did not fall.
> ∴ The stock market did not rise.

Solution   *See page S10.*   ◄

The argument in Example 3 involves three statements. Thus we use a truth table with $2^3 = 8$ rows to determine the validity of the argument.

▼ example 3   Determine the Validity of an Argument

Determine whether the following argument is valid or invalid.

> If I am going to run the marathon, then I will buy new shoes.
> If I buy new shoes, then I will not buy a television.
> ∴ If I buy a television, I will not run the marathon.

answer   Rows 3 and 4 are the only rows in which all of the premises are true.

Solution
Label the statements

> *m:*   I am going to run the marathon.
>
> *s:*   I will buy new shoes.
>
> *t:*   I will buy a television.

The symbolic form of the argument is

> $m \rightarrow s$
>
> $\underline{s \rightarrow \sim t}$
>
> $\therefore t \rightarrow \sim m$

The truth table for this argument follows.

m	*s*	*t*	First premise $m \rightarrow s$	Second premise $s \rightarrow \sim t$	Conclusion $t \rightarrow \sim m$	
T	T	T	T	F	F	row 1
T	T	F	T	T	T	row 2
T	F	T	F	T	F	row 3
T	F	F	F	T	T	row 4
F	T	T	T	F	T	row 5
F	T	F	T	T	T	row 6
F	F	T	T	T	T	row 7
F	F	F	T	T	T	row 8

The only rows in which both premises are true are rows 2, 6, 7, and 8. Because the conclusion is true in each of these rows, the argument is valid.

▼ check your progress **3**   Determine whether the following argument is valid or invalid.

> If I arrive before 8 A.M., then I will make the flight.
>
> If I make the flight, then I will give the presentation.
>
> $\therefore$ If I arrive before 8 A.M., then I will give the presentation.

Solution   *See page S10.*

Standard Forms

Some arguments can be shown to be valid if they have the same symbolic form as an argument that is known to be valid. For instance, we have shown that the argument

> $h \rightarrow m$
>
> $\underline{h \qquad}$
>
> $\therefore m$

is valid. This symbolic form is known as **modus ponens** or the **law of detachment**. All arguments that have this symbolic form are valid. Table 3.15 shows four symbolic forms and the name used to identify each form. Any argument that has a symbolic form identical to one of these symbolic forms is a valid argument.

In logic, the ability to identify standard forms of arguments is an important skill. If an argument has one of the standard forms in Table 3.15, then it is a valid argument. If an argument has one of the standard forms in Table 3.16, then it is an invalid argument. The standard forms can be thought of as laws of logic. Concerning the laws of logic, the logician Gottlob Frege (frā'gə) (1848–1925) stated, "The laws of logic are not like the laws of nature. They … are laws of the laws of nature."

TABLE 3.15 Standard Forms of Four Valid Arguments

Modus ponens	Modus tollens	Law of syllogism	Disjunctive syllogism
$p \rightarrow q$	$p \rightarrow q$	$p \rightarrow q$	$p \vee q$
p	$\sim q$	$q \rightarrow r$	$\sim p$
$\therefore q$	$\therefore \sim p$	$\therefore p \rightarrow r$	$\therefore q$

The law of syllogism can be extended to include more than two conditional premises. For example, if the premises of an argument are $a \rightarrow b$, $b \rightarrow c$, $c \rightarrow d$, …, $y \rightarrow z$, then a valid conclusion for the argument is $a \rightarrow z$. We will refer to any argument of this form with more than two conditional premises as the **extended law of syllogism**.

Table 3.16 shows two symbolic forms associated with invalid arguments. Any argument that has one of these symbolic forms is invalid.

TABLE 3.16 Standard Forms of Two Invalid Arguments

Fallacy of the converse	Fallacy of the inverse
$p \rightarrow q$	$p \rightarrow q$
q	$\sim p$
$\therefore p$	$\therefore \sim q$

▼ **example 4**   Use a Standard Form to Determine the Validity of an Argument

Use a standard form to determine whether the following argument is valid or invalid.

The program is interesting or I will watch the basketball game.
The program is not interesting.

∴ I will watch the basketball game.

Solution

Label the statements

   i: The program is interesting.
   w: I will watch the basketball game.

In symbolic form the argument is

   $i \vee w$
   $\sim i$
   $\therefore w$

This symbolic form matches the standard form known as disjunctive syllogism. Thus the argument is valid.

▼ **check your progress 4**   Use a standard form to determine whether the following argument is valid or invalid.

If I go to Florida for spring break, then I will not study.
I did not go to Florida for spring break.

∴ I studied.

Solution   *See page S11.*

Waterfall by M.C. Escher

M. C. Escher (1898–1972) created many works of art that defy logic. In this lithograph, the water completes a full cycle even though the water is always traveling downward.

Consider an argument with the following symbolic form.

$q \rightarrow r$	Premise 1
$r \rightarrow s$	Premise 2
$\sim t \rightarrow \sim s$	Premise 3
q	Premise 4
$\therefore t$	

To determine whether the argument is valid or invalid using a truth table would require a table with $2^4 = 16$ rows. It would be time consuming to construct such a table and, with the large number of truth values to be determined, we might make an error. Thus we consider a different approach that makes use of a sequence of valid arguments to arrive at a conclusion.

$q \rightarrow r$	Premise 1
$r \rightarrow s$	Premise 2
$\therefore q \rightarrow s$	Law of syllogism

$q \rightarrow s$	The previous conclusion
$s \rightarrow t$	Premise 3 expressed in an equivalent form
$\therefore q \rightarrow t$	Law of syllogism

$q \rightarrow t$	The previous conclusion
q	Premise 4
$\therefore t$	Modus ponens

This sequence of valid arguments shows that t is a valid conclusion for the original argument.

▼ **example** **5**   **Determine the Validity of an Argument**

Determine whether the following argument is valid.

If the movie was directed by Steven Spielberg (s), then I want to see it (w). The movie's production costs must exceed $50 million ($c$) or I do not want to see it. The movie's production costs were less than $50 million. Therefore, the movie was not directed by Steven Spielberg.

Solution
In symbolic form the argument is

$s \rightarrow w$	Premise 1
$c \vee \sim w$	Premise 2
$\sim c$	Premise 3
$\therefore \sim s$	Conclusion

Premise 2 can be written as $\sim w \vee c$, which is equivalent to $w \rightarrow c$. Applying the law of syllogism to Premise 1 and this equivalent form of Premise 2 produces

$s \rightarrow w$	Premise 1
$w \rightarrow c$	Equivalent form of Premise 2
$\therefore s \rightarrow c$	Law of syllogism

Combining the above conclusion $s \rightarrow c$ with Premise 3 gives us

$$s \rightarrow c \qquad \text{Conclusion from above}$$
$$\underline{\sim c} \qquad \text{Premise 3}$$
$$\therefore \sim s \qquad \text{Modus tollens}$$

This sequence of valid arguments has produced the desired conclusion, $\sim s$. Thus the original argument is valid.

▼ **check your progress 5** Determine whether the following argument is valid.

I start to fall asleep if I read a math book. I drink soda whenever I start to fall asleep. If I drink a soda, then I must eat a candy bar. Therefore, I eat a candy bar whenever I read a math book.

Hint: p whenever *q* is equivalent to $q \rightarrow p$.

Solution *See page S11.*

In the next example, we use standard forms to determine a valid conclusion for an argument.

▼ **example 6** Determine a Valid Conclusion for an Argument

Use all of the premises to determine a valid conclusion for the following argument.

We will not go to Japan ($\sim j$) or we will go to Hong Kong (h). If we visit my uncle (u), then we will go to Singapore (s). If we go to Hong Kong, then we will not go to Singapore.

Solution
In symbolic form the argument is

$$\sim j \vee h \qquad \text{Premise 1}$$
$$u \rightarrow s \qquad \text{Premise 2}$$
$$\underline{h \rightarrow \sim s} \qquad \text{Premise 3}$$
$$\therefore \, ?$$

TAKE NOTE

In Example 6 we are rewriting and reordering the statements so that the extended law of syllogism can be applied.

The first premise can be written as $j \rightarrow h$. The second premise can be written as $\sim s \rightarrow \sim u$. Therefore, the argument can be written as

$$j \rightarrow h$$
$$\sim s \rightarrow \sim u$$
$$\underline{h \rightarrow \sim s}$$
$$\therefore \, ?$$

Interchanging the second and third premises yields

$$j \rightarrow h$$
$$h \rightarrow \sim s$$
$$\underline{\sim s \rightarrow \sim u}$$
$$\therefore \, ?$$

An application of the extended law of syllogism produces

$$j \rightarrow h$$
$$h \rightarrow \sim s$$
$$\underline{\sim s \rightarrow \sim u}$$
$$\therefore j \rightarrow \sim u$$

Thus a valid conclusion for the original argument is "If we go to Japan (j), then we will not visit my uncle ($\sim u$)."

▼ **check your progress** **6**   Use all of the premises to determine a valid conclusion for the following argument.

$$\sim m \lor t$$
$$t \rightarrow \sim d$$
$$e \lor g$$
$$\underline{e \rightarrow d}$$
$$\therefore \ ?$$

Solution   *See page S11.*

MATHMATTERS   A Famous Puzzle: *Where Is the Missing Dollar?*

Most puzzles are designed to test your logical reasoning skills. The following puzzle is intriguing because of its simplicity; however, many people have found that they were unable to provide a satisfactory solution to the puzzle.

Three men decide to share the cost of a hotel room. The regular room rate is $25, but the desk clerk decides to charge them $30 because it will be easier for each man to pay one-third of $30 than it would be for each man to pay one-third of $25. Each man pays $10 and the bellhop shows them to their room.

Shortly thereafter the desk clerk starts to feel guilty and gives the bellhop 5 one dollar bills, along with instructions to return the money to the 3 men. On the way to the room the bellhop decides to give each man $1 and keep $2. After all, the three men will find it difficult to split $5 evenly.

Thus each man paid $10 dollars and received a $1 refund. After the refund, each man has each paid $9 for the room. The total amount the men have paid for the room is $3 \times \$9 = \$27$. The bellhop has $2. The $27 added to the $2 equals $29. *Where is the missing dollar?*

A solution to this puzzle is given in the Answers to Selected Exercises Appendix, on page A8, just before the Exercise Set 3.5 answers.

INSTRUCTOR NOTE

A solution to this puzzle is also given in the Answers to All Exercises Appendix, on page A17, just before the Excursion Exercises, Section 3.5 answers.

EXCURSION

Fallacies

Any argument that is not valid is called a **fallacy.** Ancient logicians enjoyed the study of fallacies and took pride in their ability to analyze and categorize different types of fallacies. In this Excursion we consider the four fallacies known as *circulus in probando,* the fallacy of experts, the fallacy of equivocation, and the fallacy of accident.

Circulus in Probando

A fallacy of *circulus in probando* is an argument that uses a premise as the conclusion. For instance, consider the following argument.

> The Chicago Bulls are the best basketball team because there is no basketball team that is better than the Chicago Bulls.

The fallacy of *circulus in probando* is also known as *circular reasoning* or *begging the question.*

Fallacy of Experts

A fallacy of experts is an argument that uses an expert (or a celebrity) to lend support to a product or an idea. Often the product or idea is outside the expert's area of expertise. The following endorsements may qualify as fallacy of experts arguments.

> Jamie Lee Curtis for Activia yogurt
>
> David Duchovny for Pedigree dog food

Fallacy of Equivocation

A fallacy of equivocation is an argument that uses a word with two interpretations in two different ways. The following argument is an example of a fallacy of equivocation.

> The highway sign read $268 fine for littering,
> so I decided fine, for $268, I will litter.

Fallacy of Accident

The following argument is an example of a fallacy of accident.

> Everyone should visit Europe.
>
> Therefore, prisoners on death row should be allowed to visit Europe.

Using more formal language, we can state the argument as follows.

> If you are a prisoner on death row (d), then you are a person (p).
> If you are a person (p), then you should be allowed to visit Europe (e).
>
> $\therefore$ If you are a prisoner on death row, then you should be allowed to visit Europe.

The symbolic form of the argument is

$$d \rightarrow p$$
$$p \rightarrow e$$
$$\therefore d \rightarrow e$$

This argument appears to be a valid argument because it has the standard form of the law of syllogism. Common sense tells us the argument is not valid, so where have we gone wrong in our analysis of the argument?

The problem occurs with the interpretation of the word "everyone." Often, when we say "everyone," we really mean "most everyone." A fallacy of accident may occur whenever we use a statement that is often true in place of a statement that is always true.

EXCURSION EXERCISES

1. Write an argument that is an example of *circulus in probando*.
2. Give an example of an argument that is a fallacy of experts.
3. Write an argument that is an example of a fallacy of equivocation.
4. Write an argument that is an example of a fallacy of accident.
5. Algebraic arguments often consist of a list of statements. In a valid algebraic argument, each statement (after the premises) can be deduced from the previous statements. The following argument that $1 = 2$ contains exactly one step that is not valid. Identify the step and explain why it is not valid.

Let
$$a = b$$ • Premise.
$$a^2 = ab$$ • Multiply each side by a.
$$a^2 - b^2 = ab - b^2$$ • Subtract b^2 from each side.
$$(a + b)(a - b) = b(a - b)$$ • Factor each side.
$$a + b = b$$ • Divide each side by $(a - b)$.
$$b + b = b$$ • Substitute b for a.
$$2b = b$$ • Collect like terms.
$$2 = 1$$ • Divide each side by b.

EXERCISE SET 3.5

(Suggested Assignment: The Enhanced WebAssign Exercises and Exercises 1, 7, 9, 21, 27, 33, 35, 39, 43, and 51)

■ In Exercises 1 to 8, use the indicated letters to write each argument in symbolic form.

1. If you can read this bumper sticker (r), you're too close (c). You can read the bumper sticker. Therefore, you're too close.

2. If Lois Lane marries Clark Kent (m), then Superman will get a new uniform (u). Superman does not get a new uniform. Therefore, Lois Lane did not marry Clark Kent.

3. If the price of gold rises (g), the stock market will fall (s). The price of gold did not rise. Therefore, the stock market did not fall.

4. I am going shopping (s) or I am going to the museum (m). I went to the museum. Therefore, I did not go shopping.

5. If we search the Internet (s), we will find information on logic (i). We searched the Internet. Therefore, we found information on logic.

6. If we check the sports results on ESPN (c), we will know who won the match (w). We know who won the match. Therefore, we checked the sports results on ESPN.

7. If the power goes off ($\sim p$), then the air conditioner will not work ($\sim a$). The air conditioner is working. Therefore, the power is not off.

8. If it snowed (s), then I did not go to my chemistry class ($\sim c$). I went to my chemistry class. Therefore, it did not snow.

■ In Exercises 9 to 24, use a truth table to determine whether the argument is valid or invalid.

9. $p \vee \sim q$
$\sim q$
$\therefore p$

10. $\sim p \wedge q$
$\sim p$
$\therefore q$

11. $p \rightarrow \sim q$
$\sim q$
$\therefore p$

12. $p \rightarrow \sim q$
p
$\therefore \sim q$

13. $\sim p \rightarrow \sim q$
$\sim p$
$\therefore \sim q$

14. $\sim p \rightarrow q$
p
$\therefore \sim q$

15. $(p \rightarrow q) \wedge (\sim p \rightarrow q)$
q
$\therefore p$

16. $(p \vee q) \wedge (p \wedge q)$
p
$\therefore q$

17. $(p \wedge \sim q) \vee (p \rightarrow q)$
$q \vee p$
$\therefore \sim p \wedge q$

18. $(p \wedge \sim q) \rightarrow (p \vee q)$
$q \rightarrow \sim p$
$\therefore p \rightarrow q$

19. $(p \wedge \sim q) \vee (p \vee r)$
r
$\therefore p \vee q$

20. $(p \rightarrow q) \rightarrow (r \rightarrow \sim q)$
p
$\therefore \sim r$

21. $p \leftrightarrow q$
$p \rightarrow r$
$\therefore \sim r \rightarrow \sim p$

22. $p \wedge r$
$p \rightarrow \sim q$
$\therefore r \rightarrow q$

23. $p \wedge \sim q$
$p \leftrightarrow r$
$\therefore q \vee r$

24. $p \rightarrow r$
$r \rightarrow q$
$\therefore \sim p \rightarrow \sim q$

■ In Exercises 25 to 30, use the indicated letters to write the argument in symbolic form. Then use a truth table to determine whether the argument is valid or invalid.

25. If you finish your homework (h), you may attend the reception (r). You did not finish your homework. Therefore, you cannot go to the reception.

26. The X Games will be held in Oceanside (o) if and only if the city of Oceanside agrees to pay \$200,000 in prize money (a). If San Diego agrees to pay \$300,000 in prize money (s), then the city of Oceanside will not agree to pay \$200,000 in prize money. Therefore, if the X Games were held in Oceanside, then San Diego did not agree to pay \$300,000 in prize money.

27. If I can't buy the house ($\sim b$), then at least I can dream about it (d). I can buy the house or at least I can dream about it. Therefore, I can buy the house.

28. If the winds are from the east (e), then we will not have a big surf ($\sim s$). We do not have a big surf. Therefore, the winds are from the east.

29. If I master college algebra (c), then I will be prepared for trigonometry (t). I am prepared for trigonometry. Therefore, I mastered college algebra.

30. If it is a blot (b), then it is not a clot ($\sim c$). If it is a zlot (z), then it is a clot. It is a blot. Therefore, it is not a zlot.

■ In Exercises 31 to 40, determine whether the argument is valid or invalid by comparing its symbolic form with the standard symbolic forms given in Tables 3.15 and 3.16. For each valid argument, state the name of its standard form.

31. If you take Art 151 in the fall, you will be eligible to take Art 152 in the spring. You were not eligible to take Art 152 in the spring. Therefore, you did not take Art 151 in the fall.

32. He will attend Stanford or Yale. He did not attend Yale. Therefore, he attended Stanford.

33. If I had a nickel for every logic problem I have solved, then I would be rich. I have not received a nickel for every logic problem I have solved. Therefore, I am not rich.

34. If it is a dog, then it has fleas. It has fleas. Therefore, it is a dog.

35. If we serve salmon, then Vicky will join us for lunch. If Vicky joins us for lunch, then Marilyn will not join us for lunch. Therefore, if we serve salmon, Marilyn will not join us for lunch.

36. If I go to college, then I will not be able to work for my Dad. I did not go to college. Therefore, I went to work for my Dad.

37. If my cat is left alone in the apartment, then she claws the sofa. Yesterday I left my cat alone in the apartment. Therefore, my cat clawed the sofa.

38. If I wish to use the new software, then I cannot continue to use this computer. I don't wish to use the new software. Therefore, I can continue to use this computer.

39. If Rita buys a new car, then she will not go on the cruise. Rita went on the cruise. Therefore, Rita did not buy a new car.

40. If Daisuke Matsuzaka pitches, then I will go to the game. I did not go to the game. Therefore, Daisuke Matsuzaka did not pitch.

■ In Exercises 41 to 46, use a sequence of valid arguments to show that each argument is valid.

41. $\sim p \rightarrow r$
$r \rightarrow t$
$\sim t$
$\therefore p$

42. $r \rightarrow \sim s$
$s \vee \sim t$
r
$\therefore \sim t$

43. If we sell the boat (s), then we will not go to the river ($\sim r$). If we don't go to the river, then we will go camping (c). If we do not buy a tent ($\sim t$), then we will not go camping. Therefore, if we sell the boat, then we will buy a tent.

44. If it is an ammonite (a), then it is from the Cretaceous period (c). If it is not from the Mesozoic era ($\sim m$), then it is not from the Cretaceous period. If it is from the Mesozoic era, then it is at least 65 million years old (s). Therefore, if it is an ammonite, then it is at least 65 million years old.

45. If the computer is not operating ($\sim o$), then I will not be able to finish my report ($\sim f$). If the office is closed (c), then the computer is not operating. Therefore, if I am able to finish my report, then the office is open.

46. If he reads the manuscript (r), he will like it (l). If he likes it, he will publish it (p). If he publishes it, then you will get royalties (m). You did not get royalties. Therefore, he did not read the manuscript.

EXTENSIONS

Critical Thinking

In Exercises 47 to 50, use all of the premises to determine a valid conclusion for the given argument.

47. $\sim(p \wedge \sim q)$
       $\underline{\quad p \quad}$
       $\therefore$?

48. $\sim s \rightarrow q$
       $\sim t \rightarrow \sim q$
       $\underline{\quad \sim t \quad}$
       $\therefore$?

49. If it is a theropod, then it is not herbivorous. If it is not herbivorous, then it is not a sauropod. It is a sauropod. Therefore, _____.

50. If you buy the car, you will need a loan. You do not need a loan or you will make monthly payments. You buy the car. Therefore, _____.

Cooperative Learning

51.  **Recreational Logic**   "Are You Smarter Than a 5th Grader?" is a popular television program that requires adult contestants to answer grade-school level questions. The show is hosted by Jeff Foxworthy.

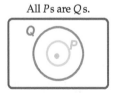

Here is an arithmetic problem that many 5th grade students can solve in less than 10 seconds after the problem has been stated.

> Start with the number 32. Add 46 to the starting number and divide the result by 2. Multiply the previous result by 3 and add 27 to that product. Take the square root of the previous sum and then quadruple that result. Multiply the latest result by 5 and divide that product by 32. Multiply the latest result by 0 and then add 24 to the result. Take half of the previous result. What is the final answer?

Explain how to solve this arithmetic problem in less than 10 seconds.

52. An Argument by Lewis Carroll   The following argument is from *Symbolic Logic* by Lewis Carroll, written in 1896. Determine whether the argument is valid or invalid.

> Babies are illogical.
> Nobody is despised who can manage a crocodile.
> Illogical persons are despised.
> Hence, babies cannot manage crocodiles.

---

section **3.6**   **Arguments and Euler Diagrams**

Arguments and Euler Diagrams

Many arguments involve sets whose elements are described using the quantifiers *all*, *some*, and *none*. The mathematician Leonhard Euler (laônhärt oi′lər) used diagrams to determine whether arguments that involved quantifiers were valid or invalid. The following figures show Euler diagrams that illustrate the four possible relationships that can exist between two sets.

All Ps are Qs.   No Ps are Qs.   Some Ps are Qs.   Some Ps are not Qs.

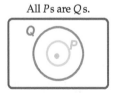 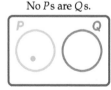 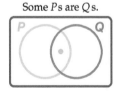 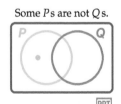

Euler diagrams

Euler used diagrams to illustrate logic concepts. Some 100 years later, John Venn extended the use of Euler's diagrams to illustrate many types of mathematics. In this section, we will construct diagrams to determine the validity of arguments. We will refer to these diagrams as Euler diagrams.

HISTORICAL NOTE

Leonhard Euler

(1707 – 1783) Euler was an exceptionally talented Swiss mathematician. He worked in many different areas of mathematics and produced more written material about mathematics than any other mathematician. His mental computational abilities were remarkable. The French astronomer and statesman Dominque François Arago wrote,

Euler calculated without apparent effort, as men breathe, or as eagles sustain themselves in the wind.

In 1776, Euler became blind; however, he continued to work in the disciplines of mathematics, physics, and astronomy. He even solved a problem that Newton had attempted concerning the motion of the moon. Euler performed all the necessary calculations in his head.

This Impressionist painting, *Dance at Bougival*, is by the French artist Pierre-Auguste Renoir.

▼ **example**   **Use an Euler Diagram to Determine the Validity of an Argument**

Use an Euler diagram to determine whether the following argument is valid or invalid.

> All college courses are fun.
>
> This course is a college course.
>
> ∴ This course is fun.

Solution

The first premise indicates that the set of college courses is a subset of the set of fun courses. We illustrate this subset relationship with an Euler diagram, as shown in Figure 3.15. The second premise tells us that "this course" is an element of the set of college courses. If we use c to represent "this course," then c must be placed inside the set of college courses, as shown in Figure 3.16.

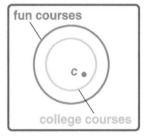

FIGURE 3.15   **FIGURE 3.16**

Figure 3.16 illustrates that c must also be an element of the set of fun courses. Thus the argument is valid.

▼ **check your progress** **1**   Use an Euler diagram to determine whether the following argument is valid or invalid.

> All lawyers drive BMWs.
>
> Susan is a lawyer.
>
> ∴ Susan drives a BMW.

Solution   *See page S11.*

If an Euler diagram can be drawn so that the conclusion does not necessarily follow from the premises, then the argument is invalid. This concept is illustrated in the next example.

▼ **example** **2**   **Use an Euler Diagram to Determine the Validity of an Argument**

Use an Euler diagram to determine whether the following argument is valid or invalid.

> Some Impressionist paintings are Renoirs.
>
> *Dance at Bougival* is an Impressionist painting.
>
> ∴ *Dance at Bougival* is a Renoir.

Solution

The Euler diagram in Figure 3.17, on page 160, illustrates the premise that some Impressionist paintings are Renoirs. Let d represent the painting *Dance at Bougival*. Figures 3.18 and 3.19 show that d can be placed in one of two regions.

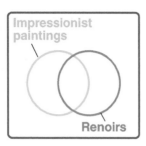

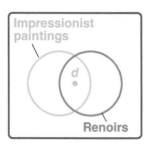

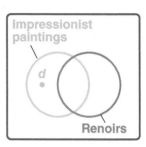

FIGURE 3.17   **FIGURE 3.18**   **FIGURE 3.19**

Although Figure 3.18 supports the argument, Figure 3.19 shows that the conclusion does not necessarily follow from the premises, and thus the argument is invalid.

▼ **check your progress 2**   Use an Euler diagram to determine whether the following argument is valid or invalid.

    No prime numbers are negative.

    The number 7 is not negative.

∴ The number 7 is a prime number.

Solution   *See page S12.*

---

question   If one particular example can be found for which the conclusion of an argument is true when its premises are true, must the argument be valid?

Some arguments can be represented by an Euler diagram that involves three sets, as shown in Example 3.

▼ **example 3**   Use an Euler Diagram to Determine the Validity of an Argument

Use an Euler diagram to determine whether the following argument is valid or invalid.

    No psychologist can juggle.

    All clowns can juggle.

∴ No psychologist is a clown.

Solution

The Euler diagram in Figure 3.20 shows that the set of psychologists and the set of jugglers are disjoint sets. Figure 3.21 shows that because the set of clowns is a subset of the set of jugglers, no psychologists *p* are elements of the set of clowns. Thus the argument is valid.

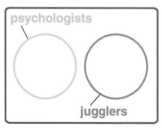

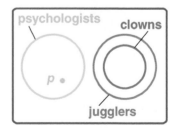

FIGURE 3.20   **FIGURE 3.21**

---

answer   No. To be a valid argument, the conclusion must be true whenever the premises are true. Just because the conclusion is true for one specific example, does not mean the argument is a valid argument.

▼ **check your progress 3** Use an Euler diagram to determine whether the following argument is valid or invalid.

> No mathematics professors are good-looking.
> All good-looking people are models.
> ∴ No mathematics professor is a model.

Solution *See page S12.*

MATHMATTERS Raymond Smullyan

Raymond Smullyan
(1919–   )

Bill Ray Photography

Raymond Smullyan is a concert pianist, a logician, a Taoist philosopher, a magician, a retired professor, and an author of many popular books on logic. Over a period of several years he has created many interesting logic problems. One of his logic problems is an enhancement of the classic logic puzzle that concerns two doors and two guards. One of the doors leads to heaven and the other door leads to hell. One of the guards always tells the truth and the other guard always lies. You do not know which guard always tells the truth and which guard always lies, and you are only allowed to ask one question of one of the guards. What one question should you ask that will allow you to determine which door leads to heaven?

Another logic puzzle that Raymond Smullyan created has been referred to as the "hardest logic puzzle ever." Information about this puzzle and the solution to the above puzzle concerning the two guards and the two doors can be found at: http://en.wikipedia.org/wiki/Raymond_Smullyan and http://en.wikipedia.org/wiki/The_Hardest_Logic_Puzzle_Ever.

Euler Diagrams and the Extended Law of Syllogism

Example 4 uses Euler diagrams to visually illustrate the extended law of syllogism from Section 3.5.

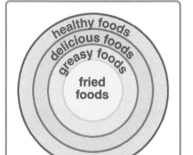

▼ **example 4** Use an Euler Diagram to Determine the Validity of an Argument

Use an Euler diagram to determine whether the following argument is valid or invalid.

> All fried foods are greasy.
> All greasy foods are delicious.
> All delicious foods are healthy.
> ∴ All fried foods are healthy.

Solution
The figure at the left illustrates that every fried food is an element of the set of healthy foods, so the argument is valid.

TAKE NOTE

Although the conclusion in Example 4 is false, the argument in Example 4 is valid.

▼ **check your progress 4** Use an Euler diagram to determine whether the following argument is valid or invalid.

> All squares are rhombi.
> All rhombi are parallelograms.
> All parallelograms are quadrilaterals.
> ∴ All squares are quadrilaterals.

Solution *See page S12.*

Using Euler Diagrams to Form Conclusions

In Example 5, we make use of an Euler diagram to determine a valid conclusion for an argument.

▼ example 5   Use an Euler Diagram to Determine a Conclusion for an Argument

Use an Euler diagram and all of the premises in the following argument to determine a valid conclusion for the argument.

> All *M*s are *N*s.
>
> No *N*s are *P*s.
>
> ∴ ?

Solution

The first premise indicates that the set of *M*s is a subset of the set of *N*s. The second premise indicates that the set of *N*s and the set of *P*s are disjoint sets. The following Euler diagram illustrates these set relationships. An examination of the Euler diagram allows us to conclude that *no Ms are Ps*.

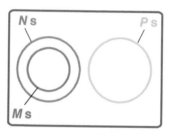

▼ check your progress 5   Use an Euler diagram and all of the premises in the following argument to determine a valid conclusion for the argument.

> Some rabbits are white.
>
> All white animals like tomatoes.
>
> ∴ ?

Solution   *See page S12.*

EXCURSION

Using Logic to Solve Cryptarithms

Many puzzles can be solved by making an assumption and then checking to see if the assumption is supported by the conditions (premises) associated with the puzzle. For instance, consider the following addition problem in which each letter represents a digit from 0 through 9 and different letters represent different digits.

TAKE NOTE

When working with cryptarithms, we assume that the leading digit of each number is a nonzero digit.

```
   T A
 + B T
 -----
 T E E
```

Note that the T in T E E is a carry from the middle column. Because the sum of any two single digits plus a previous carry of at most 1 is 19 or less, the T in T E E must be a 1. Replacing all the Ts with 1s produces:

$$\begin{array}{r} 1\,A \\ +\ B\,1 \\ \hline 1\,E\,E \end{array}$$

Now B must be an 8 or a 9, because these are the only digits that could produce a carry into the leftmost column.

Case 1: Assume B is a 9. Then A must be an 8 or smaller, and A + 1 does not produce a carry into the middle column. The sum of the digits in the middle column is 10; thus E is a 0. This presents a dilemma because the units digit of A + 1 must also be a 0, which requires A to be a 9. The assumption that B is a 9 is not supported by the conditions of the problem; thus we reject the assumption that B is a 9.

Case 2: Assume B is an 8. To produce the required carry into the leftmost column, there must be a carry from the column on the right. Thus A must be a 9, and we have the result shown below.

$$\begin{array}{r} 1\,9 \\ +\ 8\,1 \\ \hline 1\,0\,0 \end{array}$$

A check shows that this solution satisfies all the conditions of the problem.

EXCURSION EXERCISES

Solve the following cryptarithms. Assume that no leading digit is a 0. (*Source:* http://www.geocities.com/Athens/Agora/2160/puzzles.html)[3]

1.
$$\begin{array}{r} S\,O \\ +\ S\,O \\ \hline T\,O\,O \end{array}$$

2.
$$\begin{array}{r} U\,S \\ +\ A\,S \\ \hline A\,L\,L \end{array}$$

3.
$$\begin{array}{r} C\,O\,C\,A \\ +\ C\,O\,L\,A \\ \hline O\,A\,S\,I\,S \end{array}$$

4.
$$\begin{array}{r} A\,T \\ E\,A\,S\,T \\ +\ W\,E\,S\,T \\ \hline S\,O\,U\,T\,H \end{array}$$

[3] Copyright © 1998 by Jorge A C B Soares.

E X E R C I S E   S E T   3.6   (Suggested Assignment: The Enhanced WebAssign Exercises and Exercises 14, 15, 20, and 29)

■ In Exercises 1 to 20, use an Euler diagram to determine whether the argument is valid or invalid.

1.   All frogs are poetical.
Kermit is a frog.

∴ Kermit is poetical.

2.   All Oreo cookies have a filling.
All Fig Newtons have a filling.

∴ All Fig Newtons are Oreo cookies.

3.   Some plants have flowers.
All things that have flowers are beautiful.

∴ Some plants are beautiful.

4.   No squares are triangles.
Some triangles are equilateral.

∴ No squares are equilateral.

5. No rocker would do the mariachi.
All baseball fans do the mariachi.

∴ No rocker is a baseball fan.

6. Nuclear energy is not safe.
Some electric energy is safe.

∴ No electric energy is nuclear energy.

7. Some birds bite.
All things that bite are dangerous.

∴ Some birds are dangerous.

8. All fish can swim.
That barracuda can swim.

∴ That barracuda is a fish.

9. All men behave badly.
Some hockey players behave badly.

∴ Some hockey players are men.

10. All grass is green.
That ground cover is not green.

∴ That ground cover is not grass.

11. Most teenagers drink soda.
No CEOs drink soda.

∴ No CEO is a teenager.

12. Some students like history.
Vern is a student.

∴ Vern likes history.

13. No mathematics test is fun.
All fun things are worth your time.

∴ No mathematics test is worth your time.

14. All prudent people shun sharks.
No accountant is imprudent.

∴ No accountant fails to shun sharks.

15. All candidates without a master's degree will not be considered for the position of director.
All candidates who are not considered for the position of director should apply for the position of assistant.

∴ All candidates without a master's degree should apply for the position of assistant.

16. Some whales make good pets.
Some good pets are cute.
Some cute pets bite.

∴ Some whales bite.

17. All prime numbers are odd.
2 is a prime number.

∴ 2 is an odd number.

18. All Lewis Carroll arguments are valid.
Some valid arguments are syllogisms.

∴ Some Lewis Carroll arguments are syllogisms.

19. All aerobics classes are fun.
Jan's class is fun.

∴ Jan's class is an aerobics class.

20. No sane person takes a math class.
Some students that take a math class can juggle.

∴ No sane person can juggle.

■ In Exercises 21 to 26, use all of the premises in each argument to determine a valid conclusion for the argument.

21. All reuben sandwiches are good.
All good sandwiches have pastrami.
All sandwiches with pastrami need mustard.

∴ ?

22. All cats are strange.
Boomer is not strange.

∴ ?

23. All multiples of 11 end with a 5.
1001 is a multiple of 11.

∴ ?

24. If it isn't broken, then I do not fix it.
If I do not fix it, then I do not get paid.

∴ ?

25. Some horses are frisky.
All frisky horses are grey.

∴ ?

26. If we like to ski, then we will move to Vail.
If we move to Vail, then we will not buy a house.
If we do not buy a condo, then we will buy a house.

∴ ?

27. Examine the following three premises:

1. All people who have an Xbox play video games.

2. All people who play video games enjoy life.

3. Some mathematics professors enjoy life.

Now consider each of the following six conclusions. For each conclusion, determine whether the argument formed by the three premises and the conclusion is valid or invalid.

a. ∴ Some mathematics professors have an Xbox.

b. ∴ Some mathematics professors play video games.

c. ∴ Some people who play video games are mathematics professors.

d. ∴ Mathematics professors never play video games.

e. ∴ All people who have an Xbox enjoy life.

f. ∴ Some people who enjoy life are mathematics professors.

28. Examine the following three premises:

1. All people who drive pickup trucks like Garth Brooks.

2. All people who like Garth Brooks like country western music.

3. Some people who like heavy metal music like Garth Brooks.

Now consider each of the following five conclusions. For each conclusion, determine whether the argument formed by the three premises and the conclusion is valid or invalid.

a. ∴ Some people who like heavy metal music drive a pickup truck.

b. ∴ Some people who like heavy metal music like country western music.

c. ∴ Some people who like Garth Brooks like heavy metal music.

d. ∴ All people who drive a pickup truck like country western music.

e. ∴ People who like heavy metal music never drive a pickup truck.

EXTENSIONS

Critical Thinking

IN THE NEWS

The LSAT and Logic Games

A good score on the Law School Admission Test, better known as the LSAT, is viewed by many to be the most important part of getting into a top-tier law school. Rather than testing what you've already learned, it's designed to measure and project your ability to excel in law school.

Testing experts agree that the test's "logic games" section is one of the most difficult sections for students to wrap their minds around initially because it's vastly different from anything else they've seen on standardized tests.

SOURCE: U.S. News and World Report, May 28, 2010

29. LSAT Practice Problem The following exercise, created by the authors of this text, is similar to some of the questions from the logic games section of the LSAT. See the above news clip.

A cell phone provider sells seven different types of smart phones.

- Each phone has either a touch screen keyboard or a pushbutton keyboard.
- Each phone has a 3.7-inch, a 3.9-inch, or a 4.3-inch screen.

- Every phone with a 3.9-inch screen is paired with a touch screen keyboard.
- Of the seven different types of phones, most have a touch screen keyboard.
- No phone with a pushbutton keyboard is paired with a 3.7 inch screen.

Which one of the following statements CANNOT be true?

a. Five of the types of phones have 3.9-inch screens.

b. Five of the types of phones have 3.7-inch screens.

c. Four of the types of phones have a pushbutton keyboard.

d. Four of the types of phones have 4.3-inch screens.

e. Five of the types of phones have a touch screen keyboard.

Explorations

30. **Bilateral Diagrams** Lewis Carroll (Charles Dodgson) devised a *bilateral diagram* (two-part board) to analyze syllogisms. His method has some advantages over Euler diagrams and Venn diagrams. Use a library or the Internet to find information on Carroll's method of analyzing syllogisms. Write a few paragraphs that explain his method and its advantages.

CHAPTER 3 SUMMARY

The following table summarizes essential concepts in this chapter. The references given in the right-hand column list Examples and Exercises that can be used to test your understanding of a concept.

3.1 Logic Statements and Quantifiers

Statements A statement is a declarative sentence that is either true or false, but not both true and false. A simple statement is a statement that does not contain a connective.	See **Example 1** on page 112, and then try Exercises 1 to 6 on page 168.
Compound Statements A compound statement is formed by connecting simple statements with the connectives *and, or, if . . . then,* and *if and only if.*	See **Examples 3 and 4** on page 115, and then try Exercises 7 to 10 on page 168.

Truth Values The conjunction $p \wedge q$ is true if and only if both p and q are true. The disjunction $p \vee q$ is true provide p is true, q is true, or both p and q are true.	See **Example 6** on page 118, and then try Exercises 17 to 20 on page 168.
The Negation of a Quantified Statement The information in the following table can be used to write the negation of many quantified statements. All X are Y. ⟷ negation ⟷ Some X are not Y. No X are Y. ⟷ negation ⟷ Some X are Y.	See **Example 7** on page 119, and then try Exercises 11 to 16 on page 168.

3.2 Truth Tables, Equivalent Statements, and Tautologies

Construction of Truth Tables 1. If the given statement has n simple statements, then start with a standard form that has 2^n rows. Enter the truth values for each simple statement and their negations. 2. Use the truth values for each simple statement and their negations to enter the truth values under each connective within a pair of grouping symbols—parentheses ( ), brackets [ ], braces { }. If some grouping symbols are nested inside other grouping symbols, then work from the inside out. In any situation in which grouping symbols have not been used, then we use the following **order of precedence agreement**. First assign truth values to negations from left to right, followed by conjunctions from left to right, followed by disjunctions from left to right, followed by conditionals from left to right, and finally by biconditionals from left to right. 3. The truth values that are entered into the column under the connective for which truth values are assigned *last* form the truth table for the given statement.	See **Example 3** on page 125, and then try Exercises 27 to 34 on pages 168 and 169.
Equivalent Statements Two statements are equivalent if they both have the same truth value for all possible truth values of their simple statements. The notation $p \equiv q$ is used to indicate that the statements p and q are equivalent.	See **Example 4** on page 127, and then try Exercises 39 to 42 on page 169.
De Morgan's Laws for Statements For any statements p and q, $\sim(p \vee q) \equiv \sim p \wedge \sim q$ and $\sim(p \wedge q) \equiv \sim p \vee \sim q$	See **Example 5** on page 128, and then try Exercises 35 to 38 on page 169.
Tautologies and Self-Contradictions A tautology is a statement that is always true. A self-contradiction is a statement that is always false.	See **Example 6** on page 128, and then try Exercises 43 to 46 on page 169.

3.3 The Conditional and the Biconditional

Antecedent and Consequent of a Conditional In a conditional statement represented by "if p, then q" or by "if p, q," the p statement is called the antecedent and the q statement is called the consequent.	See **Example 1** on page 133, and then try Exercises 47 and 50 on page 169.
Equivalent Disjunctive Form of $p \rightarrow q$ $p \rightarrow q \equiv \sim p \vee q$ The conditional $p \rightarrow q$ is false when p is true and q is false. It is true in all other cases.	See **Examples 2 to 4** on pages 134 to 136, and then try Exercises 51 to 54 on page 169.

The Negation of $p \rightarrow q$ $\sim(p \rightarrow q) \equiv p \wedge \sim q$	See **Example 5** on page 136, and then try Exercises 55 to 58 on page 169.
The Biconditional $p \rightarrow q$ $p \leftrightarrow q \equiv [(p \rightarrow q) \wedge (q \rightarrow p)]$ The biconditional $p \leftrightarrow q$ is true only when p and q have the same truth value.	See **Example 6** on page 137, and then try Exercises 59 and 60 on page 169.

3.4 The Conditional and Related Statements

Equivalent Forms of the Conditional The conditional "if p, then q" can be stated, in English, in several equivalent forms. For example, p only if q; p implies that q; and q provided that p are all equivalent forms of if p, then q.	See **Example 1** on page 141, and then try Exercises 63 to 66 on page 169.
Statements Related to the Conditional Statement ■ The **converse** of $p \rightarrow q$ is $q \rightarrow p$. ■ The **inverse** of $p \rightarrow q$ is $\sim p \rightarrow \sim q$. ■ The **contrapositive** of $p \rightarrow q$ is $\sim q \rightarrow \sim p$.	See **Examples 2 and 3** on page 142, and then try Exercises 67 to 72 on page 169.
A Conditional Statement and Its Contrapositive A conditional and its contrapositive are equivalent statements. Therefore, if the contrapositive of a conditional statement is a true statement, then the conditional statement must also be a true statement.	See **Example 4** on page 143, and then try Exercise 74 on page 169.

3.5 Symbolic Arguments

Valid Argument An argument consists of a set of statements called premises and another statement called the conclusion. An argument is valid if the conclusion is true whenever all the premises are assumed to be true. An argument is invalid if it is not a valid argument.	See **Examples 2 and 3** on page 149, and then try Exercises 79 to 82 on page 169.
Symbolic Forms of Arguments **Standard Forms of Four Valid Arguments**	See **Examples 4 to 6** on pages 151 to 153, and then try Exercises 83 to 88 on page 170.

Modus ponens	Modus tollens	Law of syllogism	Disjunctive syllogism
$p \rightarrow q$	$p \rightarrow q$	$p \rightarrow q$	$p \vee q$
p	$\sim q$	$q \rightarrow r$	$\sim p$
$\therefore q$	$\therefore \sim p$	$\therefore p \rightarrow r$	$\therefore q$

Standard Forms of Two Invalid Arguments

Fallacy of the converse	Fallacy of the inverse
$p \rightarrow q$	$p \rightarrow q$
q	$\sim p$
$\therefore p$	$\therefore \sim q$

3.6 Arguments and Euler Diagrams

Euler Diagrams

All Ps are Qs.

No Ps are Qs.

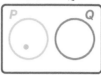

Some Ps are Qs.

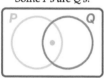

Some Ps are not Qs.

Euler diagrams can be used to determine whether arguments that involve quantifiers are valid or invalid.

Draw an Euler diagram that illustrates the conditions required by the premises of an argument.

If the conclusion of the argument must necessarily follow from all the conditions shown by the premises, then the argument is valid.

If the conclusion of the argument does not necessarily follow from the conditions shown by all the premises, then the argument is invalid.

See **Examples 1 to 4** on pages 159 to 161, and then try Exercises 89 to 92 on page 170.

CHAPTER 3 REVIEW EXERCISES

■ In Exercises 1 to 6, determine whether each sentence is a statement. Assume that a and b are real numbers.

1. How much is a ticket to London?

2. 91 is a prime number.

3. $a > b$

4. $a^2 \geq 0$

5. Lock the car.

6. Clark Kent is Superman.

■ In Exercises 7 to 10, write each sentence in symbolic form. Represent each simple statement of the sentence with the letter indicated in parentheses. Also state whether the sentence is a conjunction, a disjunction, a negation, a conditional, or a biconditional.

7. Today is Monday (m) and it is my birthday (b).

8. If x is divisible by 2 (d), then x is an even number (e).

9. I am going to the dance (g) if and only if I have a date (d).

10. All triangles (t) have exactly three sides (s).

■ In Exercises 11 to 16, write the negation of each quantified statement. Start each negation with "Some," "No," or "All."

11. Some dogs bite.

12. Every dessert at the Cove restaurant is good.

13. All winners receive a prize.

14. Some cameras do not use film.

15. No student finished the assignment.

16. At least one person enjoyed the story.

■ In Exercises 17 to 20, determine whether each statement is true or false.

17. $5 > 2$ or $5 = 2$.

18. $3 \neq 5$ and 7 is a prime number.

19. $4 \leq 7$.

20. $-3 < -1$.

■ In Exercises 21 to 26, determine the truth value of the statement given that p is true, q is false, and r is false.

21. $(p \wedge q) \vee (\sim p \vee q)$

22. $(p \rightarrow \sim q) \leftrightarrow \sim(p \vee q)$

23. $(p \wedge \sim q) \wedge (\sim r \vee q)$

24. $(r \wedge \sim p) \vee [(p \vee \sim q) \leftrightarrow (q \rightarrow r)]$

25. $[p \wedge (r \rightarrow q)] \rightarrow (q \vee \sim r)$

26. $(\sim q \vee \sim r) \rightarrow [(p \leftrightarrow \sim r) \wedge q]$

■ In Exercises 27 to 34, construct a truth table for the given statement.

27. $(\sim p \rightarrow q) \vee (\sim q \wedge p)$

28. $\sim p \leftrightarrow (q \vee p)$

29. $\sim(p \vee \sim q) \wedge (q \rightarrow p)$
30. $(p \leftrightarrow q) \vee (\sim q \wedge p)$
31. $(r \leftrightarrow \sim q) \vee (p \rightarrow q)$
32. $(\sim r \vee \sim q) \wedge (q \rightarrow p)$
33. $[p \leftrightarrow (q \rightarrow \sim r)] \wedge \sim q$
34. $\sim(p \wedge q) \rightarrow (\sim q \vee \sim r)$

■ In Exercises 35 to 38, make use of De Morgan's laws to write the given statement in an equivalent form.

35. It is not true that, Bob failed the English proficiency test and he registered for a speech course.

36. Ellen did not go to work this morning and she did not take her medication.

37. Wendy will go to the store this afternoon or she will not be able to prepare her fettuccine al pesto recipe.

38. Gina enjoyed the movie, but she did not enjoy the party.

■ In Exercises 39 to 42, use a truth table to show that the given pairs of statements are equivalent.

39. $\sim p \rightarrow \sim q$; $p \vee \sim q$
40. $\sim p \vee q$; $\sim(p \wedge \sim q)$
41. $p \vee (q \wedge \sim p)$; $p \vee q$
42. $p \leftrightarrow q$; $(p \wedge q) \vee (\sim p \wedge \sim q)$

■ In Exercises 43 to 46, use a truth table to determine whether the given statement is a tautology or a self-contradiction.

43. $p \wedge (q \wedge \sim p)$
44. $(p \wedge q) \vee (p \rightarrow \sim q)$
45. $[\sim(p \rightarrow q)] \leftrightarrow (p \wedge \sim q)$
46. $p \vee (p \rightarrow q)$

■ In Exercises 47 to 50, identify the antecedent and the consequent of each conditional statement.

47. If he has talent, he will succeed.

48. If I had a credential, I could get the job.

49. I will follow the exercise program provided I join the fitness club.

50. I will attend only if it is free.

■ In Exercises 51 to 54, write each conditional statement in its equivalent disjunctive form.

51. If she were tall, she would be on the volleyball team.

52. If he can stay awake, he can finish the report.

53. Rob will start, provided he is not ill.

54. Sharon will be promoted only if she closes the deal.

■ In Exercises 55 to 58, write the negation of each conditional statement in its equivalent conjunctive form.

55. If I get my paycheck, I will purchase a ticket.

56. The tomatoes will get big only if you provide them with plenty of water.

57. If you entered Cleggmore University, then you had a high score on the SAT exam.

58. If Ryan enrolls at a university, then he will enroll at Yale.

■ In Exercises 59 to 62, determine whether the given statement is true or false. Assume that x and y are real numbers.

59. $x = y$ if and only if $|x| = |y|$.

60. $x > y$ if and only if $x - y > 0$.

61. If $x^2 > 0$, then $x > 0$.

62. If $x^2 = y^2$, then $x = y$.

■ In Exercises 63 to 66, write each statement in "If p, then q" form.

63. Every nonrepeating, nonterminating decimal is an irrational number.

64. Being well known is a necessary condition for a politician.

65. I could buy the house provided that I could sell my condominium.

66. Being divisible by 9 is a sufficient condition for being divisible by 3.

■ In Exercises 67 to 72, write the **a.** converse, **b.** inverse, and **c.** contrapositive of the given statement.

67. If $x + 4 > 7$, then $x > 3$.

68. All recipes in this book can be prepared in less than 20 minutes.

69. If a and b are both divisible by 3, then $(a + b)$ is divisible by 3.

70. If you build it, they will come.

71. Every trapezoid has exactly two parallel sides.

72. If they like it, they will return.

73. What is the inverse of the contrapositive of $p \rightarrow q$?

74. Use the contrapositive of the following statement to determine whether the statement is true or false.

   If today is not Monday, then yesterday was not Sunday.

■ In Exercises 75 to 78, determine the original statement if the given statement is related to the original statement in the manner indicated.

75. *Converse:* If $x > 2$, then x is an odd prime number.

76. *Negation:* The senator will attend the meeting and she will not vote on the motion.

77. *Inverse:* If their manager will not contact me, then I will not purchase any of their products.

78. *Contrapositive:* If Ginny can't rollerblade, then I can't rollerblade.

■ In Exercises 79 to 82, use a truth table to determine whether the argument is valid or invalid.

79. $\dfrac{\begin{array}{c} (p \wedge \sim q) \wedge (\sim p \rightarrow q) \\ p \end{array}}{\therefore \sim q}$

80. $\dfrac{\begin{array}{c} p \rightarrow \sim q \\ q \end{array}}{\therefore \sim p}$

81. $\dfrac{\begin{array}{c} r \\ p \rightarrow \sim r \\ \sim p \rightarrow q \end{array}}{\therefore p \wedge q}$

82. $\dfrac{\begin{array}{c} (p \vee \sim r) \rightarrow (q \wedge r) \\ r \wedge p \end{array}}{\therefore p \vee q}$

■ In Exercises 83 to 88, determine whether the argument is valid or invalid by comparing its symbolic form with the symbolic forms in Tables 3.15 and 3.16, page 151.

83. We will serve either fish or chicken for lunch. We did not serve fish for lunch. Therefore, we served chicken for lunch.

84. If Mike is a CEO, then he will be able to afford to make a donation. If Mike can afford to make a donation, then he loves to ski. Therefore, if Mike does not love to ski, he is not a CEO.

85. If we wish to win the lottery, we must buy a lottery ticket. We did not win the lottery. Therefore, we did not buy a lottery ticket.

86. Robert can charge it on his MasterCard or his Visa. Robert does not use his MasterCard. Therefore, Robert charged it to his Visa.

87. If we are going to have a caesar salad, then we need to buy some eggs. We did not buy eggs. Therefore, we are not going to have a caesar salad.

88. If we serve lasagna, then Eva will not come to our dinner party. We did not serve lasagna. Therefore, Eva came to our dinner party.

■ In Exercises 89 to 92, use an Euler diagram to determine whether the argument is valid or invalid.

89. No wizard can yodel.
All lizards can yodel.

∴ No wizard is a lizard.

90. Some dogs have tails.
Some dogs are big.

∴ Some big dogs have tails.

91. All Italian villas are wonderful. It is not wise to invest in expensive villas. Some wonderful villas are expensive. Therefore, it is not wise to invest in Italian villas.

92. All logicians like to sing "It's a small world after all." Some logicians have been presidential candidates. Therefore, some presidential candidates like to sing "It's a small world after all."

CHAPTER 3 TEST

1. Determine whether each sentence is a statement.

a. Look for the cat.

b. Clark Kent is afraid of the dark.

2. Write the negation of each statement. Start each negation with "Some," "No," or "All."

a. Some trees are not green.

b. No apartments are available.

3. Determine whether each statement is true or false.

a. $5 \leq 4$

b. $-2 \geq -2$

4. Determine the truth value of each statement given that p is true, q is false, and r is true.

a. $(p \lor \sim q) \land (\sim r \land q)$

b. $(r \lor \sim p) \lor [(p \lor \sim q) \leftrightarrow (q \to r)]$

■ In Exercises 5 and 6, construct a truth table for the given statement.

5. $\sim(p \land \sim q) \lor (q \to p)$ **6.** $(r \leftrightarrow \sim q) \land (p \to q)$

7. Use one of De Morgan's laws to write the following in an equivalent form.

Elle did not eat breakfast and she did not take a lunch break.

8. What is a tautology?

9. Write $p \to q$ in its equivalent disjunctive form.

10. Determine whether the given statement is true or false. Assume that x, y, and z are real numbers.

a. $x = y$ if $|x| = |y|$.   **b.** If $x > y$, then $xz > yz$.

11. Write the **a.** converse, **b.** inverse, and **c.** contrapositive of the following statement.

If $x + 7 > 11$, then $x > 4$.

12. Write the standard form known as modus ponens.

13. Write the standard form known as the law of syllogism.

■ In Exercises 14 and 15, use a truth table to determine whether the argument is valid or invalid.

14. $(p \land \sim q) \land (\sim p \to q)$ **15.** r
          p $p \to \sim r$
          ───────────────── $\sim p \to q$
        ∴ $\sim q$ ───────────
                                                         ∴ $p \land q$

■ In Exercises 16 to 20, determine whether the argument is valid or invalid. Explain how you made your decision.

16. If we wish to win the talent contest, we must practice. We did not win the contest. Therefore, we did not practice.

17. Gina will take a job in Atlanta or she will take a job in Kansas City. Gina did not take a job in Atlanta. Therefore, Gina took a job in Kansas City.

18. No wizard can glow in the dark.
Some lizards can glow in the dark.

∴ No wizard is a lizard.

19. Some novels are worth reading.
War and Peace is a novel.

∴ *War and Peace* is worth reading.

20. If I cut my night class, then I will go to the party. I went to the party. Therefore, I cut my night class.

6

Numeration Systems and Number Theory

We start this chapter with an examination of several numeration systems. A working knowledge of these numeration systems will enable you better to understand and appreciate the advantages of our current Hindu-Arabic numeration system.

The last two sections of this chapter cover prime numbers and topics from the field of number theory. Many of the concepts in number theory are easy to comprehend but difficult, or impossible, to prove. The mathematician Karl Friedrich Gauss (1777–1855) remarked that "it is just this which gives the higher arithmetic (number theory) that magical charm which has made it the favorite science of the greatest mathematicians, not to mention its inexhaustible wealth, wherein it so greatly surpasses other parts of mathematics." Gauss referred to mathematics as "the queen of the sciences," and he considered the field of number theory "the queen of mathematics."

There are many unsolved problems in the field of number theory. One unsolved problem, dating from the year 1742, is *Goldbach's conjecture,* which states that every even number greater than 2 can be written as the sum of two prime numbers. This conjecture has yet to be proved or disproved, despite the efforts of the world's best mathematicians. The British publishing company Faber and Faber offered a $1 million prize to anyone who could provide a proof or disproof of Goldbach's conjecture between March 20, 2000, and March 20, 2002, but the prize went unclaimed and Goldbach's conjecture remains a conjecture. The company had hoped that the prize money would entice young, mathematically talented people to work on the problem. This scenario is similar to the story line in the movie *Good Will Hunting,* in which a mathematics problem posted on a bulletin board attracts the attention of a yet-to-be-discovered math genius, played by Matt Damon.

Early Numeration Systems

The Egyptian Numeration System

In mathematics, symbols that are used to represent numbers are called **numerals**. A number can be represented by many different numerals. For instance, the concept of "eightness" is represented by each of the following.

Hindu-Arabic: 8   Tally: ⅢⅢ ||| Roman: VIII

Chinese: ノ乀   Egyptian: |||||||| Babylonian: ▼▼▼▼▼▼▼▼ [PPT]

A **numeration system** consists of a set of numerals and a method of arranging the numerals to represent numbers. The numeration system that most people use today is known as the *Hindu-Arabic numeration system*. It makes use of the 10 numerals 0, 1, 2, 3, 4, 5, 6, 7, 8, and 9. Before we examine the Hindu-Arabic numeration system in detail, it will be helpful to study some of the earliest numeration systems that were developed by the Egyptians, the Romans, and the Chinese.

The Egyptian numeration system uses pictorial symbols called **hieroglyphics** as numerals. The Egyptian hieroglyphic system is an **additive system** because any given number is written by using numerals whose sum equals the number. Table 6.1 gives the Egyptian hieroglyphics for powers of 10 from 1 to 1 million.

TABLE 6.1 Egyptian Hieroglyphics for Powers of 10

Hindu-Arabic numeral	Egyptian hieroglyphic	Description of hieroglyphic
1	|	stroke
10	∩	heel bone
100	⟋	scroll
1000	⚘	lotus flower
10,000	⟋	pointing finger
100,000	⊶	fish
1,000,000	⚲	astonished person

[PPT]

To write the number 300, the Egyptians wrote the scroll hieroglyphic three times: ⟋⟋⟋. In the Egyptian hieroglyphic system, the order of the hieroglyphics is of no importance. Each of the following four Egyptian numerals represents 321.

⟋⟋⟋∩∩|,   ∩∩⟋|⟋⟋,   ⟋|∩⟋∩⟋,   [stacked hieroglyphic figure]

▼ example 1 Write a Numeral Using Egyptian Hieroglyphics

Write 3452 using Egyptian hieroglyphics.

Solution
3452 = 3000 + 400 + 50 + 2.
Thus the Egyptian numeral for 3452 is ⚘⚘⚘⟋⟋⟋⟋∩∩∩∩∩||.

▼ check your progress 1 Write 201,473 using Egyptian hieroglyphics.

Solution   *See page S20.* ◄

question Do the Egyptian hieroglyphics ୨୨∩| and ∩|୨୨ represent the same number?

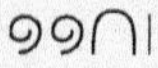

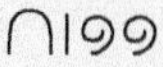

▼ example 2 Evaluate a Numeral Written Using Egyptian Hieroglyphics

Write ⚬⟋⟋⟋⚏⚏୨୨∩||| as a Hindu-Arabic numeral.

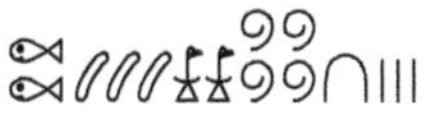

Solution
$(2 \times 100{,}000) + (3 \times 10{,}000) + (2 \times 1000) + (4 \times 100) + (1 \times 10) + (3 \times 1) = 232{,}413$

▼ check your progress 2 Write 𝓎⚬⟋⚏⚏⚏⚏୨୨∩|||| as a Hindu-Arabic numeral.

Solution   *See page S20.* ◄

One of the earliest written documents of mathematics is the Rhind papyrus (see the figure at the left). This tablet was found in Egypt in AD 1858, but it is estimated that the writings date back to 1650 BC. The Rhind papyrus contains 85 mathematical problems. Studying these problems has enabled mathematicians and historians to understand some of the mathematical procedures used in the early Egyptian numeration system.

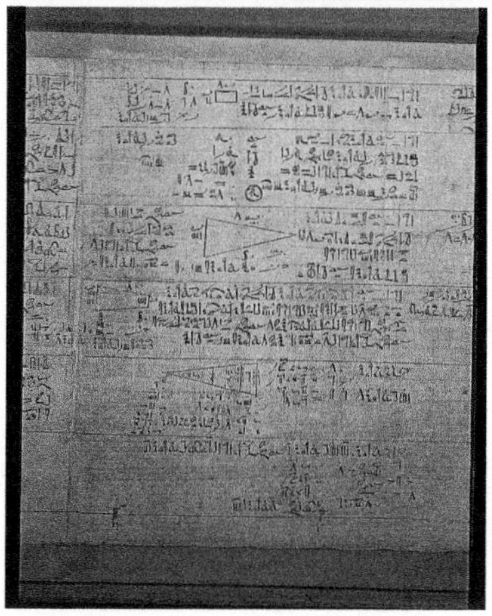

The operation of addition with Egyptian hieroglyphics is a simple grouping process. In some cases the final sum can be simplified by replacing a group of hieroglyphics by a single hieroglyphic with an equivalent numerical value. This technique is illustrated in Example 3.

▼ example 3 Use Egyptian Hieroglyphics to Find a Sum

Use Egyptian hieroglyphics to find $2452 + 1263$.

Solution
The sum is found by combining the hieroglyphics.

2452	⚏⚏୨୨୨୨∩∩∩∩ ||
+ 1263	+ ⚏୨୨∩∩∩∩∩∩|||
	⚏⚏୨୨୨୨∩∩∩ ||
	⚏ ୨୨ ∩∩∩∩∩|||

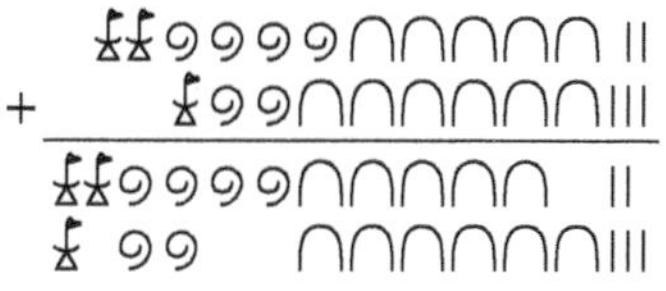

Replacing 10 heel bones with one scroll produces

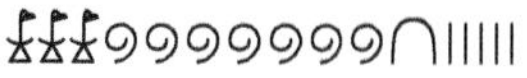

⚏⚏⚏୨୨୨୨୨୨୨∩||||| or 3715

The sum is 3715.

▼ check your progress 3 Use Egyptian hieroglyphics to find $23{,}341 + 10{,}562$.

Solution   *See page S20.* ◄

answer Yes. They both represent 211.

A portion of the Rhind papyrus

The Rhind papyrus is named after Alexander Henry Rhind, who purchased the papyrus in Egypt in AD 1858. Today the Rhind papyrus is preserved in the British Museum in London.

Art Resources, NY

In the Egyptian numeration system, subtraction is performed by removing some of the hieroglyphics from the larger numeral. In some cases it is necessary to "borrow," as shown in the next example.

▼ **example 4**   **Use Egyptian Hieroglyphics to Find a Difference**

Use Egyptian hieroglyphics to find 332,246 − 101,512.

Solution
The numerical value of one lotus flower is equivalent to the numerical value of 10 scrolls. Thus

TAKE NOTE ✓

Five scrolls cannot be removed from 2 scrolls, so one lotus flower is replaced by 10 scrolls, resulting in a total of 12 scrolls. Now 5 scrolls can be removed from 12 scrolls.

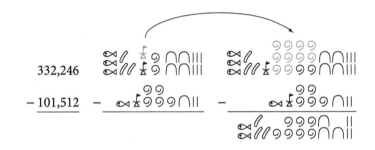

$$332,246$$

$$- 101,512$$

The difference is 230,734.

▼ **check your progress 4**   Use Egyptian hieroglyphics to find 61,432 − 45,121.

Solution   *See page S20.*

MATHMATTERS   Early Egyptian Fractions

Evidence gained from the Rhind papyrus shows that the Egyptian method of calculating with fractions was much different from the methods we use today. All Egyptian fractions (except for $\frac{2}{3}$) were represented in terms of unit fractions, which are fractions of the form $\frac{1}{n}$, for some natural number $n > 1$. The Egyptians wrote these unit fractions by placing an oval over the numeral that represented the denominator. For example,

$$\frac{\bigcirc}{|||} = \frac{1}{3} \qquad \overset{\bigcirc}{\cap|||||} = \frac{1}{15}$$

If a fraction was not a unit fraction, then the Egyptians wrote the fraction as the sum of *distinct* unit fractions. For instance,

$$\frac{2}{5} \text{ was written as the sum of } \frac{1}{3} \text{ and } \frac{1}{15}.$$

Of course, $\frac{2}{5} = \frac{1}{5} + \frac{1}{5}$, but (for some mysterious reason) the early Egyptian numeration system didn't allow repetitions. The Rhind papyrus includes a table that shows how to write fractions of the form $\frac{2}{k}$, where k is an odd number from 5 to 101, in terms of unit fractions. Some of these are listed below.

$$\frac{2}{7} = \frac{1}{4} + \frac{1}{28} \qquad \frac{2}{11} = \frac{1}{6} + \frac{1}{66} \qquad \frac{2}{19} = \frac{1}{12} + \frac{1}{76} + \frac{1}{114}$$

The Roman Numeration System

TABLE 6.2 Roman Numerals

Hindu-Arabic numeral	Roman numeral
1	I
5	V
10	X
50	L
100	C
500	D
1000	M

The Roman numeration system was used in Europe during the reign of the Roman Empire. Today we still make limited use of Roman numerals on clock faces, on the cornerstones of buildings, and in numbering the volumes of periodicals and books. Table 6.2 shows the numerals used in the Roman numeration system. If the Roman numerals are listed so that each numeral has a larger value than the numeral to its right, then the value of the Roman numeral is found by adding the values of each numeral. For example,

$$CLX = 100 + 50 + 10 = 160$$

If a Roman numeral is repeated two or three times in succession, we add to determine its numerical value. For instance, $XX = 10 + 10 = 20$ and $CCC = 100 + 100 + 100 = 300$. Each of the numerals I, X, C, and M may be repeated up to three times. The numerals V, L, and D are not repeated.

Although the Roman numeration system is an additive system, it also incorporates a subtraction property. In the Roman numeration system, the value of a numeral is determined by adding the values of the numerals from left to right. However, if the value of a numeral is less than the value of the numeral to its right, the smaller value is subtracted from the next larger value. For instance, $VI = 5 + 1 = 6$; however, $IV = 5 - 1 = 4$. In the Roman numeration system, the only numerals whose values can be subtracted from the value of the numeral to the right are I, X, and C. Also, the subtraction of these values is allowed only if the value of the numeral to the right is within two rows, as shown in Table 6.2. That is, the value of the numeral to be subtracted must be *no less than* one-tenth of the value of the numeral it is to be subtracted from. For instance, $XL = 40$ and $XC = 90$, but XD does not represent 490 because the value of X is less than one-tenth the value of D. To write 490 using Roman numerals, we write CDXC.

INSTRUCTOR NOTE

Inform your students that the C used in Roman numerals is from the Latin word for "hundred," which is *centum*. The word *century*, which means a period of 100 years, is a derivation of *centum*. The M is from the Latin word for "thousand," which is *mille*. We use the word *millennium* to designate 1000 years.

HISTORICAL NOTE

The Roman numeration system evolved over a period of several years, and thus some Roman numerals displayed on ancient structures do not adhere to the basic rules given at the right. For instance, in the Colosseum in Rome (ca. AD 80), the numeral XXVIIII appears above archway 29 instead of the numeral XXIX.

PictureQuest

▼ **A Summary of the Basic Rules Employed in the Roman Numeration System**

$I = 1, \quad V = 5, \quad X = 10, \quad L = 50, \quad C = 100, \quad D = 500, \quad M = 1000$

1. If the numerals are listed so that each numeral has a larger value than the numeral to the right, then the value of the Roman numeral is found by adding the values of the numerals.

2. Each of the numerals, I, X, C, and M may be repeated up to three times. The numerals V, L, and D are not repeated. If a numeral is repeated two or three times in succession, we add to determine its numerical value.

3. The only numerals whose values can be subtracted from the value of the numeral to the right are I, X, and C. The value of the numeral to be subtracted must be no less than one-tenth of the value of the numeral to its right.

▼ **example 5**   Evaluate a Roman Numeral

Write DCIV as a Hindu-Arabic numeral.

Solution
Because the value of D is larger than the value of C, we add their numerical values. The value of I is less than the value of V, so we subtract the smaller value from the larger value. Thus

$$DCIV = (DC) + (IV) = (500 + 100) + (5 - 1) = 600 + 4 = 604$$

 ▼ **check your progress 5**   Write MCDXLV as a Hindu-Arabic numeral.

Solution   *See page S20.*

▼ **example 6** Write a Hindu-Arabic Numeral as a Roman Numeral

Write 579 as a Roman numeral.

Solution

$$579 = 500 + 50 + 10 + 10 + 9$$

In Roman numerals, 9 is written as IX. Thus $579 = $ DLXXIX.

▼ **check your progress 6** Write 473 as a Roman numeral.

Solution *See page S20.* ◄

In the Roman numeration system, a bar over a numeral is used to denote a value 1000 times the value of the numeral. For instance,

$$\overline{V} = 5 \times 1000 = 5000 \qquad \overline{IV}LXX = (4 \times 1000) + 70 = 4070$$

TAKE NOTE ✓

The method of writing a bar over a numeral should be used only to write Roman numerals that cannot be written using the basic rules. For instance, the Roman numeral for 2003 is MMIII, not $\overline{II}III$.

▼ **example 7** Convert between Roman Numerals and Hindu-Arabic Numerals

a. Write $\overline{IV}$DLXXII as a Hindu-Arabic numeral.

b. Write 6125 as a Roman numeral.

Solution

a. $\overline{IV}$DLXXII $= (\overline{IV}) + (DLXXII)$
$\qquad\qquad\quad = (4 \times 1000) + (572)$
$\qquad\qquad\quad = 4572$

b. The Roman numeral 6 is written VI, and 125 is written as CXXV. Thus in Roman numerals, 6125 is $\overline{VI}$CXXV.

▼ **check your progress 7**

a. Write $\overline{VII}$CCLIV as a Hindu-Arabic numeral.

b. Write 8070 as a Roman numeral.

Solution *See page S20.* ◄

EXCURSION

Photolibrary

A Rosetta Tablet for the Traditional Chinese Numeration System

Most of the knowledge we have gained about early numeration systems has been obtained from inscriptions found on ancient tablets or stones. The information provided by these inscriptions has often been difficult to interpret. For several centuries, archeologists had little success in interpreting the Egyptian hieroglyphics they had discovered. Then, in

1799, a group of French military engineers discovered a basalt stone near Rosetta in the Nile delta. This stone, which we now call the Rosetta Stone, has an inscription in three scripts: Greek, Egyptian Demotic, and Egyptian hieroglyphic. It was soon discovered that all three scripts contained the same message. The Greek script was easy to translate, and from its translation, clues were uncovered that enabled scholars to translate many of the documents that up to that time had been unreadable.

Pretend that you are an archeologist. Your team has just discovered an old tablet that displays Roman numerals and traditional Chinese numerals. It also provides hints in the form of a crossword puzzle about the traditional Chinese numeration system. Study the inscriptions on the following tablet and then complete the Excursion Exercises that follow.

The Rosetta Stone

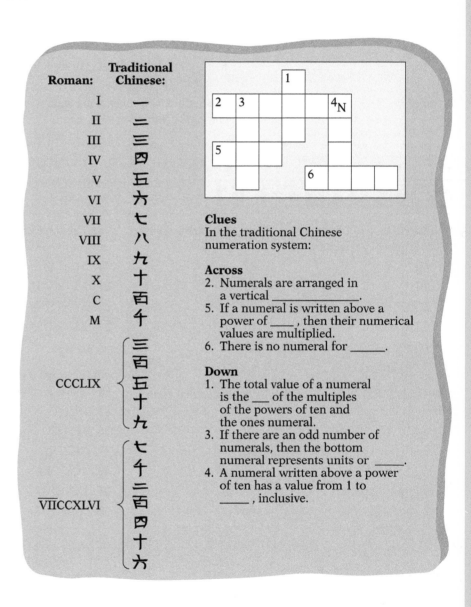

Clues
In the traditional Chinese numeration system:

Across
2. Numerals are arranged in a vertical _____.
5. If a numeral is written above a power of ____ , then their numerical values are multiplied.
6. There is no numeral for _____.

Down
1. The total value of a numeral is the ___ of the multiples of the powers of ten and the ones numeral.
3. If there are an odd number of numerals, then the bottom numeral represents units or ____.
4. A numeral written above a power of ten has a value from 1 to ____ , inclusive.

EXCURSION EXERCISES

1. Complete the crossword puzzle shown on the above tablet.

2. Write 26 as a traditional Chinese numeral.

3. Write 357 as a traditional Chinese numeral.

4. Write the Hindu-Arabic numeral given by each of the following traditional Chinese numerals.

a. 八
百
九
十
六

b. 二
千
四
百
六
十
五

5. a. How many Hindu-Arabic numerals are required to write four thousand five hundred twenty-eight?

b. How many traditional Chinese numerals are required to write four thousand five hundred twenty-eight?

6. The traditional Chinese numeration system is no longer in use. Give a reason that may have contributed to its demise.

EXERCISE SET 6.1

(Suggested Assignment: The Enhanced WebAssign Exercises and Exercises 18, 23, 25, and 31)

■ In Exercises 1 to 12, write each Hindu-Arabic numeral using Egyptian hieroglyphics.

1. 46 **2.** 82

3. 103 **4.** 157

5. 2568 **6.** 3152

7. 23,402 **8.** 15,303

9. 65,800 **10.** 43,217

11. 1,405,203 **12.** 653,271

■ In Exercises 13 to 24, write each Egyptian numeral as a Hindu-Arabic numeral.

13. 𓎆𓎆𓏤𓃟𓎩𓎩𓏤𓏤𓏤𓏤

14. 𓏤𓃟𓃟𓃟𓎩𓏤𓏤

15. 𓃟𓃟𓃟𓃟𓎩𓎩𓏤𓏤𓏤
𓃟𓃟𓃟𓃟𓎩𓎩𓏤𓏤

16. 𓃟𓎩𓎩𓎩𓎩𓏤𓏤𓏤𓏤
𓃟𓎩𓎩𓎩𓎩𓏤𓏤𓏤

17. 𓏤𓃟𓃟𓎩𓎩𓎩𓏤𓏤

18. 𓏭𓏭 𓃟𓃟𓎩𓏤
𓆈𓏭𓎆𓎆𓃟𓎩𓎩𓏤𓏤

19. 𓆈𓆈𓏭𓏭𓏤𓎩𓏤

20. 𓏭𓏭𓏭𓎆𓎆𓃟𓃟𓃟𓎩𓎩𓏤𓏤𓏤

21. 𓏭𓏭𓏭𓎆𓎆𓎆𓃟𓃟𓃟𓎩𓎩𓎩𓏤𓏤𓏤𓏤𓏤
𓏭𓏭𓏭𓎆𓎆𓃟𓃟𓃟𓃟𓎩𓎩𓎩𓏤𓏤𓏤𓏤

22. 𓅓𓆈𓏭𓏭𓎆𓎆𓎆𓃟𓃟𓃟𓃟𓃟𓎩
𓅓𓆈𓏭 𓎆𓎆𓎆𓃟𓃟𓃟𓎩𓏤𓏤𓏤

23. 𓅓𓅓𓅓𓆈𓎆𓃟𓃟𓏤𓏤𓏤
𓅓𓅓𓏭𓏭𓏭 𓎆𓃟𓃟𓏤𓏤𓏤

24. 𓅓𓆈𓆈𓎆𓃟𓃟
𓅓𓆈𓆈𓃟𓃟

■ In Exercises 25 to 32, use Egyptian hieroglyphics to find each sum or difference.

25. 51 + 43 **26.** 67 + 58

27. 231 + 435 **28.** 623 + 124

29. 83 − 51 **30.** 94 − 23

31. 254 − 198 **32.** 640 − 278

■ In Exercises 33 to 44, write each Roman numeral as a Hindu-Arabic numeral.

33. DCL **34.** MCX

35. MCDIX **36.** MDCCII

37. MCCXL **38.** MMDCIV

39. DCCCXL **40.** CDLV

41. $\overline{\text{IX}}$XLIV **42.** $\overline{\text{VII}}$DXVII

43. $\overline{\text{XI}}$CDLXI **44.** $\overline{\text{IV}}$CCXXI

■ In Exercises 45 to 56, write each Hindu-Arabic numeral as a Roman numeral.

45. 157 **46.** 231

47. 542 **48.** 783

49. 1197 **50.** 1039

51. 787 **52.** 1343

53. 683 **54.** 959

55. 6898 **56.** 4357

■ **Egyptian Multiplication** The Rhind papyrus contains problems that show a *doubling procedure* used by the Egyptians to find the product of two whole numbers. The following examples illustrate this doubling procedure. In the examples, we have used Hindu-Arabic numerals so that you can concentrate on the doubling procedure and not be distracted by the Egyptian hieroglyphics. The first example determines the product 5 × 27 by computing two successive doublings of 27 and then forming the sum of the blue numbers in the rows marked with a check. Note that the rows marked with a check show that one 27 is 27 and four 27s is 108. Thus five 27s is the sum of 27 and 108, or 135.

√ 1   27⟍ double
  2   54⟍ double
√ 4   108⟍
  5   135 ◄—— This sum is the product of 5 and 27.

In the next example, we use the Egyptian doubling procedure to find the product of 35 and 94. Because the sum of 1, 2, and 32 is 35, we add only the blue numbers in the rows marked with a check to find that 35 × 94 = 94 + 188 + 3008 = 3290.

√ 1   94 ⟍ double
√ 2   188 ⟍ double
  4   376 ⟍ double
  8   752 ⟍ double
 16   1504 ⟍ double
√ 32   3008 ⟍
 35   3290 ◄—— This sum is the product of 35 and 94.

In Exercises 57 to 64, use the Egyptian doubling procedure to find each product.

57. 8 × 63   **58.** 4 × 57

59. 7 × 29   **60.** 9 × 33

61. 17 × 35   **62.** 26 × 43

63. 23 × 108   **64.** 72 × 215

Critical Thinking

65. a. State a reason why you might prefer to use the Egyptian hieroglyphic numeration system rather than the Roman numeration system.

   b. State a reason why you might prefer to use the Roman numeration system rather than the Egyptian hieroglyphic numeration system.

66. What is the largest number that can be written using Roman numerals without using the bar over a numeral or the subtraction property?

Explorations

67. **The Ionic Greek Numeration System** The Ionic Greek numeration system assigned numerical values to the letters of the Greek alphabet. Research the Ionic Greek numeration system and write a report that explains this numeration system. Include information about some of the advantages and disadvantages of this system compared with our present Hindu-Arabic numeration system.

68. **The Method of False Position** The Rhind papyrus (see page 301) contained solutions to several mathematical problems. Some of these solutions made use of a procedure called the *method of false position*. Research the method of false position and write a report that explains this method. In your report, include a specific mathematical problem and its solution by the method of false position.

section **6.2** Place-Value Systems

Expanded Form

The most common numeration system used by people today is the Hindu-Arabic numeration system. It is called the Hindu-Arabic system because it was first developed in India (around AD 800) and then refined by the Arabs. It makes use of the 10 symbols 0, 1, 2, 3, 4, 5, 6, 7, 8, and 9. The reason for the 10 symbols, called *digits,* is related to the fact that we

HISTORICAL NOTE

Abu Ja'far Muhammad ibn Musa al'Khwarizmi

(ca. AD 790–850) Al'Khwarizmi (ăl'khwă hr'ĭz mee) produced two important texts. One of these texts advocated the use of the Hindu-Arabic numeration system. The 12th-century Latin translation of this book is called *Liber Algoritmi de Numero Indorum*, or *Al-Khwarizmi on the Hindu Art of Reckoning*. In Europe, the people who favored the adoption of the Hindu-Arabic numeration system became known as *algorists* because of Al'Khwarizmi's *Liber Algoritmi de Numero Indorum* text. The Europeans who opposed the Hindu-Arabic system were called *abacists*. They advocated the use of Roman numerals and often performed computations with the aid of an abacus.

have 10 fingers. The Hindu-Arabic numeration system is also called the *decimal system*, where the word *decimal* is a derivation of the Latin word *decem*, which means "ten."

One important feature of the Hindu-Arabic numeration system is that it is a *place-value* or *positional-value system*. This means that the numerical value of each digit in a Hindu-Arabic numeral depends on its *place* or *position* in the numeral. For instance, the 3 in 31 represents 3 tens, whereas the 3 in 53 represents 3 ones. The Hindu-Arabic numeration system is a **base ten numeration system** because the place values are the powers of 10:

$$\ldots, 10^5, 10^4, 10^3, 10^2, 10^1, 10^0$$

The place value associated with the *n*th digit of a numeral (counting from right to left) is 10^{n-1}. For instance, in the numeral 7532, the 7 is the fourth digit from the right and is in the $10^{4-1} = 10^3$, or thousands, place. The numeral 2 is the first digit from the right and is in the $10^{1-1} = 10^0$, or ones, place. The *indicated sum* of each digit of a numeral multiplied by its respective place value is called the **expanded form** of the numeral.

▼ **example 1** **Write a Numeral in Its Expanded Form**

Write 4672 in expanded form.

Solution

$$4672 = 4000 + 600 + 70 + 2$$
$$= (4 \times 1000) + (6 \times 100) + (7 \times 10) + (2 \times 1)$$

The above expanded form can also be written as

$$(4 \times 10^3) + (6 \times 10^2) + (7 \times 10^1) + (2 \times 10^0)$$

▼ **check your progress 1** Write 17,325 in expanded form.

Solution *See page S20.* ◄

If a number is written in expanded form, it can be simplified to its ordinary decimal form by performing the indicated operations. The *Order of Operations Agreement* states that we should first perform the exponentiations, then perform the multiplications, and finish by performing the additions.

▼ **example 2** **Simplify a Number Written in Expanded Form**

Simplify: $(2 \times 10^3) + (7 \times 10^2) + (6 \times 10^1) + (3 \times 10^0)$

Solution

$$(2 \times 10^3) + (7 \times 10^2) + (6 \times 10^1) + (3 \times 10^0)$$
$$= (2 \times 1000) + (7 \times 100) + (6 \times 10) + (3 \times 1)$$
$$= 2000 + 700 + 60 + 3$$
$$= 2763$$

▼ **check your progress 2** Simplify:

$$(5 \times 10^4) + (9 \times 10^3) + (2 \times 10^2) + (7 \times 10^1) + (4 \times 10^0)$$

Solution *See page S20.* ◄

In the next few examples, we make use of the expanded form of a numeral to compute sums and differences. An examination of these examples will help you better understand the computational algorithms used in the Hindu-Arabic numeration system.

▼ **example 3**   Use Expanded Forms to Find a Sum

Use expanded forms of 26 and 31 to find their sum.

Solution

$$26 = (2 \times 10) + 6$$
$$+\ 31 = (3 \times 10) + 1$$
$$(5 \times 10) + 7 = 50 + 7 = 57$$

▼ **check your progress 3**   Use expanded forms to find the sum of 152 and 234.

Solution   *See page S20.*   ◄

If the expanded form of a sum contains one or more powers of 10 that have multipliers larger than 9, then we simplify by rewriting the sum with multipliers that are less than or equal to 9. This process is known as *carrying*.

▼ **example 4**   Use Expanded Forms to Find a Sum

Use expanded forms of 85 and 57 to find their sum.

Solution

$$85 = (8 \times 10) \qquad +\ 5$$
$$+\ 57 = (5 \times 10) \qquad +\ 7$$
$$(13 \times 10) \qquad +\ 12$$
$$(10 + 3) \times 10 + 10 + 2$$
$$100 + 30 \qquad +\ 10 + 2 = 100 + 40 + 2 = 142$$

TAKE NOTE

From the expanded forms in Example 4, note that 12 is 1 ten and 2 ones. When we add columns of numbers, this is shown as "carry a 1." Because the 1 is placed in the tens column, we are actually adding 10.

$$\begin{array}{r} 1 \\ 85 \\ +\ 57 \\ \hline 142 \end{array}$$

▼ **check your progress 4**   Use expanded forms to find the sum of 147 and 329.

Solution   *See page S20.*   ◄

In the next example, we use the expanded forms of numerals to analyze the concept of "borrowing" in a subtraction problem.

▼ **example 5**   Use Expanded Forms to Find a Difference

Use the expanded forms of 457 and 283 to find 457 − 283.

Solution

$$457 = (4 \times 100) + (5 \times 10) + 7$$
$$-\ 283 = (2 \times 100) + (8 \times 10) + 3$$

From the expanded forms in Example 5, note that we "borrowed" 1 hundred as 10 tens. This explains how we show borrowing when numbers are subtracted using place value form.

$$\begin{array}{r} \overset{3}{\cancel{4}}\,^{1}57 \\ -\ 2\ 83 \\ \hline 1\ 74 \end{array}$$

At this point, this example is similar to Example 4 in Section 6.1. We cannot remove 8 tens from 5 tens, so 1 hundred is replaced by 10 tens.

$$457 = (4 \times 100) + (5 \times 10) + 7$$
$$= (3 \times 100) + (10 \times 10) + (5 \times 10) + 7$$
$$= (3 \times 100) + (15 \times 10) + 7$$

- $4 \times 100 = 3 \times 100 + 100$
  $= 3 \times 100 + 10 \times 10$

We can now remove 8 tens from 15 tens.

$$457 = (3 \times 100) + (15 \times 10) + 7$$
$$\underline{-\ 283 = (2 \times 100) +\ (8 \times 10) + 3}$$
$$= (1 \times 100) +\ (7 \times 10) + 4 = 100 + 70 + 4 = 174$$

▼ **check your progress 5**   Use expanded forms to find the difference $382 - 157$.

Solution   *See page S20.*

The Babylonian Numeration System

The Babylonian numeration system uses a base of 60. The place values in the Babylonian system are given in the following table.

TABLE 6.3   Place Values in the Babylonian Numeration System

	60^3	60^2	60^1	60^0
...	$= 216{,}000$	$= 3600$	$= 60$	$= 1$

The Babylonians recorded their numerals on damp clay using a wedge-shaped stylus. A vertical wedge shape represented 1 unit and a sideways "vee" shape represented 10 units.

$$\text{Y} \quad 1$$
$$\text{く} \quad 10$$

To represent a number smaller than 60, the Babylonians used an *additive* feature similar to that used by the Egyptians. For example, the Babylonian numeral for 32 is

くくくYY

For the number 60 and larger numbers, the Babylonians left a small space between groups of symbols to indicate a different place value. This procedure is illustrated in the following example.

▼ **example 6**   Write a Babylonian Numeral as a Hindu-Arabic Numeral

Write Y くくくY くくYYYYY as a Hindu-Arabic numeral.

Solution

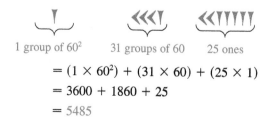

$$= (1 \times 60^2) + (31 \times 60) + (25 \times 1)$$
$$= 3600 + 1860 + 25$$
$$= 5485$$

▼ check your progress 6 Write ⟪Ⴥ ⴄⴄⴄⴄⴄ ⟪⟪Ⴥⴄⴄⴄⴄ as a Hindu-Arabic numeral.

Solution *See page S20.*

◄

question In the Babylonian numeration system, does ⴄⴄ = Ⴥ Ⴥ ?

In the next example we illustrate a division process that can be used to convert Hindu-Arabic numerals to Babylonian numerals.

▼ example 7 Write a Hindu-Arabic Numeral as a Babylonian Numeral

Write 8503 as a Babylonian numeral.

Solution
The Babylonian numeration system uses place values of

$$60^0, 60^1, 60^2, 60^3, \ldots .$$

Evaluating the powers produces

$$1, 60, 3600, 216{,}000, \ldots$$

The largest of these powers that is contained in 8503 is 3600. One method of finding how many groups of 3600 are in 8503 is to divide 3600 into 8503. Refer to the first division shown below. Now divide to determine how many groups of 60 are contained in the remainder 1303.

$$
\begin{array}{r}
2 \\
3600)\overline{8503} \\
7200 \\
\hline
1303
\end{array}
\qquad
\begin{array}{r}
21 \\
60)\overline{1303} \\
120 \\
\hline
103 \\
60 \\
\hline
43
\end{array}
$$

The above computations show that 8503 consists of 2 groups of 3600 and 21 groups of 60, with 43 left over. Thus

$$8503 = (2 \times 60^2) + (21 \times 60) + (43 \times 1)$$

As a Babylonian numeral, 8503 is written

ⴄⴄ ⟪Ⴥ ⟪⟪⟪⟪Ⴥⴄⴄ

▼ check your progress 7 Write 12,578 as a Babylonian numeral.

Solution *See page S20.*

◄

In Example 8, we find the sum of two Babylonian numerals. If a numeral for any power of 60 is larger than 59, then simplify by decreasing that numeral by 60 and increasing the numeral in the place value to its left by 1.

---

answer No. ⴄⴄ = 2, whereas Ⴥ Ⴥ = $(1 \times 60) + (1 \times 1) = 61$.

▼ **example 8** Find the Sum of Babylonian Numerals

Find the sum of the following numerals. Write the answer as a Babylonian numeral.

⟨⟨Y ⟨⟨⟨⟨YY
+ ⟨⟨⟨⟨YYY ⟨⟨⟨YYY

Solution

  ⟨⟨Y ⟨⟨⟨⟨YY
+ ⟨⟨⟨⟨YYY ⟨⟨⟨YYY
─────────────────────────
= ⟨⟨⟨⟨⟨⟨YYY ⟨⟨⟨⟨⟨⟨⟨YYYYY   • Combine the symbols for each place value.
= ⟨⟨⟨⟨⟨⟨YYYYY ⟨YYYYY   • Take away 60 from the ones place and add 1 to the 60s place.
= Y YYYYY ⟨YYYYY   • Take away 60 from the 60s place and add 1 to the 60^2 place.

⟨⟨Y ⟨⟨⟨⟨⟨YY + ⟨⟨⟨⟨YYY ⟨⟨⟨YYY = Y YYYYY ⟨YYYYY

▼ **check your progress 8** Find the sum of the following numerals. Write the answer as a Babylonian numeral.

  ⟨⟨⟨YY ⟨⟨⟨⟨⟨YYYYY
+ ⟨⟨⟨⟨YYYY ⟨⟨⟨⟨⟨YYYYYYY

Solution *See page S20.*

◀

MATHMATTERS Zero as a Placeholder and as a Number

When the Babylonian numeration system first began to develop around 1700 BC, it did not make use of a symbol for zero. The Babylonians merely used an empty space to indicate that a place value was missing. This procedure of "leaving a space" can be confusing. How big is an empty space? Is that one empty space or two empty spaces? Around 300 BC, the Babylonians started to use the symbol ♠ to indicate that a particular place value was missing. For instance, YY ♠ ⟨Y represented $(2 \times 60^2) + (11 \times 1) = 7211$. In this case the zero placeholder indicates that there are no 60s. There is evidence that although the Babylonians used the zero placeholder, they did not use the number zero.

The Mayan Numeration System

The Mayan civilization existed in the Yucatán area of southern Mexico and in Guatemala, Belize, and parts of El Salvador and Honduras. It started as far back as 9000 BC and reached its zenith during the period from AD 200 to AD 900. Among their many accomplishments, the Maya are best known for their complex hieroglyphic writing system, their sophisticated calendars, and their remarkable numeration system.

The Maya used three calendars—the solar calendar, the ceremonial calendar, and the Venus calendar. The solar calendar consisted of about 365.24 days. Of these, 360 days were divided into 18 months, each with 20 days. The Mayan numeration system was strongly influenced by this solar calendar, as evidenced by the use of the numbers 18 and 20 in determining place values. See Table 6.4.

TABLE 6.4 Place Values in the Mayan Numeration System

	18×20^3	18×20^2	18×20^1	20^1	20^0
...	$= 144{,}000$	$= 7200$	$= 360$	$= 20$	$= 1$

The Mayan numeration system was one of the first systems to use a symbol for zero as a placeholder. The Mayan numeration system used only three symbols. A dot was used to represent 1, a horizontal bar represented 5, and a conch shell represented 0. The following table shows how the Maya used a combination of these three symbols to write the whole numbers from 0 to 19. Note that each numeral contains at most four dots and at most three horizontal bars.

TABLE 6.5 Mayan Numerals

To write numbers larger than 19, the Maya used a vertical arrangement with the largest place value at the top. The following example illustrates the process of converting a Mayan numeral to a Hindu-Arabic numeral.

▼ **example 9**   Write a Mayan Numeral as a Hindu-Arabic Numeral

Write each of the following as a Hindu-Arabic numeral.

a. **b.**

Solution

a.

$10 \times 360 = 3600$
$8 \times 20 = 160$
$11 \times 1 = +11$
$\overline{\qquad 3771}$

b.

$5 \times 7200 = 36{,}000$
$0 \times 360 = \qquad 0$
$12 \times 20 = \qquad 240$
$3 \times 1 = + \quad 3$
$\overline{\qquad 36{,}243}$

▼ **check your progress 9**   Write each of the following as a Hindu-Arabic numeral.

a. **b.**

Solution   *See page S21.*

In the next example, we illustrate how the concept of place value is used to convert Hindu-Arabic numerals to Mayan numerals.

▼ **example 10** **Write a Hindu-Arabic Numeral as a Mayan Numeral**

Write 7495 as a Mayan numeral.

Solution
The place values used in the Mayan numeration system are

$$20^0, 20^1, 18 \times 20^1, 18 \times 20^2, 18 \times 20^3, \ldots$$

or

$$1, 20, 360, 7200, 144{,}000, \ldots$$

Removing 1 group of 7200 from 7495 leaves 295. No groups of 360 can be obtained from 295, so we divide 295 by the next smaller place value of 20 to find that 295 equals 14 groups of 20 with 15 left over.

$$
\begin{array}{r}
1 \\[-2pt]
\overline{7200)\,7495} \\
7200 \\ \hline
295
\end{array}
\qquad
\begin{array}{r}
14 \\[-2pt]
\overline{20)\,295} \\
20 \\ \hline
95 \\
80 \\ \hline
15
\end{array}
$$

Thus

$$7495 = (1 \times 7200) + (0 \times 360) + (14 \times 20) + (15 \times 1)$$

In Mayan numerals, 7495 is written as

▼ **check your progress 10** Write 11,480 as a Mayan numeral.

Solution *See page S21.*

EXCURSION

Subtraction via the Nines Complement and the End-Around Carry

INSTRUCTOR NOTE

The concepts in this Excursion are extended in the Excursion in Section 6.4, page 335.

In the subtraction $5627 - 2564 = 3063$, the number 5627 is called the *minuend*, 2564 is called the *subtrahend*, and 3063 is called the *difference*. In the Hindu-Arabic base ten system, subtraction can be performed by a process that involves addition and the *nines complement* of the subtrahend. The **nines complement** of a single digit n is the number $9 - n$. For instance, the nines complement of 3 is 6, the nines complement of 1 is 8, and the nines complement of 0 is 9. The nines complement of a number with more than one digit is the number that is formed by taking the nines complement of each digit. The nines complement of 25 is 74, and the nines complement of 867 is 132.

▼ **Subtraction by Using the Nines Complement and the End-Around Carry**

To subtract by using the nines complement:

1. Add the nines complement of the subtrahend to the minuend.

2. Take away 1 from the leftmost digit of the sum produced in step 1 and add 1 to the units digit. This is referred to as the end-around carry procedure.

The following example illustrates the process of subtracting 2564 from 5627 by using the nines complement.

$$
\begin{array}{r}
5627 \\
-\ 2564 \\
\end{array}
$$
Minuend
Subtrahend

$$
\begin{array}{r}
5627 \\
+\ 7435 \\
\hline
13062 \\
\end{array}
$$
Minuend
Replace the subtrahend with the nines complement of the subtrahend and add.

$$
\begin{array}{r}
1\overline{3062} \\
+\qquad 1 \\
\hline
3063 \\
\end{array}
$$
Take away 1 from the leftmost digit and add 1 to the units digit. This is the end-around carry procedure.

Thus

$$
\begin{array}{r}
5627 \\
-\ 2564 \\
\hline
3063 \\
\end{array}
$$

If the subtrahend has fewer digits than the minuend, leading zeros should be inserted in the subtrahend so that it has the same number of digits as the minuend. This process is illustrated below for $2547 - 358$.

$$
\begin{array}{r}
2547 \\
-\ \ 358 \\
\end{array}
$$
Minuend
Subtrahend

$$
\begin{array}{r}
2547 \\
-\ 0358 \\
\end{array}
$$
Insert a leading zero.

$$
\begin{array}{r}
2547 \\
+\ 9641 \\
\hline
12188 \\
\end{array}
$$
Minuend
Nines complement of subtrahend

$$
\begin{array}{r}
1\overline{2188} \\
+\qquad 1 \\
\hline
2189 \\
\end{array}
$$
Take away 1 from the leftmost digit and add 1 to the units digit.

Verify that 2189 is the correct difference.

EXCURSION EXERCISES

For Exercises 1 to 6, use the nines complement of the subtrahend and the end-around carry to find the indicated difference.

1. $724 - 351$ **2.** $2405 - 1608$

3. 91,572 − 7824
4. 214,577 − 48,231
5. 3,156,782 − 875,236
6. 54,327,105 − 7,678,235
7. Explain why the nines complement and the end-around carry procedure produce the correct answer to a subtraction problem.

EXERCISE SET 6.2

(Suggested Assignment: The Enhanced WebAssign Exercises and Exercises 16, 31, 42, 45, 47, 49, 51, and 61)

■ In Exercises 1 to 8, write each numeral in its expanded form.

1. 48
2. 93
3. 420
4. 501
5. 6803
6. 9045
7. 10,208
8. 67,482

■ In Exercises 9 to 16, simplify each expansion.

9. $(4 \times 10^2) + (5 \times 10^1) + (6 \times 10^0)$
10. $(7 \times 10^2) + (6 \times 10^1) + (3 \times 10^0)$
11. $(5 \times 10^3) + (0 \times 10^2) + (7 \times 10^1) + (6 \times 10^0)$
12. $(3 \times 10^3) + (1 \times 10^2) + (2 \times 10^1) + (8 \times 10^0)$
13. $(3 \times 10^4) + (5 \times 10^3) + (4 \times 10^2) + (0 \times 10^1) + (7 \times 10^0)$
14. $(2 \times 10^5) + (3 \times 10^4) + (0 \times 10^3) + (6 \times 10^2) + (7 \times 10^1) + (5 \times 10^0)$
15. $(6 \times 10^5) + (8 \times 10^4) + (3 \times 10^3) + (0 \times 10^2) + (4 \times 10^1) + (0 \times 10^0)$
16. $(5 \times 10^7) + (3 \times 10^6) + (0 \times 10^5) + (0 \times 10^4) + (7 \times 10^3) + (9 \times 10^2) + (0 \times 10^1) + (2 \times 10^0)$

■ In Exercises 17 to 22, use expanded forms to find each sum.

17. 35 + 41
18. 42 + 56
19. 257 + 138
20. 352 + 461
21. 1023 + 1458
22. 3567 + 2651

■ In Exercises 23 to 28, use expanded forms to find each difference.

23. 62 − 35
24. 193 − 157
25. 4725 − 1362
26. 85,381 − 64,156
27. 23,168 − 12,857
28. 59,163 − 47,956

■ In Exercises 29 to 36, write each Babylonian numeral as a Hindu-Arabic numeral.

29. 𒌋𒌋𒁹𒁹𒁹
30. 𒌋𒌋𒌋𒌋𒁹𒁹𒁹𒁹𒁹
31. 𒁹 𒌋𒌋𒌋𒁹𒁹𒁹𒁹𒁹𒁹𒁹
32. 𒌋𒁹𒁹 𒌋𒌋𒁹𒁹𒁹𒁹𒁹𒁹
33. 𒌋𒌋 𒁹𒁹 𒌋𒁹𒁹𒁹
34. 𒌋𒌋𒁹 𒌋𒁹 𒌋𒁹𒁹
35. 𒌋 𒁹𒁹𒁹 𒌋𒁹 𒁹𒁹𒁹𒁹𒁹𒁹
36. 𒌋𒌋𒁹 𒌋𒁹 𒁹 𒌋𒌋𒌋𒁹𒁹𒁹𒁹

■ In Exercises 37 to 46, write each Hindu-Arabic numeral as a Babylonian numeral.

37. 42
38. 57
39. 128
40. 540
41. 5678
42. 7821
43. 10,584
44. 12,687
45. 21,345
46. 24,567

■ In Exercises 47 to 52, find the sum of the Babylonian numerals. Write each answer as a Babylonian numeral.

47. 𒌋𒌋𒌋𒌋𒁹𒁹𒁹𒁹𒁹
\+ 𒌋𒌋𒁹𒁹𒁹

48. 𒌋𒌋𒌋𒌋𒁹𒁹𒁹𒁹𒁹𒁹𒁹𒁹𒁹
\+ 𒌋𒌋𒌋𒁹𒁹𒁹

49. 𒌋𒌋𒌋𒁹𒁹𒁹 𒌋𒌋𒌋𒌋𒁹𒁹
\+ 𒌋𒌋𒌋𒁹𒁹 𒌋𒌋𒁹

50. 𒌋𒌋𒌋𒌋𒁹𒁹 𒌋𒌋𒌋𒌋𒌋𒁹𒁹
\+ 𒌋𒌋𒌋𒁹 𒌋𒌋𒌋𒁹𒁹𒁹

51. 𒌋 𒌋𒌋𒌋𒁹 𒌋𒌋𒌋𒌋𒁹𒁹𒁹
\+ 𒁹 𒌋𒌋𒁹 𒌋𒌋𒌋𒁹𒁹

52. 𒌋𒌋 𒌋𒌋𒌋𒁹𒁹𒁹 𒌋𒌋𒌋𒁹𒁹𒁹𒁹𒁹𒁹
\+ 𒌋𒁹 𒌋𒌋𒌋𒁹𒁹 𒌋𒌋𒌋𒌋𒁹𒁹𒁹𒁹

■ In Exercises 53 to 60, write each Mayan numeral as a Hindu-Arabic numeral.

53. ····
····

54. ═══
· ·

55. ───
⊙
···

56. ···
·
⊙

57. · · ·
⊙
····
══ ·

58. ─·─
···
⊙

59. ───
⊙
───
···

60. ····
⊙
───
·

■ In Exercises 61 to 68, write each Hindu-Arabic numeral as a Mayan numeral.

61. 137   **62.** 253
63. 948   **64.** 1265
65. 1693   **66.** 2728
67. 7432   **68.** 8654

EXTENSIONS

Critical Thinking

69. a. State a reason why you might prefer to use the Babylonian numeration system instead of the Mayan numeration system.

  b. State a reason why you might prefer to use the Mayan numeration system instead of the Babylonian numeration system.

70. Explain why it might be easy to mistake the number 122 for the number 4 when 122 is written as a Babylonian numeral.

Explorations

71. A Base Three Numeration System A student has created a *base three* numeration system. The student has named this numeration system ZUT because Z, U, and T are the symbols used in this system: Z represents 0, U represents 1, and T represents 2. The place values in this system follow: ..., $3^3 = 27$, $3^2 = 9$, $3^1 = 3$, $3^0 = 1$.

Write each ZUT numeral as a Hindu-Arabic numeral.

a. TU   **b.** TZT   **c.** UZTT

Write each Hindu-Arabic numeral as a ZUT numeral.

d. 37   **e.** 87   **f.** 144

section 6.3 Different Base Systems

Converting Non–Base Ten Numerals to Base Ten

Recall that the Hindu-Arabic numeration system is a base ten system because its place values

$$..., 10^5, 10^4, 10^3, 10^2, 10^1, 10^0$$

TAKE NOTE

Recall that in the expression

$$b^n$$

b is the *base*, and n is the *exponent*.

all have 10 as their base. The Babylonian numeration system is a base sixty system because its place values

$$..., 60^5, 60^4, 60^3, 60^2, 60^1, 60^0$$

all have 60 as their base. In general, a base b (where b is a natural number greater than 1) numeration system has place values of

$$..., b^5, b^4, b^3, b^2, b^1, b^0$$

Many people think that our base ten numeration system was chosen because it is the easiest to use, but this is not the case. In reality most people find it easier to use our base ten system only because they have had a great deal of experience with the base ten system and have not had much experience with non–base ten systems. In this section, we examine some non–base ten numeration systems. To reduce the amount of memorization that would be required to learn new symbols for each of these new systems, we will (as far as possible) make use of our familiar Hindu-Arabic symbols. For instance, if we discuss a base four numeration system that requires four basic symbols, then we will use the four Hindu-Arabic symbols 0, 1, 2, and 3 and the place values

$$\ldots, 4^5, 4^4, 4^3, 4^2, 4^1, 4^0$$

The base eight, or **octal**, numeration system uses the Hindu-Arabic symbols 0, 1, 2, 3, 4, 5, 6, and 7, and the place values

$$\ldots, 8^5, 8^4, 8^3, 8^2, 8^1, 8^0$$

To differentiate between bases, we will label each non–base ten numeral with a subscript that indicates the base. For instance, 23_{four} represents a base four numeral. If a numeral is written without a subscript, then it is understood that the base is ten. Thus 23 written without a subscript is understood to be the base ten numeral 23.

To convert a non–base ten numeral to base ten, we write the numeral in its expanded form, as shown in the following example.

TAKE NOTE

Because 23_{four} is *not* equal to the base ten number 23, it is important *not* to read 23_{four} as "twenty-three." To avoid confusion, read 23_{four} as "two three base four."

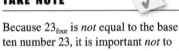

▼ **example 1**   **Convert to Base Ten**

Convert 2314_{five} to base ten.

Solution
In the base five numeration system, the place values are

$$\ldots, 5^4, 5^3, 5^2, 5^1, 5^0$$

The expanded form of 2314_{five} is

$$2314_{\text{five}} = (2 \times 5^3) + (3 \times 5^2) + (1 \times 5^1) + (4 \times 5^0)$$
$$= (2 \times 125) + (3 \times 25) + (1 \times 5) + (4 \times 1)$$
$$= 250 + 75 + 5 + 4$$
$$= 334$$

Thus $2314_{\text{five}} = 334$.

▼ **check your progress 1**   Convert 3156_{seven} to base ten.

Solution   *See page S21.*

question   Does the notation 26_{five} make sense?

In base two, which is called the **binary numeration system**, the place values are the powers of two.

$$\ldots, 2^7, 2^6, 2^5, 2^4, 2^3, 2^2, 2^1, 2^0$$

The binary numeration system uses only the two digits 0 and 1. These *bi*nary dig*its* are often called **bits**. To convert a base two numeral to base ten, write the numeral in its expanded form and then evaluate the expanded form.

answer   No. The expression 26_{five} is a meaningless expression because there is no 6 in base five.

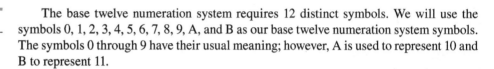

▼ example **2**   Convert to Base Ten

Convert 10110111_{two} to base ten.

Solution

$$
\begin{aligned}
10110111_{two} &= (1 \times 2^7) + (0 \times 2^6) + (1 \times 2^5) + (1 \times 2^4) + (0 \times 2^3) \\
&\quad + (1 \times 2^2) + (1 \times 2^1) + (1 \times 2^0) \\
&= (1 \times 128) + (0 \times 64) + (1 \times 32) + (1 \times 16) + (0 \times 8) \\
&\quad + (1 \times 4) + (1 \times 2) + (1 \times 1) \\
&= 128 + 0 + 32 + 16 + 0 + 4 + 2 + 1 \\
&= 183
\end{aligned}
$$

▼ check your progress **2**   Convert 111000101_{two} to base ten.

Solution   *See page S21.*

The base twelve numeration system requires 12 distinct symbols. We will use the symbols 0, 1, 2, 3, 4, 5, 6, 7, 8, 9, A, and B as our base twelve numeration system symbols. The symbols 0 through 9 have their usual meaning; however, A is used to represent 10 and B to represent 11.

▼ example **3**   Convert to Base Ten

Convert $B37_{twelve}$ to base ten.

Solution

In the base twelve numeration system, the place values are

$$\dots, 12^4, 12^3, 12^2, 12^1, 12^0$$

Thus

$$
\begin{aligned}
B37_{twelve} &= (11 \times 12^2) + (3 \times 12^1) + (7 \times 12^0) \\
&= 1584 + 36 + 7 \\
&= 1627
\end{aligned}
$$

▼ check your progress **3**   Convert $A5B_{twelve}$ to base ten.

Solution   *See page S21.*

Computer programmers often write programs that use the base sixteen numeration system, which is also called the **hexadecimal system**. This system uses the symbols 0, 1, 2, 3, 4, 5, 6, 7, 8, 9, A, B, C, D, E, and F. Table 6.6, on page 320, shows that A represents 10, B represents 11, C represents 12, D represents 13, E represents 14, and F represents 15.

▼ example **4**   Convert to Base Ten

Convert $3E8_{sixteen}$ to base ten.

Solution

In the base sixteen numeration system, the place values are

$$\dots, 16^4, 16^3, 16^2, 16^1, 16^0$$

TABLE 6.6 Decimal and Hexadecimal Equivalents

Base ten decimal	Base sixteen hexadecimal
0	0
1	1
2	2
3	3
4	4
5	5
6	6
7	7
8	8
9	9
10	A
11	B
12	C
13	D
14	E
15	F

Thus

$$3E8_{sixteen} = (3 \times 16^2) + (14 \times 16^1) + (8 \times 16^0)$$
$$= 768 + 224 + 8$$
$$= 1000$$

▼ **check your progress** **4** Convert $C24F_{sixteen}$ to base ten.

Solution *See page S21.*

Converting from Base Ten to Another Base

The most efficient method of converting a number written in base ten to another base makes use of a *successive division process*. For example, to convert 219 to base four, divide 219 by 4 and write the quotient 54 and the remainder 3, as shown below. Now divide the quotient 54 by the base to get a new quotient of 13 and a new remainder of 2. Continuing the process, divide the quotient 13 by 4 to get a new quotient of 3 and a remainder of 1. Because our last quotient, 3, is less than the base, 4, we stop the division process. The answer is given by the last quotient, 3, and the remainders, shown in red in the following diagram. That is, $219 = 3123_{four}$.

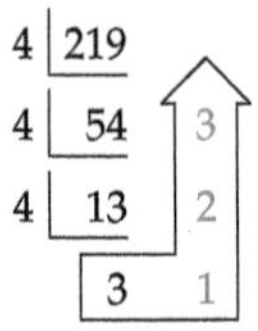

You can understand how the successive division process converts a base ten numeral to another base by analyzing the process. The first division shows there are 54 fours in 219, with **3 ones** left over. The second division shows that there are 13 sixteens (two successive divisions by 4 is the same as dividing by 16) in 219, and the remainder 2 indicates that there are **2 fours** left over. The last division shows that there are 3 sixty-fours (three successive divisions by 4 is the same as dividing by 64) in 219, and the remainder 1 indicates that there is **1 sixteen** left over. In mathematical notation these results are written as follows.

$$219 = (3 \times 64) + (1 \times 16) + (2 \times 4) + (3 \times 1)$$
$$= (3 \times 4^3) + (1 \times 4^2) + (2 \times 4^1) + (3 \times 4^0)$$
$$= 3123_{four}$$

▼ **example** **5** Convert a Base Ten Numeral to Another Base

Convert 5821 to **a.** base three and **b.** base sixteen.

Solution

a.

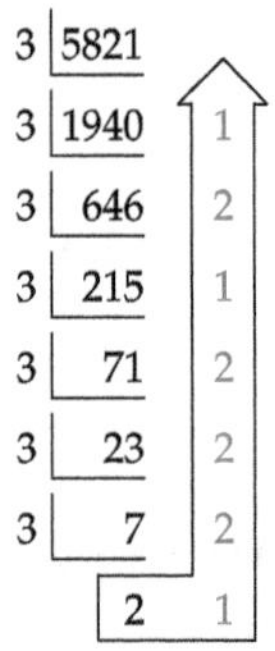

$$5821 = 21222121_{three}$$

b.

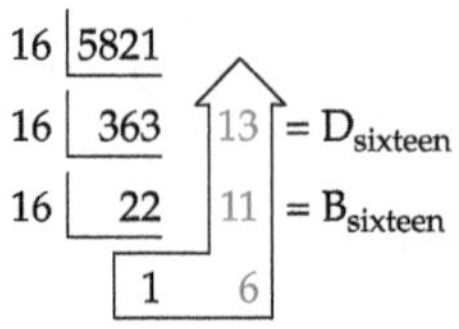

$$5821 = 16BD_{sixteen}$$

▼ check your progress **5**　Convert 1952 to **a.** base five and **b.** base twelve.

Solution　*See page S21.*

MATHMATTERS　Music by the Numbers

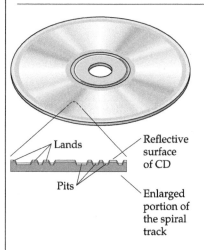

Lands

Reflective
surface
of CD

Pits

Enlarged
portion of
the spiral
track

The binary numeration system is used to encode music on a CD (compact disc). The figure at the left shows the surface of a CD, which consists of flat regions called *lands* and small indentations called *pits*. As a laser beam tracks along a spiral path, the beam is reflected to a sensor when it shines on a land, but it is not reflected to the sensor when it shines on a pit. The sensor interprets a reflection as a 1 and no reflection as a 0. As the CD is playing, the sensor receives a series of 1s and 0s, which the CD player converts to music. On a typical CD, the spiral path that the laser follows loops around the disc over 20,000 times and contains about 650 megabytes of data. A **byte** is 8 bits, so this amounts to 5,200,000,000 bits, each of which is represented by a pit or a land.

TABLE 6.7　Octal and Binary Equivalents

Octal	Binary
0	000
1	001
2	010
3	011
4	100
5	101
6	110
7	111

PPT

Converting Directly between Computer Bases

Although computers compute internally by using base two (binary system), humans generally find it easier to compute with a larger base. Fortunately, there are easy conversion techniques that can be used to convert a base two numeral directly to a base eight (octal) numeral or a base sixteen (hexadecimal) numeral. Before we explain the techniques, it will help to become familiar with the information in Table 6.7, which shows the eight octal symbols and their binary equivalents.

To convert from octal to binary, just replace each octal numeral with its 3-bit binary equivalent.

▼ example **6**　Convert Directly from Base Eight to Base Two

Convert 5724_{eight} directly to binary form.

Solution

$$\begin{array}{cccc} 5 & 7 & 2 & 4_{\text{eight}} \\ \| & \| & \| & \| \\ 101 & 111 & 010 & 100_{\text{two}} \end{array}$$

$5724_{\text{eight}} = 101111010100_{\text{two}}$

▼ check your progress **6**　Convert 63210_{eight} directly to binary form.

Solution　*See page S21.*

Because every group of three binary bits is equivalent to an octal symbol, we can convert from binary directly to octal by breaking a binary numeral into groups of three (from right to left) and replacing each group with its octal equivalent.

▼ **example 7** **Convert Directly from Base Two to Base Eight**

Convert 11100101_{two} directly to octal form.

Solution
Starting from the right, break the binary numeral into groups of three. Then replace each group with its octal equivalent.

This zero was inserted to make a group of three.

$$
\begin{array}{ccc}
011 & 100 & 101_{two} \\
\parallel & \parallel & \parallel \\
3 & 4 & 5_{eight}
\end{array}
$$

$11100101_{two} = 345_{eight}$

▼ **check your progress 7** Convert 111010011100_{two} directly to octal form.

Solution *See page S21.*

TABLE 6.8 Hexadecimal and Binary Equivalents

Hexadecimal	Binary
0	0000
1	0001
2	0010
3	0011
4	0100
5	0101
6	0110
7	0111
8	1000
9	1001
A	1010
B	1011
C	1100
D	1101
E	1110
F	1111

Table 6.8 shows the hexadecimal symbols and their binary equivalents. To convert from hexadecimal to binary, replace each hexadecimal symbol with its four-bit binary equivalent.

▼ **example 8** **Convert Directly from Base Sixteen to Base Two**

Convert $BAD_{sixteen}$ directly to binary form.

Solution

$$
\begin{array}{ccc}
B & A & D_{sixteen} \\
\parallel & \parallel & \parallel \\
1011 & 1010 & 1101_{two}
\end{array}
$$

$BAD_{sixteen} = 101110101101_{two}$

▼ **check your progress 8** Convert $C5A_{sixteen}$ directly to binary form.

Solution *See page S21.*

Because every group of four binary bits is equivalent to a hexadecimal symbol, we can convert from binary to hexadecimal by breaking the binary numeral into groups of four (from right to left) and replacing each group with its hexadecimal equivalent.

▼ **example 9** **Convert Directly from Base Two to Base Sixteen**

Convert 10110010100011_{two} directly to hexadecimal form.

Solution
Starting from the right, break the binary numeral into groups of four. Replace each group with its hexadecimal equivalent.

Insert two zeros to make a group of four.

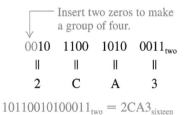

$$
\begin{array}{cccc}
0010 & 1100 & 1010 & 0011_{two} \\
\parallel & \parallel & \parallel & \parallel \\
2 & C & A & 3
\end{array}
$$

$10110010100011_{two} = 2CA3_{sixteen}$

▼ **check your progress 9**   Convert 101000111010010_{two} directly to hexadecimal form.

Solution   *See page S21.*

The Double-Dabble Method

There is a shortcut that can be used to convert a base two numeral to base ten. The advantage of this shortcut, called the *double-dabble method,* is that you can start at the left of the numeral and work your way to the right without first determining the place value of each bit in the base two numeral.

INSTRUCTOR NOTE

Many students enjoy using the double-dabble method. It is best to demonstrate this method first with small base two numerals, such as 10100_{two}.

▼ **example 10**   Apply the Double-Dabble Method

Use the double-dabble method to convert 1011001_{two} to base ten.

Solution
Start at the left with the first 1 and move to the right. Every time you pass by a 0, double your current number. Every time you pass by a 1, dabble. Dabbling is accomplished by doubling your current number and adding 1.

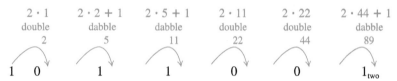

$2 \cdot 1$	$2 \cdot 2 + 1$	$2 \cdot 5 + 1$	$2 \cdot 11$	$2 \cdot 22$	$2 \cdot 44 + 1$
double	dabble	dabble	double	double	dabble
2	5	11	22	44	89
1 0	1	1	0	0	1_{two}

As we pass by the final 1 in the units place, we dabble 44 to get 89. Thus $1011001_{two} = 89$.

▼ **check your progress 10**   Use the double-dabble method to convert 1110010_{two} to base ten.

Solution   *See page S21.*

EXCURSION

Information Retrieval via a Binary Search

To complete this Excursion, you must first construct a set of 31 cards that we refer to as a deck of *binary cards*. Templates for constructing the cards are available at our website, www.cengage.com/math/aufmann, under the file name Binary Cards. Use a computer to print the templates onto a medium-weight card stock similar to that used for playing cards.

We are living in the information age, but information is not useful if it cannot be retrieved when you need it. The binary numeration system is vital to the retrieval of information. To illustrate the connection between retrieval of information and the binary system, examine the card in the following figure. The card is labeled with the base ten numeral 20, and the holes and notches at the top of the card represent 20 in binary notation. A hole is used to indicate a 1, and a notch is used to indicate a 0. In the figure, the card has holes in the third and fifth binary-place-value positions (counting from right to left) and notches cut out of the first, second, and fourth positions.

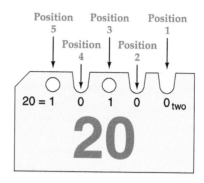

After you have constructed your deck of binary cards, take a few seconds to shuffle the deck. To find the card labeled with the numeral 20, complete the following process.

1. Use a thin dowel (or the tip of a sharp pencil) to lift out the cards that have a hole in the fifth position. *Keep* these cards and set the other cards off to the side.

2. From the cards that are *kept,* use the dowel to lift out the cards with a hole in the fourth position. Set these cards off to the side.

3. From the cards that are *kept,* use the dowel to lift out the cards that have a hole in the third position. *Keep* these cards and place the others off to the side.

4. From the cards that are *kept,* use the dowel to lift out the cards with a hole in the second position. Set these cards off to the side.

5. From the cards that are *kept,* use the dowel to lift out the card that has a hole in the first position. Set this card off to the side.

The card that remains is the card labeled with the numeral 20. You have just completed a binary search.

EXCURSION EXERCISES

The binary numeration system can also be used to implement a *sorting* procedure. To illustrate, shuffle your deck of cards. Use the dowel to lift out the cards that have a hole in the rightmost position. Place these cards, *face up,* behind the other cards. Now use the dowel to lift out the cards that have a hole in the next position to the left. Place these cards, face up, behind the other cards. Continue this process of lifting out the cards in the next position to the left and placing them behind the other cards until you have completed the process for all five positions.

1. Examine the numerals on the cards. What do you notice about the order of the numerals? Explain why they are in this order.

2. If you wanted to sort 1000 cards from smallest to largest value by using the binary sort procedure, how many positions (where each position is either a hole or a notch) would be required at the top of each card? How many positions are needed to sort 10,000 cards?

3. Explain why the above sorting procedure cannot be implemented with base three cards.

EXERCISE SET **6.3** (Suggested Assignment: The Enhanced WebAssign Exercises and Exercises 14, 22, 32, 57, and 58)

■ In Exercises 1 to 10, convert the given numeral to base ten.

1. 243_{five}
2. 145_{seven}
3. 67_{nine}
4. 573_{eight}
5. 3154_{six}
6. 735_{eight}
7. 13211_{four}
8. 102022_{three}
9. $B5_{sixteen}$
10. $4A_{twelve}$

■ In Exercises 11 to 20, convert the given base ten numeral to the indicated base.

11. 267 to base five
12. 362 to base eight
13. 1932 to base six
14. 2024 to base four
15. 15,306 to base nine
16. 18,640 to base seven
17. 4060 to base two
18. 5673 to base three
19. 283 to base twelve
20. 394 to base sixteen

■ In Exercises 21 to 28, use expanded forms to convert the given base two numeral to base ten.

21. 1101_{two}
22. 10101_{two}
23. 11011_{two}
24. 101101_{two}
25. 1100100_{two}
26. 11110101000_{two}
27. 10001011_{two}
28. 110110101_{two}

■ In Exercises 29 to 34, use the double-dabble method to convert the given base two numeral to base ten.

29. 101001_{two}
30. 1110100_{two}
31. 1011010_{two}
32. 10001010_{two}
33. 10100111010_{two}
34. 10000000100_{two}

■ In Exercises 35 to 46, convert the given numeral to the indicated base.

35. 34_{six} to base eight
36. 71_{eight} to base five
37. 878_{nine} to base four
38. 546_{seven} to base six
39. 1110_{two} to base five
40. 21200_{three} to base six

41. 3440_{eight} to base nine
42. 1453_{six} to base eight
43. $56_{sixteen}$ to base eight
44. 43_{twelve} to base six
45. $A4_{twelve}$ to base sixteen
46. $C9_{sixteen}$ to base twelve

■ In Exercises 47 to 56, convert the given numeral *directly* (without first converting to base ten) to the indicated base.

47. 352_{eight} to base two
48. $A4_{sixteen}$ to base two
49. 11001010_{two} to base eight
50. 111011100101_{two} to base sixteen
51. 101010001_{two} to base sixteen
52. 56721_{eight} to base two
53. $BEF3_{sixteen}$ to base two
54. $6A7B8_{sixteen}$ to base two
55. $BA5CF_{sixteen}$ to base two
56. 47134_{eight} to base two

57. **The Triple-Whipple-Zipple Method** There is a procedure that can be used to convert a base three numeral directly to base ten without using the expanded form of the numeral. Write an explanation of this procedure, which we will call the *triple-whipple-zipple* method. *Hint:* The method is an extension of the double-dabble method.

58. Determine whether the following statements are true or false.

a. A number written in base two is divisible by 2 if and only if the number ends with a 0.

b. In base six, the next counting number after 55_{six} is 100_{six}.

c. In base sixteen, the next counting number after $3BF_{sixteen}$ is $3C0_{sixteen}$.

EXTENSIONS

Critical Thinking

The D'ni Numeration System In the computer game *Riven*, a D'ni numeration system is used. Although the D'ni numeration system is a base twenty-five numeration system with 25 distinct numerals, you really need to memorize only the first 5 numerals, which are shown below.

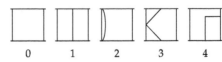

0   1   2   3   4

The basic D'ni numerals

If two D'ni numerals are placed side by side, then the numeral on the left is in the twenty-fives place and the numeral on the right is in the ones place. Thus

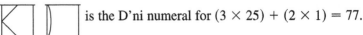 is the D'ni numeral for $(3 \times 25) + (2 \times 1) = 77$.

59. Convert the following D'ni numeral to base ten.

60. Convert the following D'ni numeral to base ten.

Rotating any of the D'ni numerals for 1, 2, 3, and 4 by a 90° counterclockwise rotation produces a numeral with a value five times its original value. For instance, rotating the numeral for 1 produces [symbol], which is the D'ni numeral for 5, and

rotating the numeral for 2 produces [symbol], which is the D'ni numeral for 10.

61. Write the D'ni numeral for 15.

62. Write the D'ni numeral for 20.

In the D'ni numeration system explained above, many numerals are obtained by rotating a basic numeral and then overlaying it on one of the basic numerals. For instance, if you rotate the D'ni numeral for 1, you get the numeral for 5. If you then overlay the numeral for 5 on the numeral for 1, you get the numeral for $5 + 1 = 6$.

5 overlayed on 1 produces 6.

63. Write the D'ni numeral for 8.

64. Write the D'ni numeral for 22.

65. Convert the following D'ni numeral to base ten.

66. Convert the following D'ni numeral to base ten.

67. a. State one advantage of the hexadecimal numeration system over the decimal numeration system.

b. State one advantage of the decimal numeration system over the hexadecimal numeration system.

68. a. State one advantage of the D'ni numeration system over the decimal numeration system.

b. State one advantage of the decimal numeration system over the D'ni numeration system.

Explorations

69. **The ASCII Code** ASCII, pronounced *ask-key,* is an acronym for the American Standard Code for Information Interchange. In this code, each of the characters that can be typed on a computer keyboard is represented by a number. For instance, the letter A is assigned the number 65, which when written as an 8-bit binary numeral is 01000001. Research the topic of ASCII. Write a report about ASCII and its applications.

70. **The Postnet Code** The U.S. Postal Service uses a *Postnet code* to write zip codes + 4 on envelopes. The Postnet code is a bar code that is based on the binary numeration system. Postnet code is very useful because it can be read by a machine. Write a few paragraphs that explain how to convert a zip code + 4 to its Postnet code. What is the Postnet code for your zip code + 4?

Erin Q. Smith
1836 First Avenue
Escondido, CA
92027-4405

s e c t i o n **6.4** **Arithmetic in Different Bases**

Addition in Different Bases

Most computers and calculators make use of the base two (binary) numeration system to perform arithmetic computations. For instance, if you use a calculator to find the sum of 9 and 5, the calculator first converts the 9 to 1001_{two} and the 5 to 101_{two}. The calculator uses electronic circuitry called *binary adders* to find the sum of 1001_{two} and 101_{two} as 1110_{two}. The calculator then converts 1110_{two} to base ten and displays the sum 14. All of the conversions and the base two addition are done internally in a fraction of a second, which gives the user the impression that the calculator performed the addition in base ten.

The following examples illustrate how to perform arithmetic in different bases. We first consider the operation of addition in the binary numeration system. Table 6.9 is an addition table for base two. It is similar to the base ten addition table that you memorized in elementary school, except that it is much smaller because base two involves only the bits 0 and 1. The numerals shown in red in Table 6.9 illustrate that $1_{two} + 1_{two} = 10_{two}$.

TABLE 6.9 A Binary Addition Table

	Second addend	
+	0	1
0	0	1
1	1	10

First addend / Sums

▼ **example 1** **Add Base Two Numerals**

Find the sum of 11110_{two} and 1011_{two}.

Solution
Arrange the numerals vertically, keeping the bits of the same place value in the same column.

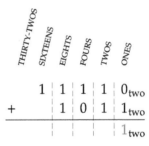

$$
\begin{array}{cccccc}
 & 1 & 1 & 1 & 1 & 0_{two} \\
+ & & 1 & 0 & 1 & 1_{two} \\
\hline
 & & & & & 1_{two}
\end{array}
$$

Start by adding the bits in the ones column: $0_{two} + 1_{two} = 1_{two}$. Then move left and add the bits in the twos column. When the sum of the bits in a column exceeds 1, the addition will involve carrying, as shown below.

TAKE NOTE

In this section, assume that the small numerals, used to indicate a carry, are written in the same base as the numerals in the given problem.

$$
\begin{array}{cccccc}
 & & & & 1 & \\
 & 1 & 1 & 1 & 1 & 0_{two} \\
+ & & 1 & 0 & 1 & 1_{two} \\
\hline
 & & & & 0 & 1_{two}
\end{array}
$$

Add the bits in the twos column.
$1_{two} + 1_{two} = 10_{two}$
Write the 0 in the twos column and carry the 1 to the fours column.

$$
\begin{array}{cccccc}
 & & & 1 & 1 & \\
 & 1 & 1 & 1 & 1 & 0_{two} \\
+ & & 1 & 0 & 1 & 1_{two} \\
\hline
 & & & 0 & 0 & 1_{two}
\end{array}
$$

Add the bits in the fours column.
$(1_{two} + 1_{two}) + 0_{two} = 10_{two} + 0_{two} = 10_{two}$
Write the 0 in the fours column and carry the 1 to the eights column.

$$\begin{array}{r} \overset{1\ \ 1\ \ 1\ \ 1}{1\ \ 1\ \ 1\ \ 1\ \ 0}_{two} \\ +\ \ \ \ \ 1\ \ 0\ \ 1\ \ 1_{two} \\ \hline 1\ \ 0\ \ 1\ \ 0\ \ 0\ \ 1_{two} \end{array}$$

Add the bits in the eights column.
$(1_{two} + 1_{two}) + 1_{two} = 10_{two} + 1_{two} = 11_{two}$
Write a 1 in the eights column and carry a 1 to the
sixteens column. Continue to add the bits in each column
to the left of the eights column.

The sum of 11110_{two} and 1011_{two} is 101001_{two}.

▼ **check your progress 1**   Find the sum of 11001_{two} and 1101_{two}.

Solution   *See page S21.*

TABLE 6.10 A Base
Four Addition Table

+	0	1	2	3
0	0	1	2	3
1	1	2	3	10
2	2	3	10	11
3	3	10	11	12

There are four symbols in base four, namely 0, 1, 2, and 3. Table 6.10 shows a base
four addition table that lists all the sums that can be produced by adding two base four
digits. The numerals shown in red in Table 6.10 illustrate that $2_{four} + 3_{four} = 11_{four}$.

In the next example, we compute the sum of two numbers written in base four.

▼ **example 2**   **Add Base Four Numerals**

Find the sum of 23_{four} and 13_{four}.

Solution

Arrange the numbers vertically, keeping the numerals of the same place value in the
same column.

$$\begin{array}{r} \text{\tiny SIXTEENS}\ \ \text{\tiny FOURS}\ \ \text{\tiny ONES} \\ \overset{1}{2}\ \ 3_{four} \\ +\ \ 1\ \ 3_{four} \\ \hline 2_{four} \end{array}$$

Add the numerals in the ones column.
Table 6.10 shows that $3_{four} + 3_{four} = 12_{four}$.
Write the 2 in the ones column and carry the 1 to the fours column.

$$\begin{array}{r} \overset{1\ \ \ 1}{2}\ \ 3_{four} \\ +\ \ 1\ \ 3_{four} \\ \hline 1\ \ 0\ \ 2_{four} \end{array}$$

Add the numerals in the fours column:
$(1_{four} + 2_{four}) + 1_{four} = 3_{four} + 1_{four} = 10_{four}.$
Write the 0 in the fours column and carry the 1 to the sixteens
column. Bring down the 1 that was carried to the sixteens column to
form the sum 102_{four}.

The sum of 23_{four} and 13_{four} is 102_{four}.

▼ **check your progress 2**   Find $32_{four} + 12_{four}$.

Solution   *See page S21.*

In the previous examples, we used a table to determine the necessary sums. However,
it is generally quicker to find a sum by computing the base ten sum of the numerals in each
column and then converting each base ten sum back to its equivalent in the given base.
The next two examples illustrate this summation technique.

▼ **example 3** Add Base Six Numerals

Find $25_{six} + 32_{six} + 42_{six}$.

Solution

Arrange the numbers vertically, keeping the numerals of the same place value in the same column.

$$
\begin{array}{ccc}
\text{\scriptsize THIRTY-SIXES} & \text{\scriptsize SIXES} & \text{\scriptsize ONES} \\
 & \overset{1}{2} & 5_{six} \\
 & 3 & 2_{six} \\
+ & 4 & 2_{six} \\
\hline
 & & 3_{six}
\end{array}
$$

Add the numerals in the ones column:
$5_{six} + 2_{six} + 2_{six} = 5 + 2 + 2 = 9$.
Convert 9 to base six. ($9 = 13_{six}$)
Write the 3 in the ones column and carry the 1 to the sixes column.

$$
\begin{array}{ccc}
\overset{1}{} & \overset{1}{2} & 5_{six} \\
 & 3 & 2_{six} \\
+ & 4 & 2_{six} \\
\hline
1 & 4 & 3_{six}
\end{array}
$$

Add the numerals in the sixes column and convert the sum to base six.
$1_{six} + 2_{six} + 3_{six} + 4_{six} = 1 + 2 + 3 + 4 = 10 = 14_{six}$
Write the 4 in the sixes column and carry the 1 to the thirty-sixes column. Bring down the 1 that was carried to the thirty-sixes column to form the sum 143_{six}.

$25_{six} + 32_{six} + 42_{six} = 143_{six}$

▼ **check your progress 3** Find $35_{seven} + 46_{seven} + 24_{seven}$.

Solution *See page S21.*

In the next example, we solve an addition problem that involves a base greater than ten.

▼ **example 4** Add Base Twelve Numerals

Find $A97_{twelve} + 8BA_{twelve}$.

Solution

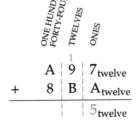

$$
\begin{array}{ccc}
\text{\scriptsize ONE HUNDRED} & \text{\scriptsize TWELVES} & \text{\scriptsize ONES} \\
\text{\scriptsize FORTY-FOURS} & & \\
A & \overset{1}{9} & 7_{twelve} \\
+\quad 8 & B & A_{twelve} \\
\hline
 & & 5_{twelve}
\end{array}
$$

Add the numerals in the ones column.
$7_{twelve} + A_{twelve} = 7 + 10 = 17$
Convert 17 to base twelve. ($17 = 15_{twelve}$)
Write the 5 in the ones column and carry the 1 to the twelves column.

$$\begin{array}{r} \overset{1\;\;\;\;1}{A \;|\; 9 \;|\; 7_{twelve}} \\ +\quad 8 \;|\; B \;|\; A_{twelve} \\ \hline 9 \;|\; 5_{twelve} \end{array}$$

Add the numerals in the twelves column.
$1_{twelve} + 9_{twelve} + B_{twelve} = 1 + 9 + 11 = 21 = 19_{twelve}$
Write the 9 in the twelves column and carry the 1 to the one hundred forty-fours column.

$$\begin{array}{r} \overset{1\;\;\;1\;\;\;1}{A \;|\; 9 \;|\; 7_{twelve}} \\ +\quad 8 \;|\; B \;|\; A_{twelve} \\ \hline 1 \;\; 7 \;|\; 9 \;|\; 5_{twelve} \end{array}$$

Add the numerals in the one hundred forty-fours column.
$1_{twelve} + A_{twelve} + 8_{twelve} = 1 + 10 + 8 = 19 = 17_{twelve}$
Write the 7 in the one hundred forty-fours column and carry the 1 to the one thousand seven hundred twenty-eights column. Bring down the 1 that was carried to the one thousand seven hundred twenty-eights column to form the sum 1795_{twelve}.

$A97_{twelve} + 8BA_{twelve} = 1795_{twelve}$

▼ **check your progress** **4** Find $AC4_{sixteen} + 6E8_{sixteen}$.

Solution   *See page S21.*

Subtraction in Different Bases

To subtract two numbers written in the same base, begin by arranging the numbers vertically, keeping numerals that have the same place value in the same column. It will be necessary to borrow whenever a numeral in the subtrahend is greater than its corresponding numeral in the minuend. Every number that is borrowed will be a power of the base.

▼ **example** **5**   **Subtract Base Seven Numerals**

Find $463_{seven} - 124_{seven}$.

Solution
Arrange the numbers vertically, keeping the numerals of the same place value in the same column.

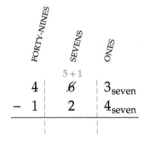

Because $4_{seven} > 3_{seven}$, it is necessary to borrow from the 6 in the sevens column.
(6 sevens =
  5 sevens + 1 seven)

Borrow 1 seven from the sevens column and add $7 = 10_{seven}$ to the 3_{seven} in the ones column.

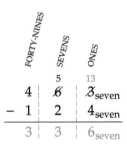

Subtract the numerals in each column. The 6_{seven} in the ones column was produced by the following arithmetic:
$13_{seven} - 4_{seven} = 10 - 4$
$= 6$
$= 6_{seven}$

$463_{seven} - 124_{seven} = 336_{seven}$

▼ **check your progress** **5** Find $365_{nine} - 183_{nine}$.

Solution   *See page S22.*

▼ **example 6**   Subtract Base Sixteen Numerals

Find $7AB_{sixteen} - 3E4_{sixteen}$.

Solution

Table 6.11 shows the hexadecimal numerals and their decimal equivalents. Because $B_{sixteen}$ is greater than $4_{sixteen}$, there is no need to borrow to find the difference in the ones column. However, $A_{sixteen}$ is less than $E_{sixteen}$, so it is necessary to borrow to find the difference in the sixteens column.

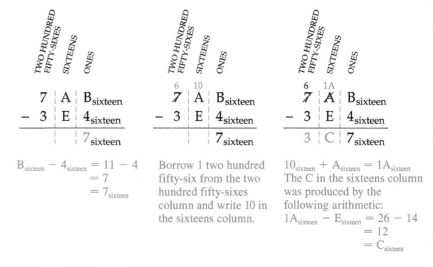

$$7AB_{sixteen} - 3E4_{sixteen} = 3C7_{sixteen}$$

▼ **check your progress 6**   Find $83A_{twelve} - 467_{twelve}$.

Solution   *See page S22.*   ◀

Multiplication in Different Bases

To perform multiplication in bases other than base ten, it is often helpful to first write a multiplication table for the given base. Table 6.12 shows a multiplication table for base four. The numbers shown in red in the table illustrate that $2_{four} \times 3_{four} = 12_{four}$. You can verify this result by converting the numbers to base ten, multiplying in base ten, and then converting back to base four. Here is the actual arithmetic.

$$2_{four} \times 3_{four} = 2 \times 3 = 6 = 12_{four}$$

question   What is $5_{six} \times 4_{six}$?

▼ **example 7**   Multiply Base Four Numerals

Use the base four multiplication table to find $3_{four} \times 123_{four}$.

Solution

Align the numerals of the same place value in the same column. Use Table 6.12 to multiply 3_{four} times each numeral in 123_{four}. If any of these multiplications produces a two-numeral product, then write down the numeral on the right and carry the numeral on the left.

answer $5_{six} \times 4_{six} = 20 = 32_{six}$.

TABLE 6.11 Decimal and Hexadecimal Equivalents

Base ten decimal	Base sixteen hexadecimal
0	0
1	1
2	2
3	3
4	4
5	5
6	6
7	7
8	8
9	9
10	A
11	B
12	C
13	D
14	E
15	F

TABLE 6.12 A Base Four Multiplication Table

×	0	1	2	3
0	0	0	0	0
1	0	1	2	3
2	0	2	10	12
3	0	3	12	21

First diagram (columns: SIXTY-FOURS, SIXTEENS, FOURS, ONES):

$$
\begin{array}{r}
\overset{2}{} \\
1 \;\; 2 \;\; 3_{\text{four}} \\
\times \quad\quad 3_{\text{four}} \\
\hline
1_{\text{four}}
\end{array}
$$

$3_{\text{four}} \times 3_{\text{four}} = 21_{\text{four}}$
Write the 1 in the ones column and carry the 2.

Second diagram:

$$
\begin{array}{r}
\overset{2}{} \;\; \overset{2}{} \\
1 \;\; 2 \;\; 3_{\text{four}} \\
\times \quad\quad 3_{\text{four}} \\
\hline
0 \;\; 1_{\text{four}}
\end{array}
$$

$3_{\text{four}} \times 2_{\text{four}} = 12_{\text{four}}$
$12_{\text{four}} + 2_{\text{four}}^{\text{(the carry)}} = 20_{\text{four}}$
Write the 0 in the fours column and carry the 2.

Third diagram:

$$
\begin{array}{r}
\overset{1}{} \;\; \overset{2}{} \;\; \overset{2}{} \\
1 \;\; 2 \;\; 3_{\text{four}} \\
\times \quad\quad 3_{\text{four}} \\
\hline
1 \;\; 1 \;\; 0 \;\; 1_{\text{four}}
\end{array}
$$

$3_{\text{four}} \times 1_{\text{four}} = 3_{\text{four}}$
$3_{\text{four}} + 2_{\text{four}}^{\text{(the carry)}} = 11_{\text{four}}$
Write a 1 in the sixteens column and carry a 1 to the sixty-fours column. Bring down the 1 that was carried to the sixty-fours column to form the product 1101_{four}.

$3_{\text{four}} \times 123_{\text{four}} = 1101_{\text{four}}$

▼ **check your progress 7** Find $2_{\text{four}} \times 213_{\text{four}}$.

Solution *See page S22.*

Writing all of the entries in a multiplication table for a large base such as base twelve can be time-consuming. In such cases you may prefer to multiply in base ten and then convert each product back to the given base. The next example illustrates this multiplication method.

▼ **example 8** **Multiply Base Twelve Numerals**

Find $53_{\text{twelve}} \times 27_{\text{twelve}}$.

Solution
Align the numerals of the same place value in the same column. Start by multiplying each numeral of the multiplicand (53_{twelve}) by the ones numeral of the multiplier (27_{twelve}).

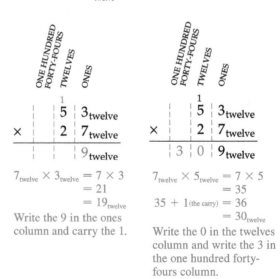

$7_{\text{twelve}} \times 3_{\text{twelve}} = 7 \times 3$
$= 21$
$= 19_{\text{twelve}}$
Write the 9 in the ones column and carry the 1.

$7_{\text{twelve}} \times 5_{\text{twelve}} = 7 \times 5$
$= 35$
$35 + 1_{\text{(the carry)}} = 36$
$= 30_{\text{twelve}}$
Write the 0 in the twelves column and write the 3 in the one hundred forty-fours column.

Now multiply each numeral of the multiplicand by the twelves numeral of the multiplier.

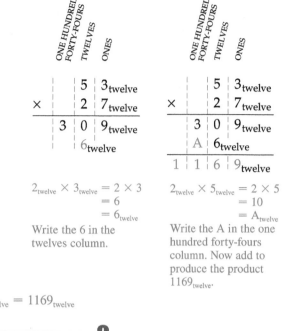

$$2_{\text{twelve}} \times 3_{\text{twelve}} = 2 \times 3$$
$$= 6$$
$$= 6_{\text{twelve}}$$
Write the 6 in the
twelves column.

$$2_{\text{twelve}} \times 5_{\text{twelve}} = 2 \times 5$$
$$= 10$$
$$= A_{\text{twelve}}$$
Write the A in the one
hundred forty-fours
column. Now add to
produce the product
1169_{twelve}.

$$53_{\text{twelve}} \times 27_{\text{twelve}} = 1169_{\text{twelve}}$$

▼ check your progress 8   Find $25_{\text{eight}} \times 34_{\text{eight}}$.

Solution   *See page S22.*

MATHMATTERS   The Fields Medal

The International
Mathematical Union

The Fields Medal

A Nobel Prize is awarded each year in the categories of chemistry, physics, physiology, medicine, literature, and peace. However, no award is given in mathematics. Why Alfred Nobel chose not to provide an award in the category of mathematics is unclear. There has been some speculation that Nobel had a personal conflict with the mathematician Gosta Mittag-Leffler.

The Canadian mathematician John Charles Fields (1863–1932) felt that a prestigious award should be given in the area of mathematics. Fields helped establish the Fields Medal, which was first given to Lars Valerian Ahlfors and Jesse Douglas in 1936. The International Congress of Mathematicians had planned to give two Fields Medals every four years after 1936, but because of World War II, the next Fields Medals were not given until 1950.

It was Fields' wish that the Fields Medal recognize both existing work and the promise of future achievement. Because of this concern for future achievement, the International Congress of Mathematicians decided to restrict those eligible for the Fields Medal to mathematicians under the age of 40.

Division in Different Bases

To perform a division in a base other than base ten, it is helpful to first make a list of a few multiples of the divisor. This procedure is illustrated in the following example.

▼ **example 9**   **Divide Base Seven Numerals**

Find $253_{seven} \div 3_{seven}$.

Solution

First list a few multiples of the divisor 3_{seven}.

$3_{seven} \times 0_{seven} = 3 \times 0 = 0 = 0_{seven}$ $\qquad 3_{seven} \times 4_{seven} = 3 \times 4 = 12 = 15_{seven}$

$3_{seven} \times 1_{seven} = 3 \times 1 = 3 = 3_{seven}$ $\qquad 3_{seven} \times 5_{seven} = 3 \times 5 = 15 = 21_{seven}$

$3_{seven} \times 2_{seven} = 3 \times 2 = 6 = 6_{seven}$ $\qquad 3_{seven} \times 6_{seven} = 3 \times 6 = 18 = 24_{seven}$

$3_{seven} \times 3_{seven} = 3 \times 3 = 9 = 12_{seven}$

Because $3_{seven} \times 6_{seven} = 24_{seven}$ is slightly less than 25_{seven}, we pick 6 as our first numeral in the quotient when dividing 25_{seven} by 3_{seven}.

$$
\begin{array}{r}
6 \\
3_{seven} \overline{) 2\ 5\ 3_{seven}} \\
2\ 4 \\
\hline
1
\end{array}
\qquad
\begin{array}{l}
3_{seven} \times 6_{seven} = 24_{seven} \\
\text{Subtract } 24_{seven} \text{ from } 25_{seven}.
\end{array}
$$

quotient

$$
\begin{array}{r}
6\ 3_{seven}\\
3_{seven} \overline{) 2\ 5\ 3_{seven}} \\
2\ 4 \\
\hline
1\ 3\\
1\ 2\\
\hline
1
\end{array}
\qquad
\begin{array}{l}
\\
\\
\text{Bring down the 3.}\\
3_{seven} \times 3_{seven} = 12_{seven}\\
\text{Subtract } 12_{seven} \text{ from } 13_{seven}.
\end{array}
$$

remainder

Thus $253_{seven} \div 3_{seven} = 63_{seven}$ with a remainder of 1_{seven}.

▼ **check your progress 9**   Find $324_{five} \div 3_{five}$.

Solution   *See page S22.*

In a base two division problem, the only multiples of the divisor that are used are zero times the divisor and one times the divisor.

▼ **example 10**   **Divide Base Two Numerals**

Find $101011_{two} \div 11_{two}$.

Solution

The divisor is 11_{two}. The multiples of the divisor that may be needed are $11_{two} \times 0_{two} = 0_{two}$ and $11_{two} \times 1_{two} = 11_{two}$. Also note that because $10_{two} - 1_{two} = 2 - 1 = 1 = 1_{two}$, we know that

$$
\begin{array}{r}
10_{two}\\
-\ 1_{two}\\
\hline
1_{two}
\end{array}
$$

INSTRUCTOR NOTE

Many students find division with non–base ten numerals to be difficult. If you are short on time, you may find that it is best to skip division with non–base ten numerals.

Instructors who decide to cover this topic may find it helpful to first review division using a base ten example.

$$
\begin{array}{r}
1\ 1\ 1\ 0_{\text{two}} \\
11_{\text{two}}\overline{)1\ 0\ 1\ 0\ 1\ 1_{\text{two}}} \\
\underline{1\ 1} \\
1\ 0\ 0 \\
\underline{1\ 1} \\
1\ 1 \\
\underline{1\ 1} \\
0\ 1 \\
\underline{0} \\
1
\end{array}
$$

Therefore, $101011_{\text{two}} \div 11_{\text{two}} = 1110_{\text{two}}$ with a remainder of 1_{two}.

▼**check your progress 10** Find $1110011_{\text{two}} \div 10_{\text{two}}$.

Solution *See page S22.*

EXCURSION

Subtraction in Base Two via the Ones Complement and the End-Around Carry

INSTRUCTOR NOTE

The concepts in this Excursion are an extension of the concepts in the Excursion in Section 6.2, page 314.

Computers and calculators are often designed so that the number of required circuits is minimized. Instead of using separate circuits to perform addition and subtraction, engineers make use of an *end-around carry procedure* that uses addition to perform subtraction. The end-around carry procedure also makes use of the ones complement of a number. In base two, the ones complement of 0 is 1 and the ones complement of 1 is 0. Thus the ones complement of any base two number can be found by changing each 1 to a 0 and each 0 to a 1.

▼ **Subtraction Using the Ones Complement and the End-Around Carry**

To subtract a base two number from a larger base two number:

1. Add the ones complement of the subtrahend to the minuend.

2. Take away 1 from the leftmost bit of the sum and add 1 to the units bit. This is referred to as the end-around carry procedure.

The following example illustrates the process of subtracting 1001_{two} from 1101_{two} using the ones complement and the end-around carry procedure.

$$
\begin{array}{r}
1101_{\text{two}} \\
-\ \ 1001_{\text{two}} \\
\end{array}
\qquad
\begin{array}{l}
\text{Minuend} \\
\text{Subtrahend}
\end{array}
$$

$$
\begin{array}{r}
1101_{\text{two}} \\
+\ \ 0110_{\text{two}} \\
\hline
10011_{\text{two}}
\end{array}
\qquad
\begin{array}{l}
\text{Replace the subtrahend with the ones} \\
\text{complement of the subtrahend and add.}
\end{array}
$$

$$10011_{two}$$
$$+ \quad 1_{two}$$
$$\overline{\quad 100_{two}}$$

Take away 1 from the leftmost bit and add 1 to the ones bit. This is the end-around carry procedure.

$$1101_{two} - 1001_{two} = 100_{two}$$

If the subtrahend has fewer bits than the minuend, leading zeros should be inserted in the subtrahend so that it has the same number of bits as the minuend. This process is illustrated below for the subtraction $1010110_{two} - 11001_{two}$.

$$1010110_{two}$$
$$- \quad 11001_{two}$$

Minuend

Subtrahend

$$1010110_{two}$$
$$- \quad 0011001_{two}$$

Insert two leading zeros.

$$1010110_{two}$$
$$+ \quad 1100110_{two}$$
$$\overline{10111100_{two}}$$

Ones complement of subtrahend

$$10111100_{two}$$
$$+ \quad 1_{two}$$
$$\overline{\quad 111101_{two}}$$

Take away 1 from the leftmost bit and add 1 to the ones bit.

$$1010110_{two} - 11001_{two} = 111101_{two}$$

EXCURSION EXERCISES

Use the ones complement of the subtrahend and the end-around carry method to find each difference. State each answer as a base two numeral.

1. $1110_{two} - 1001_{two}$

2. $101011_{two} - 100010_{two}$

3. $101001010_{two} - 1011101_{two}$

4. $111011100110_{two} - 101010100_{two}$

5. $1111101011_{two} - 1001111_{two}$

6. $1110010101100_{two} - 100011110_{two}$

EXERCISE SET 6.4

(Suggested Assignment: The Enhanced WebAssign Exercises and Exercises 3, 6, 21, 24, 30, 33, 44, 49, 50, and 55)

■ In Exercises 1 to 12, find each sum in the same base as the given numerals.

1. $204_{five} + 123_{five}$

2. $323_{four} + 212_{four}$

3. $5625_{seven} + 634_{seven}$

4. $1011_{two} + 101_{two}$

5. $110101_{two} + 10011_{two}$

6. $11001010_{two} + 1100111_{two}$

7. $8B5_{twelve} + 578_{twelve}$

8. $379_{sixteen} + 856_{sixteen}$

9. $C489_{sixteen} + BAD_{sixteen}$

10. $221_{three} + 122_{three}$

11. $435_{six} + 245_{six}$

12. $5374_{eight} + 615_{eight}$

■ In Exercises 13 to 24, find each difference in the same base as the given numerals.

13. $434_{five} - 143_{five}$

14. $534_{six} - 241_{six}$

15. $7325_{eight} - 563_{eight}$

16. $6148_{nine} - 782_{nine}$

17. $11010_{two} - 1011_{two}$

18. $111001_{two} - 10101_{two}$

19. $11010100_{two} - 1011011_{two}$

20. $9C5_{sixteen} - 687_{sixteen}$

21. $43A7_{twelve} - 289_{twelve}$

22. $BAB2_{twelve} - 475_{twelve}$

23. $762_{nine} - 367_{nine}$

24. $3223_{four} - 133_{four}$

■ In Exercises 25 to 38, find each product in the same base as the given numerals.

25. $3_{\text{six}} \times 145_{\text{six}}$

26. $5_{\text{seven}} \times 542_{\text{seven}}$

27. $2_{\text{three}} \times 212_{\text{three}}$

28. $4_{\text{five}} \times 4132_{\text{five}}$

29. $5_{\text{eight}} \times 7354_{\text{eight}}$

30. $11_{\text{two}} \times 11011_{\text{two}}$

31. $10_{\text{two}} \times 101010_{\text{two}}$

32. $101_{\text{two}} \times 110100_{\text{two}}$

33. $25_{\text{eight}} \times 453_{\text{eight}}$

34. $43_{\text{six}} \times 1254_{\text{six}}$

35. $132_{\text{four}} \times 1323_{\text{four}}$

36. $43_{\text{twelve}} \times 895_{\text{twelve}}$

37. $5_{\text{sixteen}} \times BAD_{\text{sixteen}}$

38. $23_{\text{sixteen}} \times 798_{\text{sixteen}}$

■ In Exercises 39 to 49, find each quotient and remainder in the same base as the given numerals.

39. $132_{\text{four}} \div 2_{\text{four}}$

40. $124_{\text{five}} \div 2_{\text{five}}$

41. $231_{\text{four}} \div 3_{\text{four}}$

42. $672_{\text{eight}} \div 5_{\text{eight}}$

43. $5341_{\text{six}} \div 4_{\text{six}}$

44. $11011_{\text{two}} \div 10_{\text{two}}$

45. $101010_{\text{two}} \div 11_{\text{two}}$

46. $1011011_{\text{two}} \div 100_{\text{two}}$

47. $457_{\text{twelve}} \div 5_{\text{twelve}}$

48. $832_{\text{sixteen}} \div 7_{\text{sixteen}}$

49. $234_{\text{five}} \div 12_{\text{five}}$

50. If $232_x = 92$, find the base x.

51. If $143_x = 10200_{\text{three}}$, find the base x.

52. If $46_x = 101010_{\text{two}}$, find the base x.

53. Consider the addition $384 + 245$.

   a. Use base ten addition to find the sum.

   b. Convert 384 and 245 to base two.

   c. Find the base two sum of the base two numbers you found in part b.

   d. Convert the base two sum from part c to base ten.

   e. How does the answer to part a compare with the answer to part d?

54. Consider the subtraction $457 - 318$.

   a. Use base ten subtraction to find the difference.

   b. Convert 457 and 318 to base two.

   c. Find the base two difference of the base two numbers you found in part b.

   d. Convert the base two difference from part c to base ten.

   e. How does the answer to part a compare with the answer to part d?

55. Consider the multiplication 247×26.

   a. Use base ten multiplication to find the product.

   b. Convert 247 and 26 to base two.

   c. Find the base two product of the base two numbers you found in part b.

   d. Convert the base two product from part c to base ten.

   e. How does the answer to part a compare with the answer to part d?

EXTENSIONS

Critical Thinking

56. Explain the error in the following base eight subtraction.

$$\begin{array}{r} 751_{\text{eight}} \\ -\ 126_{\text{eight}} \\ \hline 625_{\text{eight}} \end{array}$$

57. Determine the base used in the following multiplication.

$$314_{\text{base } x} \times 24_{\text{base } x} = 11202_{\text{base } x}$$

58. The base ten number 12 is an even number. In base seven, 12 is written as 15_{seven}. Is 12 an odd number in base seven?

59. Explain why there is no numeration system with a base of 1.

60. A Cryptarithm In the following base four addition problem, each letter represents one of the numerals 0, 1, 2, or 3. No two different letters represent the same numeral. Determine which numeral is represented by each letter.

$$\begin{array}{r} \text{N O} \\ +\ \text{A T} \\ \hline \text{N O T} \end{array}_{\text{four}}$$

61. A Cryptarithm In the following base six addition problem, each letter represents one of the numerals 0, 1, 2, 3, 4, or 5. No two different letters represent the same numeral. Determine which numeral is represented by each letter.

$$\begin{array}{r} \text{M A} \\ +\ \text{A S} \\ \hline \text{M O M} \end{array}_{\text{six}}$$

Explorations

62. Negative Base Numerals It is possible to use a negative number as the base of a numeration system. For instance, the negative base four numeral $32_{\text{negative four}}$ represents the number
$$3 \times (-4)^1 + 2 \times (-4)^0 = -12 + 2 = -10.$$

   a. Convert each of the following negative base numerals to base 10:

     $143_{\text{negative five}}$

     $74_{\text{negative nine}}$

     $10110_{\text{negative two}}$

   b. Write -27 as a negative base five numeral.

   c. Write 64 as a negative base three numeral.

   d. Write 112 as a negative base ten numeral.

section 6.5   Prime Numbers

Prime Numbers

Number theory is a mathematical discipline that is primarily concerned with the properties that are exhibited by the natural numbers. The mathematician Carl Friedrich Gauss established many theorems in number theory. As we noted in the chapter opener, Gauss called mathematics the queen of the sciences and number theory the queen of mathematics. Many topics in number theory involve the concept of a *divisor* or *factor* of a natural number.

▼ Divisor of a Natural Number

The natural number a is a **divisor** or **factor** of the natural number b, provided there exists a natural number j such that $aj = b$.

In less formal terms, a natural number a is a divisor of the natural number b provided $b \div a$ has a remainder of 0. For instance, 10 has divisors of 1, 2, 5, and 10 because each of these numbers divides into 10 with a remainder of 0.

▼ example 1   Find Divisors

Determine all of the natural number divisors of each number.

a. 6   **b.** 42   **c.** 17

Solution

a. Divide 6 by 1, 2, 3, 4, 5, and 6. The division of 6 by 1, 2, 3, and 6 each produces a natural number quotient and a remainder of 0. Thus 1, 2, 3, and 6 are divisors of 6. Dividing 6 by 4 and 6 by 5 does not produce a remainder of 0. Therefore 4 and 5 are not divisors of 6.

b. The only natural numbers from 1 to 42 that divide into 42 with a remainder of 0 are 1, 2, 3, 6, 7, 14, 21, and 42. Thus the divisors of 42 are 1, 2, 3, 6, 7, 14, 21, and 42.

c. The only natural number divisors of 17 are 1 and 17.

▼ check your progress 1   Determine all of the natural number divisors of each number.

a. 9   **b.** 11   **c.** 24

Solution   *See page S22.*

It is worth noting that every natural number greater than 1 has itself as a factor and 1 as a factor. If a natural number greater than 1 has only 1 and itself as factors, then it is a very special number known as a *prime number*.

▼ Prime Numbers and Composite Numbers

A **prime number** is a natural number greater than 1 that has exactly two factors (divisors): itself and 1.
A **composite number** is a natural number greater than 1 that is not a prime number.

TAKE NOTE

The natural number 1 is neither a prime number nor a composite number.

The 10 smallest prime numbers are 2, 3, 5, 7, 11, 13, 17, 19, 23, and 29. Each of these numbers has only itself and 1 as factors. The 10 smallest composite numbers are 4, 6, 8, 9, 10, 12, 14, 15, 16, and 18.

▼ example 2 **Classify a Number as a Prime Number or a Composite Number**

Determine whether each number is a prime number or a composite number.

a. 41   **b.** 51   **c.** 119

Solution

a. The only divisors of 41 and 1 are 41. Thus 41 is a prime number.

b. The divisors of 51 are 1, 3, 17, and 51. Thus 51 is a composite number.

c. The divisors of 119 are 1, 7, 17, and 119. Thus 119 is a composite number.

▼ check your progress 2 Determine whether each number is a prime number or a composite number.

a. 47   **b.** 171   **c.** 91

Solution *See page S22.*   ◄

question Are all prime numbers odd numbers?

Divisibility Tests

To determine whether one number is divisible by a smaller number, we often apply a **divisibility test**, which is a procedure that enables one to determine whether the smaller number is a divisor of the larger number without actually dividing the smaller number into the larger number. Table 6.13 provides divisibility tests for the numbers 2, 3, 4, 5, 6, 8, 9, 10, and 11.

TABLE 6.13 Base Ten Divisibility Tests

A number is divisible by the following divisor if:	Divisibility test	Example
2	The number is an even number.	846 is divisible by 2 because 846 is an even number.
3	The sum of the digits of the number is divisible by 3.	531 is divisible by 3 because $5 + 3 + 1 = 9$ is divisible by 3.
4	The last two digits of the number form a number that is divisible by 4.	1924 is divisible by 4 because the last two digits form the number 24, which is divisible by 4.
5	The number ends with a 0 or a 5.	8785 is divisible by 5 because it ends with 5.
6	The number is divisible by 2 and by 3.	972 is divisible by 6 because it is divisible by 2 and also by 3.
8	The last three digits of the number form a number that is divisible by 8.	19,168 is divisible by 8 because the last three digits form the number 168, which is divisible by 8.
9	The sum of the digits of the number is divisible by 9.	621,513 is divisible by 9 because the sum of the digits is 18, which is divisible by 9.
10	The last digit is 0.	970 is divisible by 10 because it ends with 0.
11	Start at one end of the number and compute the sum of every other digit. Next compute the sum of the remaining digits. If the difference of these sums is divisible by 11, then the original number is divisible by 11.	4807 is divisible by 11 because the difference of the sum of the digits shown in blue ($8 + 7 = 15$) and the sum of the remaining digits shown in red ($4 + 0 = 4$) is $15 - 4 = 11$, which is divisible by 11.

PPT

answer No. The even number 2 is a prime number.

▼ example **3** Apply Divisibility Tests

Use divisibility tests to determine whether 16,278 is divisible by the following numbers.

a. 2   **b.** 3   **c.** 5   **d.** 8   **e.** 11

Solution

a. Because 16,278 is an even number, it is divisible by 2.

b. The sum of the digits of 16,278 is 24, which is divisible by 3. Therefore, 16,278 is divisible by 3.

c. The number 16,278 does not end with a 0 or a 5. Therefore, 16,278 is not divisible by 5.

d. The last three digits of 16,278 form the number 278, which is not divisible by 8. Thus 16,278 is not divisible by 8.

e. The sum of the digits with even place-value powers is $1 + 2 + 8 = 11$. The sum of the digits with odd place-value powers is $6 + 7 = 13$. The difference of these sums is $13 - 11 = 2$. This difference is not divisible by 11, so 16,278 is not divisible by 11.

▼ check your progress **3**   Use divisibility tests to determine whether 341,565 is divisible by each of the following numbers.

a. 3   **b.** 4   **c.** 10   **d.** 11

Solution   *See page S22.*   ◀

Prime Factorization

The **prime factorization** of a composite number is a factorization that contains only prime numbers. Many proofs in number theory make use of the following important theorem.

> ▼ **The Fundamental Theorem of Arithmetic**
>
> Every composite number can be written as a unique product of prime numbers (disregarding the order of the factors).

To find the prime factorization of a composite number, rewrite the number as a product of two smaller natural numbers. If these smaller numbers are both prime numbers, then you are finished. If either of the smaller numbers is not a prime number, then rewrite it as a product of smaller natural numbers. Continue this procedure until all factors are primes. In Example 4 we make use of a *tree diagram* to organize the factorization process.

▼ example **4**   Find the Prime Factorization of a Number

Determine the prime factorization of the following numbers.

a. 84   **b.** 495   **c.** 4004

Solution

a. The following tree diagrams show two different ways of finding the prime factorization of 84, which is $2 \cdot 2 \cdot 3 \cdot 7 = 2^2 \cdot 3 \cdot 7$. Each number in the tree is equal to the product of the two smaller numbers below it. The numbers (in red) at the extreme ends of the branches are the prime factors.

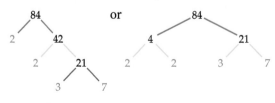

$$84 = 2^2 \cdot 3 \cdot 7$$

TAKE NOTE

The TI-83/84 program listed on page 135 can be used to find the prime factorization of a given natural number less than 10 billion.

TAKE NOTE

The following compact division procedure can also be used to determine the prime factorization of a number.

```
2 | 4004
2 | 2002
7 | 1001
11 | 143
      13
```

In this procedure, we use only prime number divisors, and we continue until the last quotient is a prime number. The prime factorization is the product of all the prime numbers, which are shown in red.

b.

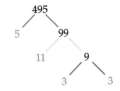

$$495 = 3^2 \cdot 5 \cdot 11$$

c.

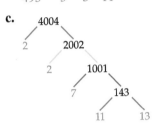

$$4004 = 2^2 \cdot 7 \cdot 11 \cdot 13$$

▼ **check your progress 4** Determine the prime factorization of the following numbers.

a. 315   **b.** 273   **c.** 1309

Solution   *See page S23.*

MATHMATTERS   Srinivasa Ramanujan

Srinivasa Ramanujan
(1887–1920)

On January 16, 1913, the 26-year-old Srinivasa Ramanujan (Rä-mä′noo-jûn) sent a letter from Madras, India, to the illustrious English mathematician G.H. Hardy. The letter requested that Hardy give his opinion about several mathematical ideas that Ramanujan had developed. In the letter Ramanujan explained, "I have not trodden through the conventional regular course which is followed in a University course, but I am striking out a new path for myself." Much of the mathematics was written using unconventional terms and notation; however, Hardy recognized (after many detailed readings and with the help of other mathematicians at Cambridge University) that Ramanujan was "a mathematician of the highest quality, a man of altogether exceptional originality and power."

On March 17, 1914, Ramanujan set sail for England, where he joined Hardy in a most unusual collaboration that lasted until Ramanujan returned to India in 1919. The following famous story is often told to illustrate the remarkable mathematical genius of Ramanujan.

INSTRUCTOR NOTE

The number 1729 is also a Carmichael number, a number that is not prime but satisfies Fermat's little theorem (see Exercise 35, Section 6.6). Thus 1729 provides a counterexample to the converse of Fermat's little theorem.

After Hardy had taken a taxicab to visit Ramanujan, he made the remark that the license plate number for the taxi was "1729, a rather dull number." Ramanujan immediately responded by saying that 1729 was a most interesting number, because it is the smallest natural number that can be expressed in two different ways as the sum of two cubes.

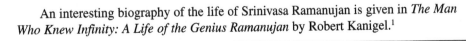

$$1^3 + 12^3 = 1729 \quad \text{and} \quad 9^3 + 10^3 = 1729$$

An interesting biography of the life of Srinivasa Ramanujan is given in *The Man Who Knew Infinity: A Life of the Genius Ramanujan* by Robert Kanigel.[1]

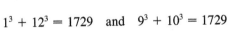

[1] Kanigel, Robert. *The Man Who Knew Infinity: A Life of the Genius Ramanujan*. New York: Simon & Schuster, 1991.

To determine whether a natural number is a prime number, it is necessary to consider only divisors from 2 up to the square root of the number because every composite number n has at least one divisor less than or equal to $\sqrt{n}$. The proof of this statement is outlined in Exercise 77 of this section.

It is possible to determine whether a natural number n is a prime number by checking each natural number from 2 up to the largest integer not greater than $\sqrt{n}$ to see whether each is a divisor of n. If none of these numbers is a divisor of n, then n is a prime number. For large values of n, this division method is generally time consuming and tedious. The Greek astronomer and mathematician Eratosthenes (ca. 276–192 BC) recognized that multiplication is generally easier than division, and he devised a method that makes use of multiples to determine every prime number in a list of natural numbers. Today we call this method the *sieve of Eratosthenes.*

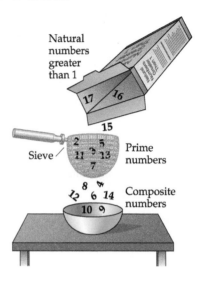

In article 329 of *Disquisitiones Arithmeticae,* Gauss wrote:

"The problem of distinguishing prime numbers from composite numbers and of resolving the latter into their prime factors is known to be one of the most important and useful in arithmetic.... The dignity of the science itself seems to require that every possible means be explored for the solution of a problem so elegant and so celebrated." (*Source: The Little Book of Big Primes* by Paulo Ribenboim. New York: Springer-Verlag, 1991)

To sift prime numbers, first make a list of consecutive natural numbers. In Table 6.14, we have listed the consecutive counting numbers from 2 to 100.

- Cross out every multiple of 2 larger than 2. The next smallest remaining number in the list is 3. Cross out every multiple of 3 larger than 3.
- Call the next smallest remaining number in the list k. Cross out every multiple of k larger than k. Repeat this step for all $k < \sqrt{100}$.

TABLE 6.14 The Sieve Method of Finding Primes

	2	3	4	5	6	7	8	9	10
11	12	13	14	15	16	17	18	19	20
21	22	23	24	25	26	27	28	29	30
31	32	33	34	35	36	37	38	39	40
41	42	43	44	45	46	47	48	49	50
51	52	53	54	55	56	57	58	59	60
61	62	63	64	65	66	67	68	69	70
71	72	73	74	75	76	77	78	79	80
81	82	83	84	85	86	87	88	89	90
91	92	93	94	95	96	97	98	99	100

The numbers in blue that are not crossed out are prime numbers. Table 6.14 shows that there are 25 prime numbers smaller than 100.

Over 2000 years ago, Euclid proved that the set of prime numbers is an infinite set. Euclid's proof is an *indirect proof* or a *proof by contradiction.* Essentially, his proof shows that for any finite list of prime numbers, we can create a number T, as described below, such that any prime factor of T can be shown to be a prime number that is not in the list. Thus there must be an infinite number of primes because it is not possible for all of the primes to be in any finite list.

Euclid's Proof   Assume that *all* of the prime numbers are contained in the list $p_1, p_2, p_3, \ldots, p_r$. Let $T = (p_1 \cdot p_2 \cdot p_3 \cdots p_r) + 1$. Either T is a prime number or T has a prime divisor. If T is a prime, then it is a prime that is not in our list and we have reached a contradiction. If T is not prime, then one of the primes $p_1, p_2, p_3, \ldots, p_r$ must be a divisor of T. However, the number T is not divisible by any of the primes $p_1, p_2, p_3, \ldots, p_r$ because each p_i divides $p_1 \cdot p_2 \cdot p_3 \cdots p_r$ but does not divide 1. Hence any prime divisor of T, say p, is a prime number that is not in the list $p_1, p_2, p_3, \ldots, p_r$. So p is yet another prime number, and $p_1, p_2, p_3, \ldots, p_r$ is not a complete list of all the prime numbers.

We conclude this section with two quotations about prime numbers. The first is by the illustrious mathematician Paul Erdös (1913–1996), and the second by the mathematics professor Don B. Zagier of the Max-Planck Institute, Bonn, Germany.

> It will be millions of years before we'll have any understanding, and even then it won't be a complete understanding, because we're up against the infinite.— P. Erdös, about prime numbers in *Atlantic Monthly*, November 1987, p. 74. *Source:* http://www.mlahanas.de/Greeks/Primes.htm.

In a 1975 lecture, D. Zagier commented,

> There are two facts about the distribution of prime numbers of which I hope to convince you so overwhelmingly that they will be permanently engraved in your hearts. The first is that, despite their simple definition and role as the building blocks of the natural numbers, the prime numbers belong to the most arbitrary and ornery objects studied by mathematicians: they grow like weeds among the natural numbers, seeming to obey no other law than that of chance, and nobody can predict where the next one will sprout. The second fact is even more astonishing, for it states just the opposite: that the prime numbers exhibit stunning regularity, that there are laws governing their behavior, and that they obey these laws with almost military precision. *Source:* "The First 50 Million Prime Numbers" by Don Zagler.

Don B. Zagier

EXCURSION

The Distribution of the Primes

Many mathematicians have searched without success for a mathematical formula that can be used to generate the sequence of prime numbers. We know that the prime numbers form an infinite sequence. However, the distribution of the prime numbers within the sequence of natural numbers is very complicated. The ratio of prime numbers to composite numbers appears to become smaller and smaller as larger and larger numbers are considered. In general, the number of consecutive composite numbers that come between two prime numbers tends to increase as the size of the numbers becomes larger; however, this increase is erratic and appears to be unpredictable.

In this Excursion, we refer to a list of two or more consecutive composite numbers as a **prime desert**. For instance, 8, 9, 10 is a prime desert because it consists of three consecutive composite numbers. The longest prime desert shown in Table 6.14 is the seven consecutive composite numbers 90, 91, 92, 93, 94, 95, and 96. A formula that involves *factorials* can be used to form prime deserts of any finite length.

▼ *n* factorial

If *n* is a natural number, then *n*!, which is read "*n* factorial," is defined as

$$n! = n \cdot (n - 1) \cdot \cdots \cdot 3 \cdot 2 \cdot 1$$

As an example of a factorial, consider $4! = 4 \cdot 3 \cdot 2 \cdot 1 = 24$.

The sequence

$$4! + 2, 4! + 3, 4! + 4$$

is a prime desert of the three composite numbers 26, 27, and 28. Figure 6.1 below shows a prime desert of 10 consecutive composite numbers. Figure 6.2 shows a procedure that can be used to produce a prime desert of *n* composite numbers, where *n* is any natural number greater than 2.

$$
\left.
\begin{aligned}
11! &+ 2 \\
11! &+ 3 \\
11! &+ 4 \\
11! &+ 5 \\
11! &+ 6 \\
&\ \ \vdots \\
11! &+ 10 \\
11! &+ 11
\end{aligned}
\right\}
$$
A prime desert of 10 consecutive composite numbers

$$
\left.
\begin{aligned}
(n + 1)! &+ 2 \\
(n + 1)! &+ 3 \\
(n + 1)! &+ 4 \\
(n + 1)! &+ 5 \\
(n + 1)! &+ 6 \\
&\ \ \vdots \\
(n + 1)! &+ n \\
(n + 1)! &+ (n + 1)
\end{aligned}
\right\}
$$
A prime desert of *n* consecutive composite numbers

FIGURE 6.1   **FIGURE 6.2**

A prime desert of length 1 million is shown by the sequence

$$1{,}000{,}001! + 2;\ 1{,}000{,}001! + 3;\ 1{,}000{,}001! + 4;\ \ldots;\ 1{,}000{,}001! + 1{,}000{,}001$$

It appears that the distribution of prime numbers is similar to the situation wherein a mathematical gardener plants an infinite number of grass seeds on a windy day. Many of the grass seeds fall close to the gardener, but are blown down the street and into the next neighborhood. There are gaps where no grass seeds are within 1 mile of each other. Farther down the road there are gaps where no grass seeds are within 10 miles of each other. No matter how far the gardener travels and how long it has been since the last grass seed was spotted, the gardener knows that more grass seeds will appear.

EXCURSION EXERCISES

1. Explain how you know that each of the numbers

   $$1{,}000{,}001! + 2;\ 1{,}000{,}001! + 3;\ 1{,}000{,}001! + 4;\ \ldots;\ 1{,}000{,}001! + 1{,}000{,}001$$

   is a composite number.

2. Use factorials to generate the numbers in a prime desert of 12 consecutive composite numbers. Now use a calculator to evaluate each number in this prime desert.

3. Use factorials and "..." notation to represent a prime desert of

   a. 20 consecutive composite numbers.

   b. 500,000 consecutive composite numbers.

   c. 7 billion consecutive composite numbers.

EXERCISE SET 6.5 (Suggested Assignment: The Enhanced WebAssign Exercises and Exercises 10, 20, 27, 35, 41, 42, 43, and 73)

■ In Exercises 1 to 10, determine all natural number divisors of the given number.

1. 20   **2.** 32

3. 65   **4.** 75

5. 41   **6.** 79

7. 110   **8.** 150

9. 385   **10.** 455

■ In Exercises 11 to 20, determine whether each number is a prime number or a composite number.

11. 21   **12.** 31

13. 37   **14.** 39

15. 101   **16.** 81

17. 79   **18.** 161

19. 203   **20.** 211

■ In Exercises 21 to 28, use the divisibility tests in Table 6.13 to determine whether the given number is divisible by each of the following: 2, 3, 4, 5, 6, 8, 9, and 10.

21. 210   **22.** 314

23. 51   **24.** 168

25. 2568   **26.** 3525

27. 4190   **28.** 6123

■ In Exercises 29 to 40, write the prime factorization of the number.

29. 18   **30.** 48

31. 120   **32.** 380

33. 425   **34.** 625

35. 1024   **36.** 1410

37. 6312   **38.** 3155

39. 18,234   **40.** 19,345

41. Use the sieve of Eratosthenes procedure to find all prime numbers from 2 to 200. *Hint:* Because $\sqrt{200} \approx 14.1$, you need to continue the sieve procedure up to $k = 13$. Note: You do not need to consider $k = 14$ because 14 is not a prime number.

42. Use your list of prime numbers from Exercise 41 to find the number of prime numbers from:

    a. 2 to 50   **b.** 51 to 100

    c. 101 to 150   **d.** 151 to 200

43. Twin Primes If the natural numbers n and $n + 2$ are both prime numbers, then they are said to be **twin primes**. For example, 11 and 13 are twin primes. It is not known whether the set of twin primes is an infinite set or a finite set. Use the list of primes from Exercise 41 to write all twin primes less than 200.

44. Twin Primes Find a pair of twin primes between 200 and 300. See Exercise 43.

45. Twin Primes Find a pair of twin primes between 300 and 400. See Exercise 43.

46. ✎ **A Prime Triplet** If the natural numbers n, $n + 2$, and $n + 4$ are all prime numbers, then they are said to be **prime triplets**. Write a few sentences that explain why the prime triplets 3, 5, and 7 are the only prime triplets.

47. Goldbach's Conjecture In 1742, Christian Goldbach conjectured that every even number greater than 2 can be written as the sum of two prime numbers. Many mathematicians have tried to prove or disprove this conjecture without succeeding. Show that *Goldbach's conjecture* is true for each of the following even numbers.

    a. 24   **b.** 50

    c. 86   **d.** 144

    e. 210   **f.** 264

48. ✎ **Perfect Squares** The square of a natural number is called a **perfect square**. Pick six perfect squares. For each perfect square, determine the number of distinct natural-number factors of the perfect square. Make a conjecture about the number of distinct natural-number factors of any perfect square.

Every prime number has a divisibility test. Many of these divisibility tests are slight variations of the following divisibility test for 7.

A divisibility test for 7 To determine whether a given base ten number is divisible by 7, double the ones digit of the given number. Find the difference between this number and the number formed by omitting the ones digit from the given number. If necessary, repeat this procedure until you obtain a small final difference.

If the final difference is divisible by 7, then the given number is also divisible by 7.

If the final difference is not divisible by 7, then the given number is not divisible by 7.

Example Use the above divisibility test to determine whether 301 is divisible by 7.

Solution The double of the ones digit is 2. Subtracting 2 from 30, which is the number formed by omitting the ones digit from the original number, yields 28. Because 28 is divisible by 7, the original number 301 is divisible by 7.

■ In Exercises 49 to 56, use the divisibility test for 7, given on page 345, to determine whether each number is divisible by 7.

49. 182

50. 203

51. 1001

52. 2403

53. 11,561

54. 13,842

55. 204,316

56. 789,327

A divisibility test for 13 To determine whether a given base ten number is divisible by 13, multiply the ones digit of the given number by 4. Find the sum of this multiple of 4 and the number formed by omitting the ones digit from the given number. If necessary, repeat this procedure until you obtain a small final sum.

If the final sum is divisible by 13, then the given number is divisible by 13.

If the final sum is not divisible by 13, then the given number is not divisible by 13.

Example Use the divisibility test for 13 to determine whether 1079 is divisible by 13.

Solution Four times the ones digit is 36. The number formed by omitting the ones digit is 107. The sum of 36 and 107 is 143. Now repeat the procedure on 143. Four times the ones digit is 12. The sum of 12 and 14, which is the number formed by omitting the ones digit, is 26. Because 26 is divisible by 13, the original number 1079 is divisible by 13.

■ In Exercises 57 to 64, use the divisibility test for 13 to determine whether each number is divisible by 13.

57. 91

58. 273

59. 1885

60. 8931

61. 14,507

62. 22,184

63. 13,351

64. 85,657

EXTENSIONS

Critical Thinking

65. Factorial Primes A prime number of the form $n! \pm 1$ is called a **factorial prime**. Recall that the notation $n!$ is called n factorial and represents the product of all natural numbers from 1 to n. For example, $4! = 4 \cdot 3 \cdot 2 \cdot 1 = 24$. Factorial primes are of interest to mathematicians because they often signal the end or the beginning of a lengthy string of consecutive composite numbers. See the Excursion on page 343.

   a. Find the smallest value of n such that $n! + 1$ and $n! - 1$ are twin primes.

   b. Find the smallest value of n for which $n! + 1$ is a composite number and $n! - 1$ is a prime number.

66. Primorial Primes The notation $p\#$ represents the product of all the prime numbers less than or equal to the prime number p. For instance,

$$3\# = 2 \cdot 3 = 6$$

$$5\# = 2 \cdot 3 \cdot 5 = 30$$

$$11\# = 2 \cdot 3 \cdot 5 \cdot 7 \cdot 11 = 2310$$

A **primorial prime** is a prime number of the form $p\# \pm 1$. For instance, $3\# + 1 = 2 \cdot 3 + 1 = 7$ and $3\# - 1 = 2 \cdot 3 - 1 = 5$ are both primorial primes. Large primorial primes are often examined in the search for a pair of large twin primes.

   a. Find the smallest prime number p, where $p \geq 7$, such that $p\# + 1$ and $p\# - 1$ are twin primes.

   b. Find the smallest prime number p, such that $p\# + 1$ is a prime number but $p\# - 1$ is a composite number.

67. A Divisibility Test for 17 Determine a divisibility test for 17. *Hint*: One divisibility test for 17 is similar to the divisibility test for 7 on the previous page in that it involves the last digit of the given number and the operation of subtraction.

68. A Divisibility Test for 19 Determine a divisibility test for 19. *Hint*: One divisibility test for 19 is similar to the divisibility test for 13, shown above between Exercises 56 and 57, in that it involves the last digit of the given number and the operation of addition.

Number of Divisors of a Composite Number The following method can be used to determine the number of divisors of a composite number. First find the prime factorization (in exponential form) of the composite number. Add 1 to each exponent in the prime factorization and then compute the product of these exponents. This product is equal to the number of divisors of the composite number. To illustrate that this procedure yields the correct result, consider the composite number 12, which has the six divisors 1, 2, 3, 4, 6, and 12. The prime factorization of 12 is $2^2 \cdot 3^1$. Adding 1 to each of the exponents produces the numbers 3 and 2. The product of 3 and 2 is 6, which agrees with the result obtained by listing all of the divisors.

■ In Exercises 69 to 74, determine the number of divisors of each composite number.

69. 60

70. 84

71. 297

72. 288

73. 360

74. 875

Explorations

75. Kummer's Proof In the 1870s, the mathematician Eduard Kummer used a proof similar to the following to show that there exist an infinite number of prime numbers. Supply the missing reasons in parts a and b.

a. *Proof* Assume there exist only a finite number of prime numbers, say p_1, p_2, p_3, ... , p_r. Let $N = p_1p_2p_3\cdots p_r > 2$. The natural number $N - 1$ has at least one common prime factor with N. Why?

b. Call the common prime factor from part a p_i. Now p_i divides N and p_i divides $N - 1$. Thus p_i divides their difference: $N - (N - 1) = 1$. Why?

This leads to a contradiction, because no prime number is a divisor of 1. Hence Kummer concluded that the original assumption was incorrect and there must exist an infinite number of prime numbers.

76. *Theorem* If a number of the form $111\ldots1$ is a prime number, then the number of 1s in the number is a prime number. For instance,

$$11 \text{ and } 1{,}111{,}111{,}111{,}111{,}111{,}111$$

are prime numbers, and the number of 1s in each number (2 in the first number and 19 in the second number) is a prime number.

a. What is the converse of the above theorem? *Hint:* The converse of "If p then q," where p and q are statements, is "If q then p." Is the converse of a theorem always true?

b. The number $111 = 3 \cdot 37$, so 111 is not a prime number. Explain why this does not contradict the above theorem.

77. State the missing reasons in parts a, b, and c of the following proof.

Theorem Every composite number n has at least one divisor less than or equal to $\sqrt{n}$.

a. *Proof* Assume that a is a divisor of n. Then there exists a natural number j such that $aj = n$. Why?

b. Now a and j cannot both be greater than $\sqrt{n}$, because this would imply that $aj > \sqrt{n}\sqrt{n}$. However, $\sqrt{n}\sqrt{n}$ simplifies to n, which equals aj. What contradiction does this lead to?

c. Thus either a or j must be less than or equal to $\sqrt{n}$. Because j is also a divisor of n, the proof is complete. How do we know that j is a divisor of n?

78. **The RSA Algorithm** In 1977, Ron Rivest, Adi Shamir, and Leonard Adleman invented a method for encrypting information. Their method is known as the RSA algorithm. Today the RSA algorithm is used by the Internet Explorer Web browser, as well as by VISA and MasterCard to ensure secure electronic credit card transactions. The RSA algorithm involves large prime numbers. Research the RSA algorithm and write a report about some of the reasons why this algorithm has become one of the most popular of all the encryption algorithms.

79. **Theorems and Conjectures** Use the Internet or a text on prime numbers to determine whether each of the following statements is an established theorem or a conjecture. (*Note: n* represents a natural number).

a. There are infinitely many twin primes.

b. There are infinitely many primes of the form $n^2 + 1$.

c. There is always a prime number between n and $2n$ for $n \geq 2$.

d. There is always a prime number between n^2 and $(n + 1)^2$.

e. Every odd number greater than 5 can be written as the sum of three primes.

f. Every positive even number can be written as the difference of two primes.

---

section 6.6 Topics from Number Theory

Perfect, Deficient, and Abundant Numbers

The ancient Greek mathematicians personified the natural numbers. For instance, they considered the odd natural numbers as male and the even natural numbers as female. They also used the concept of a *proper factor* to classify a natural number as *perfect, deficient,* or *abundant.*

> **▼ Proper Factors**
>
> The **proper factors** of a natural number consist of all the natural number factors of the number other than the number itself.

For instance, the proper factors of 10 are 1, 2, and 5. The proper factors of 16 are 1, 2, 4, and 8. The proper factors of a number are also called the **proper divisors** of the number.

> **▼ Perfect, Deficient, and Abundant Numbers**
>
> A natural number is
> - **perfect** if it is equal to the sum of its proper factors.
> - **deficient** if it is greater than the sum of its proper factors.
> - **abundant** if it is less than the sum of its proper factors.

POINT OF INTEREST

Six is a number perfect in itself, and not because God created the world in 6 days; rather the contrary is true. God created the world in six days because this number is perfect, and it would remain perfect, even if the work of the six days did not exist.
—*St. Augustine (354–430)*

Marin Mersenne (1588–1648)

 **▼ example 1** Classify a Number as Perfect, Deficient, or Abundant

Determine whether the following numbers are perfect, deficient, or abundant.

a. 6   **b.** 20   **c.** 25

Solution

a. The proper factors of 6 are 1, 2, and 3. The sum of these proper factors is $1 + 2 + 3 = 6$. Because 6 is equal to the sum of its proper divisors, 6 is a perfect number.

b. The proper factors of 20 are 1, 2, 4, 5, and 10. The sum of these proper factors is $1 + 2 + 4 + 5 + 10 = 22$. Because 20 is less than the sum of its proper factors, 20 is an abundant number.

c. The proper factors of 25 are 1 and 5. The sum of these proper factors is $1 + 5 = 6$. Because 25 is greater than the sum of its proper factors, 25 is a deficient number.

▼ check your progress 1 Determine whether the following numbers are perfect, deficient, or abundant.

a. 24   **b.** 28   **c.** 35

Solution *See page S23.*

Mersenne Numbers and Perfect Numbers

As a French monk in the religious order known as the Minims, Marin Mersenne (mər-sĕn′) devoted himself to prayer and his studies, which included topics from number theory. Mersenne took it upon himself to collect and disseminate mathematical information to scientists and mathematicians through Europe. Mersenne was particularly interested in prime numbers of the form $2^n - 1$, where n is a prime number. Today numbers of the form $2^n - 1$, where n is a prime number, are known as **Mersenne numbers**.

Some Mersenne numbers are prime and some are composite. For instance, the Mersenne numbers $2^2 - 1$ and $2^3 - 1$ are prime numbers, but the Mersenne number $2^{11} - 1 = 2047$ is not prime because $2047 = 23 \cdot 89$.

▼ **example** **2**   Determine Whether a Mersenne Number Is a Prime Number

Determine whether the Mersenne number $2^5 - 1$ is a prime number.

Solution

$2^5 - 1 = 31$, and 31 is a prime number. Thus $2^5 - 1$ is a Mersenne prime.

▼ **check your progress** **2**   Determine whether the Mersenne number $2^7 - 1$ is a prime number.

Solution   *See page S23.*   ◄

The ancient Greeks knew that the first four perfect numbers were 6, 28, 496, and 8128. In fact, proposition 36 from Volume IX of Euclid's *Elements* states a procedure that uses Mersenne primes to produce a perfect number.

INSTRUCTOR NOTE

Inform your students that the number $2^n - 1$ can be a prime number only if n is a prime number. However, the fact that n is a prime number does not guarantee that $2^n - 1$ is a prime number. For instance, if $n = 11$, then $2^n - 1 = 2047 = 23 \cdot 89$, which is not a prime number.

▼ **Euclid's Procedure for Generating a Perfect Number**

If n and $2^n - 1$ are both prime numbers, then $2^{n-1}(2^n - 1)$ is a perfect number.

Euclid's procedure shows how every Mersenne prime can be used to produce a perfect number. For instance:

$$\text{If } n = 2, \text{ then } 2^{2-1}(2^2 - 1) = 2(3) = 6.$$
$$\text{If } n = 3, \text{ then } 2^{3-1}(2^3 - 1) = 4(7) = 28.$$
$$\text{If } n = 5, \text{ then } 2^{5-1}(2^5 - 1) = 16(31) = 496.$$
$$\text{If } n = 7, \text{ then } 2^{7-1}(2^7 - 1) = 64(127) = 8128.$$

INSTRUCTOR NOTE

The fifth and sixth perfect numbers both end with 6. Thus the 6, 8 alternating pattern for the ones digits of perfect numbers does not hold for the first seven perfect numbers.

The fifth perfect number was not discovered until the year 1461. It is $2^{12}(2^{13} - 1) = 33,550,336$. The sixth and seventh perfect numbers were discovered in 1588 by P. A. Cataldi. In exponential form, they are $2^{16}(2^{17} - 1)$ and $2^{18}(2^{19} - 1)$. It is interesting to observe that the ones digits of the first five perfect numbers alternate: 6, 8, 6, 8, 6. Evaluate $2^{16}(2^{17} - 1)$ and $2^{18}(2^{19} - 1)$ to determine if this alternating pattern continues for the first seven perfect numbers.

▼ **example** **3**   Use a Given Mersenne Prime Number to Write a Perfect Number

In 1750, Leonhard Euler proved that $2^{31} - 1$ is a Mersenne prime. Use Euclid's theorem to write the perfect number associated with this prime.

Solution

Euler's theorem states that if n and $2^n - 1$ are both prime numbers, then $2^{n-1}(2^n - 1)$ is a perfect number. In this example, $n = 31$, which is a prime number. We are given that $2^{31} - 1$ is a prime number, so the perfect number we seek is $2^{30}(2^{31} - 1)$.

▼ **check your progress** **3**   In 1883, I. M. Pervushin proved that $2^{61} - 1$ is a Mersenne prime. Use Euclid's theorem to write the perfect number associated with this prime.

Solution   *See page S23.*   ◄

question   Must a perfect number produced by Euclid's perfect number–generating procedure be an even number?

---

answer   Yes. The 2^{n-1} factor of $2^{n-1}(2^n - 1)$ ensures that this product will be an even number.

The search for Mersenne primes still continues. In fact, in 1999 the Electronic Frontier Foundation offered a reward of $100,000 to the first person or group to discover a prime number with at least 10 million digits.

As of August 2011, only 47 Mersenne primes (and 47 perfect numbers) had been discovered. The largest of these 47 Mersenne primes is $2^{43112609} - 1$. This gigantic Mersenne prime number is also the largest known prime number as of August 2011, and it was the first prime number to be discovered with at least 10 million digits. Its discoverers received the $100,000 reward offered by the Electronic Frontier Foundation. More information on Mersenne primes is given in Table 6.15 below and in the Math Matters on page 351.

TABLE 6.15 Selected Known Mersenne Primes as of August 2011 (Listed from Smallest to Largest)

Rank	Mersenne prime	Discovery date	Discoverer
1	$2^2 - 1 = 3$	BC	Ancient Greek mathematicians
2	$2^3 - 1 = 7$	BC	Ancient Greek mathematicians
3	$2^5 - 1 = 31$	BC	Ancient Greek mathematicians
4	$2^7 - 1 = 127$	BC	Ancient Greek mathematicians
5	$2^{13} - 1 = 8191$	1456	Anonymous
6	$2^{17} - 1 = 131071$	1588	Cataldi
7	$2^{19} - 1 = 524287$	1588	Cataldi
8	$2^{31} - 1 = 2147483647$	1772	Euler
9	$2^{61} - 1 = 2305843009213693951$	1883	Pervushin
⋮	⋮	⋮	⋮
38	$2^{6972593} - 1$	June 1999	GIMPS* / Hajratwala
39	$2^{13466917} - 1$	Nov. 2001	GIMPS / Cameron
40	$2^{20996011} - 1$	Nov. 2003	GIMPS / Shafer
41?	$2^{24036583} - 1$	May 2004	GIMPS / Findley
42?	$2^{25964951} - 1$	Feb. 2005	GIMPS / Nowak
43?	$2^{30402457} - 1$	Dec. 2005	GIMPS / Cooper and Boone
44?	$2^{32582657} - 1$	Sept. 2006	GIMPS / Cooper and Boone
45?	$2^{37156667} - 1$	Sept. 2008	GIMPS / Elvenich
46?	$2^{42643801} - 1$	April 2009	GIMPS / Strindmo
47?	$2^{43112609} - 1$	Aug. 2008	GIMPS / Smith

Source of data: The *Mersenne prime* website at: http://en.wikipedia.org/wiki/Mersenne_prime.
*GIMPS is a project involving volunteers who use computer software to search for Mersenne prime numbers. See the Math Matters on page 351.

In the last seven rows of Table 6.15, a question mark has been inserted after the rank number because as of August 2011, it is not known if there are other Mersenne prime numbers larger than $2^{20996011} - 1$ but less than $2^{43112609} - 1$ that have yet to be discovered.

MATHMATTERS The Great Internet Mersenne Prime Search

From 1952 to 1996, large-scale computers were used to find Mersenne primes. However, during recent years, personal computers working in parallel have joined in the search. This search has been organized by an Internet organization known as the Great Internet Mersenne Prime Search (GIMPS).

Members of this group use the Internet to download a Mersenne prime program that runs on their computers. Each member is assigned a range of numbers to check for Mersenne primes. A search over a specified range of numbers can take several weeks, but because the program is designed to run in the background, you can still use your computer to perform its regular duties. As of August 2011, a total of 13 Mersenne primes had been discovered by members of the GIMPs.

One of the current goals of GIMPS is to find a prime number that has at least 100 million digits. In fact, the Electronic Frontier Foundation is offering a $150,000 reward to the first person or group to discover a 100 million–digit prime number. If you are just using your computer to run a screen saver, why not join in the search? You can get the Mersenne prime program and additional information at

http://www.mersenne.org

Who knows, maybe you will discover a new Mersenne prime and share in the reward money. Of course, we know that you really just want to have your name added to Table 6.15.

$150,000 REWARD

TO THE PERSON OR GROUP THAT FIRST DISCOVERS A PRIME NUMBER WITH AT LEAST 100 MILLION DIGITS!

Offered by the Electronic Frontier Foundation

The Number of Digits in b^x

To determine the number of digits in a Mersenne number, mathematicians make use of the following formula, which involves finding the *greatest integer* of a number. **The greatest integer** of a number k is the greatest integer less than or equal to k. For instance, the greatest integer of 5 is 5 and the greatest integer of 7.8 is 7.

> ▼ **The Number of Digits in b^x**
>
> The number of digits in the number b^x, where b is a natural number less than 10 and x is a natural number, is the greatest integer of $(x \log b) + 1$.

CALCULATOR NOTE

To evaluate 19 log 2 on a TI-83/84 calculator, press

19 [LOG] 2 [)] [ENTER]

On a scientific calculator, press

19 [×] 2 [LOG] [=]

▼ **example** **4** Determine the Number of Digits in a Mersenne Number

Find the number of digits in the Mersenne prime number $2^{19} - 1$.

Solution
First consider just 2^{19}. The base b is 2. The exponent x is 19.

$$(x \log b) + 1 = (19 \log 2) + 1$$
$$\approx 5.72 + 1$$
$$= 6.72$$

The greatest integer of 6.72 is 6. Thus 2^{19} has six digits. If 2^{19} were a power of 10, then $2^{19} - 1$ would have one fewer digit than 2^{19}. However, 2^{19} is not a power of 10, so $2^{19} - 1$ has the same number of digits as 2^{19}. Thus the Mersenne prime number $2^{19} - 1$ also has six digits.

▼ **check your progress** **4** Find the number of digits in the Mersenne prime number $2^{2976221} - 1$.

Solution *See page S23.* ◄

Euclid's perfect number–generating formula produces only *even* perfect numbers. Do odd perfect numbers exist? As of August 2011, no odd perfect number had been discovered and no one had been able to prove either that odd perfect numbers exist or that they do not exist. The question of the existence of an odd perfect number is one of the major unanswered questions of number theory.

Fermat's Last Theorem

In 1637, the French mathematician Pierre de Fermat wrote in the margin of a book:

> It is impossible to divide a cube into two cubes, or a fourth power into two fourth powers, or in general any power greater than the second into two like powers, and I have a truly marvelous demonstration of it. But this margin will not contain it.

This problem, which became known as **Fermat's last theorem**, can also be stated in the following manner.

HISTORICAL NOTE

Pierre de Fermat

(1601–1665) In the mathematical community, Pierre de Fermat (fĕ hr'məh) is known as the Prince of Amateurs because although he spent his professional life as a councilor and judge, he spent his leisure time working on mathematics. As an amateur mathematician, Fermat did important work in analytic geometry and calculus, but he is remembered today for his work in the area of number theory. Fermat stated theorems he had developed, but seldom did he provide the actual proofs. He did not want to waste his time showing the details required of a mathematical proof and then spend additional time defending his work once it was scrutinized by other mathematicians.

By the twentieth century, all but one of Fermat's proposed theorems had been proved by other mathematicians. The remaining unproved theorem became known as Fermat's last theorem.

▼ **Fermat's Last Theorem**

There are no natural numbers x, y, z, and n that satisfy $x^n + y^n = z^n$, where n is greater than 2.

Fermat's last theorem has attracted a great deal of attention over the last three centuries. The theorem has become so well known that it is simply called "FLT." Here are some of the reasons for its popularity:

- Very little mathematical knowledge is required to understand the statement of FLT.
- FLT is an extension of the well-known Pythagorean theorem $x^2 + y^2 = z^2$, which has several natural number solutions. Two such solutions are $3^2 + 4^2 = 5^2$ and $5^2 + 12^2 = 13^2$.
- It seems so simple. After all, while reading the text *Arithmetica* by Diophantus, Fermat wrote that he had discovered a *truly marvelous proof* of FLT. The only reason that Fermat gave for not providing his proof was that the margin of *Arithmetica* was too narrow to contain it.

Many famous mathematicians have worked on FLT. Some of these mathematicians tried to disprove FLT by searching for natural numbers x, y, z, and n that satisfied $x^n + y^n = z^n$, $n > 2$.

▼ **example** **5** Check a Possible Solution to Fermat's Last Theorem

Determine whether $x = 6$, $y = 8$, and $n = 3$ satisfies the equation $x^n + y^n = z^n$, where z is a natural number.

Solution

Substituting 6 for x, 8 for y, and 3 for n in $x^n + y^n = z^n$ yields

$$6^3 + 8^3 = z^3$$
$$216 + 512 = z^3$$
$$728 = z^3$$

The real solution of $z^3 = 728$ is $\sqrt[3]{728} \approx 8.99588289$, which is not a natural number. Thus $x = 6$, $y = 8$, and $n = 3$ does not satisfy the equation $x^n + y^n = z^n$, where z is a natural number.

▼ **check your progress** **5**   Determine whether $x = 9$, $y = 11$, and $n = 4$ satisfies the equation $x^n + y^n = z^n$, where z is a natural number.

Solution   *See page S23.*   ◄

In the eighteenth century, the great mathematician Leonhard Euler was able to make some progress on a proof of FLT. He adapted a technique that he found in Fermat's notes about another problem. In this problem, Fermat gave an outline of how to prove that the special case $x^4 + y^4 = z^4$ has no natural number solutions. Using a similar procedure, Euler was able to show that $x^3 + y^3 = z^3$ also has no natural number solutions. Thus all that was left was to show that $x^n + y^n = z^n$ has no solutions with n greater than 4.

In the nineteenth century, additional work on FLT was done by Sophie Germain, Augustin Louis Cauchy, and Gabriel Lame. Each of these mathematicians produced some interesting results, but FLT still remained unsolved.

In 1983, the German mathematician Gerd Faltings used concepts from differential geometry to prove that the number of solutions to FLT must be finite. Then, in 1988, the Japanese mathematician Yoichi Miyaoka claimed he could show that the number of solutions to FLT was not only finite, but that the number of solutions was zero, and thus he had proved FLT. At first Miyaoka's work appeared to be a valid proof, but after a few weeks of examination, a flaw was discovered. Several mathematicians looked for a way to repair the flaw, but eventually they came to the conclusion that although Miyaoka had developed some interesting mathematics, he had failed to establish the validity of FLT.

In 1993, Andrew Wiles of Princeton University made a major advance toward a proof of FLT. Wiles first became familiar with FLT when he was only 10 years old. At his local public library, Wiles first learned about FLT in the book *The Last Problem* by Eric Temple Bell. In reflecting on his first thoughts about FLT, Wiles recalled

> It looked so simple, and yet all the great mathematicians in history couldn't solve it. Here was a problem that I, a ten-year-old, could understand and I knew from that moment that I would never let it go. I had to solve it.[2]

Wiles took a most unusual approach to solving FLT. Whereas most contemporary mathematicians share their ideas and coordinate their efforts, Wiles decided to work alone. After seven years of working in the attic of his home, Wiles was ready to present his work. In June of 1993, Wiles gave a series of three lectures at the Isaac Newton Institute in Cambridge, England. After showing that FLT was a corollary of his major theorem, Wiles's concluding remark was "I think I'll stop here." Many mathematicians in the audience felt that Wiles had produced a valid proof of FLT, but a formal verification by several mathematical referees was required before Wiles's work could be classified as an official proof. The verification process was lengthy and complex. Wiles's written work was about 200 pages in length, it covered several different areas of mathematics, and it used hundreds of sophisticated logical arguments and a great many mathematical calculations. Any mistake could result in an invalid proof. Thus it was not too surprising when a flaw was discovered in late 1993. The flaw did not necessarily imply that Wiles's proof could not be repaired, but it did indicate that it was not a valid proof in its present form.

It appeared that once again the proof of FLT had eluded a great effort by a well-known mathematician. Several months passed and Wiles was still unable to fix the flaw. It was a most depressing period for Wiles. He felt that he was close to solving one of the world's hardest mathematical problems, yet all of his creative efforts failed to turn the

POINT OF INTEREST

Andrew Wiles

Fermat's last theorem had been labeled by some mathematicians as the world's hardest mathematical problem. After solving Fermat's last theorem, Andrew Wiles made the following remarks: "Having solved this problem there's certainly a sense of loss, but at the same time there is this tremendous sense of freedom. I was so obsessed by this problem that for eight years I was thinking about it all the time—when I woke up in the morning to when I went to sleep at night. That's a long time to think about one thing. That particular odyssey is now over. My mind is at rest."

---

[2] Singh, Simon. *Fermat's Enigma: The Quest to Solve the World's Greatest Mathematical Problem.* New York: Walker Publishing Company, Inc., 1997, p. 6.

INSTRUCTOR NOTE

NOVA has produced a wonderful video called *The Proof* that chronicles Wiles's efforts to prove Fermat's last theorem. This video can be ordered by calling 1-800-949-8670.

flawed proof into a valid proof. Several mathematicians felt that it was not possible to repair the flaw, but Wiles did not give up. Finally, in late 1994, Wiles had an insight that eventually led to a valid proof. The insight required Wiles to seek additional help from Richard Taylor, who had been a student of Wiles. On October 15, 1994, Wiles and Taylor presented to the world a proof that has now been judged to be a valid proof of FLT. Their proof is certainly not the "truly marvelous proof" that Fermat said he had discovered. But it has been deemed a wonderful proof that makes use of several new mathematical procedures and concepts.

EXCURSION

A Sum of the Divisors Formula

Consider the numbers 10, 12, and 28 and the sums of the proper factors of these numbers.

$$10: 1 + 2 + 5 = 8 \qquad 12: 1 + 2 + 3 + 4 + 6 = 16$$

$$28: 1 + 2 + 4 + 7 + 14 = 28$$

From the above sums we see that 10 is a deficient number, 12 is an abundant number, and 28 is a perfect number. The goal of this Excursion is to find a method that will enable us to determine whether a number is deficient, abundant, or perfect without having to first find all of its proper factors and then compute their sum.

In the following example, we use the number 108 and its prime factorization $2^2 \cdot 3^3$ to illustrate that every factor of 108 can be written as a product of the powers of its prime factors. Table 6.16 includes all the proper factors of 108 (the numbers in blue) plus the factor $2^2 \cdot 3^3$, which is 108 itself. The sum of each column is shown at the bottom (the numbers in red).

TABLE 6.16 Every Factor of 108 Expressed as a Product of Powers of Its Prime Factors

	1	$1 \cdot 3$	$1 \cdot 3^2$	$1 \cdot 3^3$
	2	$2 \cdot 3$	$2 \cdot 3^2$	$2 \cdot 3^3$
	2^2	$2^2 \cdot 3$	$2^2 \cdot 3^2$	$2^2 \cdot 3^3$
Sum	7	$7 \cdot 3$	$7 \cdot 3^2$	$7 \cdot 3^3$

The sum of *all* the factors of 108 is the sum of the numbers in the bottom row.

$$\text{Sum of all factors of } 108 = 7 + 7 \cdot 3 + 7 \cdot 3^2 + 7 \cdot 3^3$$

$$= 7(1 + 3 + 3^2 + 3^3) = 7(40) = 280$$

To find the sum of just the *proper* factors, we must subtract 108 from 280, which gives us 172. Thus 108 is an abundant number.

We now look for a pattern for the sum of all the factors. Note that

Sum of left column   Sum of top row

$$1 + 2 + 2^2 \qquad\qquad 1 + 3 + 3^2 + 3^3$$

$$7 \cdot 40$$

This result suggests that the sum of the factors of a number can be found by finding the sum of all the prime power factors of each prime factor and then computing the product of those sums. Because we are interested only in the sum of the proper factors, we subtract the original number. Although we have not proved this result, it is a true statement and can save much time and effort. For instance, the sum of the proper factors of 3240 can be found as follows:

$$3240 = 2^3 \cdot 3^4 \cdot 5$$

Compute the sum of *all* the prime power factors of each prime factor.

$$1 + 2 + 2^2 + 2^3 = 15 \qquad 1 + 3 + 3^2 + 3^3 + 3^4 = 121 \qquad 1 + 5 = 6$$

The sum of the proper factors of 3240 is $(15)(121)(6) - 3240 = 7650$. Thus 3240 is an abundant number.

EXCURSION EXERCISES

Use the above technique to find the sum of the proper factors of each number and then state whether the number is deficient, abundant, or perfect.

1. 200   **2.** 262   **3.** 325   **4.** 496

5. Use deductive reasoning to prove that every prime number is deficient.

6. Use inductive reasoning to decide whether every multiple of 6 greater than 6 is abundant.

EXERCISE SET 6.6

(Suggested Assignment: The Enhanced WebAssign Exercises and Exercises 31, 33, 37, 39, and 41)

■ In Exercises 1 to 16, determine whether each number is perfect, deficient, or abundant.

1. 18   **2.** 32

3. 91   **4.** 51

5. 19   **6.** 144

7. 204   **8.** 128

9. 610   **10.** 508

11. 291   **12.** 1001

13. 176   **14.** 122

15. 260   **16.** 258

■ In Exercises 17 to 20, determine whether each Mersenne number is a prime number.

17. $2^3 - 1$ **18.** $2^5 - 1$

19. $2^7 - 1$ **20.** $2^{13} - 1$

21. In 1876, E. Lucas proved, without the aid of a computer, that $2^{127} - 1$ is a Mersenne prime. Use Euclid's theorem to write the perfect number associated with this prime.

22. In 1952, R. M. Robinson proved, with the aid of a computer, that $2^{521} - 1$ is a Mersenne prime. Use Euclid's theorem to write the perfect number associated with this prime.

■ In Exercises 23 to 28, determine the number of digits in the given Mersenne prime.

23. $2^{17} - 1$ **24.** $2^{132049} - 1$

25. $2^{1398269} - 1$ **26.** $2^{3021377} - 1$

27. $2^{6972593} - 1$ **28.** $2^{20996011} - 1$

> **IN THE NEWS**
>
> ### *Time Magazine* Picks the Discovery of Two New Mersenne Primes as the 29th Best Invention of 2008
>
> Searching for larger and larger Mersenne primes is the unofficial national sport of mathematicians. This year two large Mersenne primes were discovered. The largest of these prime numbers has almost 13 million digits. It was discovered by a team directed by Edson Smith at UCLA (University of California, Los Angeles).
>
> SOURCE: Best Inventions of 2008, *Time Magazine*, October 29, 2008

The two Mersenne prime numbers that were discovered in 2008 follow.

$$2^{37156667} - 1 \quad \text{and} \quad 2^{43112609} - 1$$

29. Determine the number of digits in $2^{37156667} - 1$.

30. Determine the number of digits in $2^{43112609} - 1$.

31. Verify that $x = 9$, $y = 15$, and $n = 5$ do not yield a solution to the equation $x^n + y^n = z^n$ where z is a natural number.

32. Verify that $x = 7$, $y = 19$, and $n = 6$ do not yield a solution to the equation $x^n + y^n = z^n$ where z is a natural number.

33. Determine whether each of the following statements is a true statement, a false statement, or a conjecture.

a. If n is a prime number, then $2^n - 1$ is also a prime number.

b. Fermat's last theorem is called his last theorem because we believe that it was the last theorem he proved.

c. All perfect numbers of the form $2^{n-1}(2^n - 1)$ are even numbers.

d. Every perfect number is an even number.

34. Prove that $4078^n + 3433^n = 12{,}046^n$ cannot be a solution to the equation $x^n + y^n = z^n$ where n is a natural number. *Hint:* Examine the ones digits of the powers.

35. *Fermat's Little Theorem*   A theorem known as *Fermat's little theorem* states, "If n is a prime number and a is any natural number, then $a^n - a$ is divisible by n." Verify Fermat's little theorem for

a. $n = 7$ and $a = 12$.

b. $n = 11$ and $a = 8$.

36. **Amicable Numbers**   The Greeks considered the pair of numbers 220 and 284 to be *amicable* or *friendly* numbers because the sum of the proper divisors of one of the numbers is the other number.

The sum of the proper factors of 220 is

$$1 + 2 + 4 + 5 + 10 + 11 + 20 + 22 +$$
$$44 + 55 + 110 = 284$$

The sum of the proper factors of 284 is

$$1 + 2 + 4 + 71 + 142 = 220$$

Determine whether

a. 60 and 84 are amicable numbers.

b. 1184 and 1210 are amicable numbers.

37. **A Sum of Cubes Property**   The perfect number 28 can be written as $1^3 + 3^3$. The perfect number 496 can be written as $1^3 + 3^3 + 5^3 + 7^3$. Verify that the next perfect number, 8128, can also be written as the sum of the cubes of consecutive odd natural numbers, starting with 1^3.

38. **A Sum of the Digits Theorem**   If you sum the digits of any even perfect number (except 6), then sum the digits of the resulting number, and repeat this process until you get a single digit, that digit will be 1.

As an example, consider the perfect number 28. The sum of its digits is 10. The sum of the digits of 10 is 1.

Verify the previous theorem for each of the following perfect numbers.

a. 496   **b.** 8128

c. 33,550,336   **d.** 8,589,869,056

39. **A Sum of Reciprocals Theorem**   The sum of the reciprocals of all the positive divisors of a perfect number is always 2.

Verify the above theorem for each of the following perfect numbers.

a. 6   **b.** 28

E X T E N S I O N S

Critical Thinking

40. **The Smallest Odd Abundant Number**   Determine the smallest odd abundant number. *Hint:* It is greater than 900 but less than 1000.

41. **Fermat Numbers**   Numbers of the form $2^{2^m} + 1$, where m is a whole number, are called *Fermat numbers*. Fermat believed that all Fermat numbers were prime. Prove that Fermat was wrong.

Explorations

42. **Semiperfect Numbers**   Any number that is the sum of *some or all* of its proper divisors is called a **semiperfect number**. For instance, 12 is a semiperfect number because it has 1, 2, 3, 4, and 6 as proper factors, and $12 = 1 + 2 + 3 + 6$.

The first twenty-five semiperfect numbers are 6, 12, 18, 20, 24, 28, 30, 36, 40, 42, 48, 54, 56, 60, 66, 72,

78, 80, 84, 88, 90, 96, 100, 102, and 104. It has been established that every natural number multiple of a semiperfect number is semiperfect and that a semiperfect number cannot be a deficient number.

a. Use the definition of a semiperfect number to verify that 20 is a semiperfect number.

b. Explain how to verify that 200 is a semiperfect number without examining its proper factors.

43. **Weird Numbers**   Any number that is an abundant number but not a semiperfect number (see Exercise 42) is called a **weird number**. Find the only weird number less than 100. *Hint:* The abundant numbers less than 100 are 12, 18, 20, 24, 30, 36, 40, 42, 48, 54, 56, 60, 66, 70, 72, 78, 80, 84, 88, 90, and 96.

44. **A False Prediction**   In 1811, Peter Barlow wrote in his text *Theory of Numbers* that the eighth perfect number, $2^{30}(2^{31} - 1) = 2{,}305{,}843{,}008{,}139{,}952{,}128$, which was discovered by Leonhard Euler in 1772, "is the greatest perfect number known at present, and probably the greatest that ever will be discovered; for as they are merely curious, without being useful, it is not likely that any person will attempt to find one beyond it." (*Source:* http://en.wikipedia.org/wiki/2147483647)

The current search for larger and larger perfect numbers shows that Barlow's prediction did not come true. Search the Internet for answers to the question "Why do people continue the search for large perfect numbers (or large prime numbers)?" Write a brief summary of your findings.

CHAPTER 6   SUMMARY

The following table summarizes essential concepts in this chapter. The references given in the right-hand column list Examples and Exercises that can be used to test your understanding of a concept.

6.1 Early Numeration Systems

The Egyptian Numeration System The Egyptian numeration system is an additive system that uses symbols called hieroglyphics as numerals.	See **Examples 1 to 4** on pages 300 to 302, and then try Exercises 1 to 4 on page 359.

Hindu-Arabic numeral	Egyptian hieroglyphic	Description of hieroglyphic
1	|	stroke
10	∩	heel bone
100	୧	scroll
1000	⚘	lotus flower
10,000	∂	pointing finger
100,000	⋈	fish
1,000,000	⚲	atonished person

The Roman Numeration System The Roman numeration system is an additive system that also incorporates a subtraction property. Basic Rules Employed in the Roman Numeration System I = 1, V = 5, X = 10, L = 50, C = 100, D = 500, M = 1000 **1.** If the numerals are listed so that each numeral has a larger value than the numeral to the right, then the value of the Roman numeral is found by adding the values of the numerals. **2.** Each of the numerals, I, X, C, and M may be repeated up to three times. The numerals V, L, and D are not repeated. If a numeral is repeated two or three times in succession, we add to determine its numerical value. **3.** The only numerals whose values can be subtracted from the value of the numeral to the right are I, X, and C. The value of the numeral to be subtracted must be no less than one-tenth of the value of the numeral to its right.	See **Examples 5 to 7** on pages 303 and 304, and then try Exercises 5 to 12 on page 359.

6.2 Place-Value Systems

The Hindu-Arabic Numeration System The Hindu-Arabic numeration system is a base ten place-value system. The representation of a numeral as the indicated sum of each digit of the numeral multiplied by its respective place value is called the expanded form of the numeral.	See **Examples 1 to 5** on pages 308 and 309, and then try Exercises 13 to 16 on page 360.

The Babylonian Numeration System The Babylonian numeration system is a base sixty place-value system. A vertical wedge shape represents one unit, and a sideways "vee" shape represents 10 units. ▼ 1 ◀ 10	See **Examples 6 to 8** on pages 310 to 312, and then try Exercises 17 to 24 on page 360.
The Mayan Numeration System The numerals and the place values used in the Mayan numeration system are shown below. 0 1 2 3 4 5 6 7 8 9 10 11 12 13 14 15 16 17 18 19 Place Values in the Mayan Numeration System 18×20^3 = 144,000 18×20^2 = 7200 18×20^1 = 360 20^1 = 20 20^0 = 1	See **Examples 9 and 10** on pages 313 and 314, and then try Exercises 25 to 32 on page 360.

6.3 Different Base Systems

Convert Non–Base Ten Numerals to Base Ten In general, a base b numeration system has place values of $\ldots, b^5, b^4, b^3, b^2, b^1, b^0$. To convert a non–base ten numeral to base ten, we write the numeral in expanded form and simplify.	See **Examples 1 to 4** on pages 318 and 319, and then try Exercises 33 to 36 on page 360.
Convert from Base Ten to Another Base The method of converting a number written in base ten to another base makes use of the successive division process described on page 320.	See **Example 5** on page 320, and then try Exercises 37 to 40 on page 360.
Converting Directly between Computer Bases To convert directly from: ■ octal to binary, replace each octal numeral with its 3-bit binary equivalent. ■ binary to octal, break the binary numeral into groups of three, from right to left, and replace each group with its octal equivalent. ■ hexadecimal to binary, replace each hexadecimal numeral with its 4-bit binary equivalent. ■ binary to hexadecimal, break the binary numeral into groups of four, from right to left, and replace each group with its hexadecimal equivalent.	See **Examples 6 to 9** on pages 321 and 322, and then try Exercises 45 to 52 on page 360.
The Double-Dabble Method The double-dabble method is a procedure that can be used to convert a base two numeral to base ten. The advantage of this method is that you can start at the left of the base two numeral and work your way to the right without first determining the place value of each bit in the base two numeral.	See **Example 10** on page 323, and then try Exercises 53 to 56 on page 360.

6.4 Arithmetic in Different Bases

Arithmetic in Different Bases This section illustrates how to perform arithmetic operations in different base systems.	See **Examples 4, 5, 8, and 9** on pages 329 to 334, and then try Exercises 57 to 64 on page 360.

Divisor of a Number The natural number a is a divisor or a factor of the natural number b provided there exists a natural number j such that $aj = b$.	See **Example 1** on page 338, and then try Exercises 65 and 66 on page 360.
Prime Numbers and Composite Numbers A prime number is a natural number greater than 1 that has exactly two factors: itself and 1. A composite number is a natural number greater than 1 that is not a prime number.	See **Example 2** on page 339, and then try Exercises 67 to 70 on page 360.
The Fundamental Theorem of Arithmetic Every composite number can be written as a unique product of prime numbers (disregarding the order of the factors).	See **Example 4** on page 340, and then try Exercises 71 to 74 on page 360.

6.6 Topics from Number Theory

Perfect, Deficient, and Abundant Numbers A natural number is <ul><li>**perfect** if it is equal to the sum of its proper factors.</li><li>**deficient** if it is greater than the sum of its proper factors.</li><li>**abundant** if it is less than the sum of its proper factors.</li></ul>	See **Example 1** on page 348, and then try Exercises 75 to 78 on page 360.
Euclid's Procedure for Generating a Perfect Number A Mersenne number is a number of the form $2^n - 1$, where n is a prime number. Some Mersenne numbers are prime and some are composite. If n is a prime number and the Mersenne number $2^n - 1$ is a prime number, then $2^{n-1}(2^n - 1)$ is a perfect number.	See **Example 3** on page 349, and then try Exercises 79 and 80 on page 361.
The Number of Digits in b^x The number of digits in the number b^x, where b is a natural number less than 10 and x is a natural number, is the greatest integer of $(x \log b) + 1$.	See **Example 4** on page 351, and then try Exercises 85 and 86 on page 361.
Fermat's Last Theorem There are no natural numbers x, y, z, and n that satisfy $x^n + y^n = z^n$ where n is greater than 2.	See **Example 5** on page 352, and then try Exercise 80 on page 361.

CHAPTER 6   REVIEW EXERCISES

1. Write 4,506,325 using Egyptian hieroglyphics.
2. Write 3,124,043 using Egyptian hieroglyphics.
3. Write the Egyptian hieroglyphic

as a Hindu-Arabic numeral.
4. Write the Egyptian hieroglyphic

as a Hindu-Arabic numeral.

■ In Exercises 5 to 8, write each Roman numeral as a Hindu-Arabic numeral.

5. CCCXLIX
6. DCCLXXIV
7. $\overline{\text{IX}}$DCXL
8. $\overline{\text{XCII}}$CDXLIV

■ In Exercises 9 to 12, write each Hindu-Arabic numeral as a Roman numeral.

9. 567
10. 823
11. 2489
12. 1335

■ In Exercises 13 and 14, write each Hindu-Arabic numeral in expanded form.

13. 432   **14.** 456,327

■ In Exercises 15 and 16, simplify each expanded form.

15. $(5 \times 10^6) + (3 \times 10^4) + (8 \times 10^3) +$
$(2 \times 10^2) + (4 \times 10^0)$

16. $(3 \times 10^5) + (8 \times 10^4) + (7 \times 10^3) +$
$(9 \times 10^2) + (6 \times 10^1)$

■ In Exercises 17 to 20, write each Babylonian numeral as a Hindu-Arabic numeral.

17. ⟨𝖸𝖸𝖸 ⟨⟨𝖸

18. ⟨⟨𝖸𝖸𝖸𝖸𝖸𝖸 ⟨⟨⟨⟨𝖸𝖸𝖸

19. ⟨⟨𝖸 ⟨𝖸𝖸𝖸𝖸 𝖸

20. ⟨⟨𝖸𝖸𝖸𝖸 ⟨𝖸𝖸𝖸𝖸𝖸𝖸 ⟨⟨⟨𝖸𝖸𝖸

■ In Exercises 21 to 24, write each Hindu-Arabic numeral as a Babylonian numeral.

21. 721   **22.** 1080
23. 12,543   **24.** 19,281

■ In Exercises 25 to 28, write each Mayan numeral as a Hindu-Arabic numeral.

25. · · · ·
· · · ·
═══

26. · · ·
· ·   ·

27. ·
⟨👁⟩

28. · · ·
═══
⟨👁⟩

■ In Exercises 29 to 32, write each Hindu-Arabic numeral as a Mayan numeral.

29. 522   **30.** 346
31. 1862   **32.** 1987

■ In Exercises 33 to 36, convert each numeral to base ten.

33. 45_{six} **34.** 172_{nine}
35. $E3_{\text{sixteen}}$ **36.** $1BA_{\text{twelve}}$

■ In Exercises 37 to 40, convert each base ten numeral to the indicated base.

37. 45 to base three   **38.** 123 to base seven
39. 862 to base eleven   **40.** 3021 to base twelve

■ In Exercises 41 to 44, convert each numeral to the indicated base.

41. 346_{nine} to base six   **42.** 1532_{six} to base eight
43. 275_{twelve} to base nine   **44.** $67A_{\text{sixteen}}$ to base twelve

■ In Exercises 45 to 52, convert each numeral directly (without first converting to base ten) to the indicated base.

45. 11100_{two} to base eight
46. 1010100_{two} to base eight
47. 1110001101_{two} to base sixteen
48. 11101010100_{two} to base sixteen
49. 25_{eight} to base two
50. 1472_{eight} to base two
51. $4A_{\text{sixteen}}$ to base two
52. $C72_{\text{sixteen}}$ to base two

■ In Exercises 53 to 56, use the double-dabble method to convert each base two numeral to base ten.

53. 110011010_{two} **54.** 100010101_{two}
55. 10000010001_{two} **56.** 11001010000_{two}

■ In Exercises 57 to 64, perform the indicated operation. Write the answers in the same base as the given numerals.

57. $235_{\text{six}} + 144_{\text{six}}$ **58.** $673_{\text{eight}} + 345_{\text{eight}}$
59. $672_{\text{nine}} - 135_{\text{nine}}$ **60.** $1332_{\text{four}} - 213_{\text{four}}$
61. $25_{\text{eight}} \times 542_{\text{eight}}$ **62.** $43_{\text{five}} \times 3421_{\text{five}}$
63. $1010101_{\text{two}} \div 11_{\text{two}}$ **64.** $321_{\text{four}} \div 12_{\text{four}}$

■ In Exercises 65 and 66, use divisibility tests to determine whether the given number is divisible by 2, 3, 4, 5, 6, 8, 9, 10, or 11.

65. 1485   **66.** 4268

■ In Exercises 67 to 70, determine whether the given number is a prime number or a composite number.

67. 501   **68.** 781
69. 689   **70.** 1003

■ In Exercises 71 to 74, determine the prime factorization of the given number.

71. 45   **72.** 54
73. 153   **74.** 285

■ In Exercises 75 to 78, determine whether the given number is perfect, deficient, or abundant.

75. 28   **76.** 81
77. 144   **78.** 200

■ In Exercises 79 and 80, use Euclid's perfect number–generating procedure to write the perfect number associated with the given Mersenne prime.

79. $2^{61} - 1$ **80.** $2^{1279} - 1$

■ In Exercises 81 to 84, use the Egyptian doubling procedure to find the given product.

81. 8×46 **82.** 9×57
83. 14×83 **84.** 21×143

85. Find the number of digits in the Mersenne number $2^{132049} - 1$.

86. Find the number of digits in the Mersenne number $2^{2976221} - 1$.

87. How many odd perfect numbers had been discovered as of August 2011?

88. Determine whether there is a natural number z such that
$$2^3 + 17^3 = z^3$$

C H A P T E R 6   TEST

1. Write 3124 using Egyptian hieroglyphics.

2. Write the Egyptian hieroglyphic

as a Hindu-Arabic numeral.

3. Write the Roman numeral MCDXLVII as a Hindu-Arabic numeral.

4. Write 2609 as a Roman numeral.

5. Write 67,485 in expanded form.

6. Simplify:
$$(5 \times 10^5) + (3 \times 10^4) + (2 \times 10^2)$$
$$+ (8 \times 10^1) + (4 \times 10^0)$$

7. Write the Babylonian numeral

❮ ❮❮𐤂 ❮𐤂𐤂𐤂𐤂

as a Hindu-Arabic numeral.

8. Write 9675 as a Babylonian numeral.

9. Write the Mayan numeral

· · ·
═══
———

as a Hindu-Arabic numeral.

10. Write 502 as a Mayan numeral.

11. Convert 3542_{six} to base ten.

12. Convert 2148 to **a.** base eight and **b.** base twelve.

13. Convert 4567_{eight} to binary form.

14. Convert $101010110111_{\text{two}}$ to hexadecimal form.

■ In Exercises 15 to 18, perform the indicated operation. Write the answers in the same base as the given numerals.

15. $34_{\text{five}} + 23_{\text{five}}$

16. $462_{\text{eight}} - 147_{\text{eight}}$

17. $101_{\text{two}} \times 101110_{\text{two}}$

18. $431_{\text{seven}} \div 5_{\text{seven}}$

19. Determine the prime factorization of 230.

20. Determine whether 1001 is a prime number or a composite number.

21. Use divisibility tests to determine whether 1,737,285,147 is divisible by **a.** 2, **b.** 3, or **c.** 5.

22. Use divisibility tests to determine whether 19,531,333,276 is divisible by **a.** 4, **b.** 6, or **c.** 11.

23. Determine whether 96 is perfect, deficient, or abundant.

24. Use Euclid's perfect number–generating procedure to write the perfect number associated with the Mersenne prime $2^{17} - 1$.

Mathematical Systems

It's a dark, rainy night as you pull up to the drive at your home. You press a remote in your car that opens your garage door and you drive in. You press the remote again and the garage door closes.

Unbeknownst to you, there is a crook lurking on the street with a radio scanner who picks up the signal from your remote. The next time you're not home, the crook plans on using the saved signal to open your garage door and burglarize your home. However, when the crook arrives and sends the signal, the door doesn't open.

Now suppose you want to make a purchase on the Internet using a credit card. You may have noticed that the typical http:// that precedes a web address is replaced by https://. The "s" at the end indicates a *secure* website. This means that someone who may be trying to steal credit card information cannot intercept the information you send.

Although a garage door opener and a secure website may seem to be quite different, the mathematics behind each of these is similar. Both are based on modular arithmetic, one of the topics of this chapter.

Modular arithmetic, in turn, is part of a branch of mathematics called group theory, another topic in this chapter. Group theory is used in a variety of many seemingly unrelated subjects such as the structure of a diamond, wallpaper patterns, quantum physics, and the 12-tone chromatic scale in music.

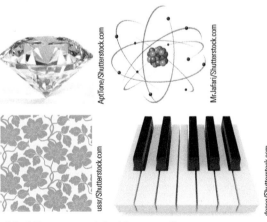

All of these are related to group theory.

Modular Arithmetic

Introduction to Modular Arithmetic

POINT OF INTEREST

The abbreviation A.M. comes from the Latin *ante* (before) *meridiem* (midday). The abbreviation P.M. comes from the Latin *post* (after) *meridiem* (midday).

FIGURE 8.1A

FIGURE 8.1B

FIGURE 8.2

Many clocks have the familiar 12-hour design. We designate whether the time is before noon or after noon by using the abbreviations A.M. and P.M. A reference to 7:00 A.M. means 7 hours after 12:00 midnight; a reference to 7:00 P.M. means 7 hours after 12:00 noon. In both cases, once 12 is reached on the clock, we begin again with 1.

If we want to determine a time in the future or in the past, it is necessary to consider whether we have passed 12 o'clock. To determine the time 8 hours after 3 o'clock, we add 3 and 8. Because we did not pass 12 o'clock, the time is 11 o'clock (Figure 8.1A). However, to determine the time 8 hours after 9 o'clock, we must take into consideration that once we have passed 12 o'clock, we begin again with 1. Therefore, 8 hours after 9 o'clock is 5 o'clock, as shown in Figure 8.1B.

We will use the symbol $\oplus$ to denote addition on a 12-hour clock. Using this notation,

$$3 \oplus 8 = 11 \quad \text{and} \quad 9 \oplus 8 = 5$$

on a 12-hour clock.

We can also perform subtraction on a 12-hour clock. If the time now is 10 o'clock, then 7 hours ago the time was 3 o'clock, which is the difference between 10 and 7 ($10 - 7 = 3$). However, if the time now is 3 o'clock, then, using Figure 8.2, we see that 7 hours ago it was 8 o'clock. If we use the symbol $\ominus$ to denote subtraction on a 12-hour clock, we can write

$$10 \ominus 7 = 3 \quad \text{and} \quad 3 \ominus 7 = 8$$

▼ **example** **1** **Perform Clock Arithmetic**

Evaluate each of the following, where $\oplus$ and $\ominus$ indicate addition and subtraction, respectively, on a 12-hour clock.

a. $8 \oplus 7$ **b.** $7 \oplus 12$ **c.** $8 \ominus 11$ **d.** $2 \ominus 8$

Solution
Calculate using a 12-hour clock.

a. $8 \oplus 7 = 3$ **b.** $7 \oplus 12 = 7$ **c.** $8 \ominus 11 = 9$ **d.** $2 \ominus 8 = 6$

▼ **check your progress** **1** Evaluate each of the following using a 12-hour clock.

a. $6 \oplus 10$ **b.** $5 \oplus 9$ **c.** $7 \ominus 11$ **d.** $5 \ominus 10$

Solution   *See page S27.* ◀

A similar example involves day-of-the-week arithmetic. If we associate each day of the week with a number, as shown at the left, then 6 days after Friday is Thursday and 16 days after Monday is Wednesday. Symbolically, we write

$$5 \boxplus 6 = 4 \quad \text{and} \quad 1 \boxplus 16 = 3$$

Monday = 1	Friday = 5
Tuesday = 2	Saturday = 6
Wednesday = 3	Sunday = 7
Thursday = 4	

Note: We are using the $\boxplus$ symbol for days-of-the-week arithmetic to differentiate from the $\oplus$ symbol for clock arithmetic.

Another way to determine the day of the week is to note that when the sum $5 + 6 = 11$ is divided by 7, the number of days in a week, the remainder is 4, the number associated with Thursday. When $1 + 16 = 17$ is divided by 7, the remainder is 3, the number associated with Wednesday. This works because the days of the week repeat every 7 days.

The same method can be applied to 12-hour-clock arithmetic. From Example 1a, when $8 + 7 = 15$ is divided by 12, the number of hours on a 12-hour clock, the remainder is 3, the time 7 hours after 8 o'clock.

Situations such as these that repeat in cycles are represented mathematically by using **modular arithmetic**, or **arithmetic modulo** *n*.

▼ **Modulo *n***

Two integers *a* and *b* are said to be **congruent modulo *n***, where *n* is a natural number, if $\dfrac{a - b}{n}$ is an integer. In this case, we write $a \equiv b \bmod n$. The number *n* is called the **modulus**. The statment $a \equiv b \bmod n$ is called a **congruence**.

▼ **example 2**   Determine Whether a Congruence Is True

Determine whether the congruence is true.

a. $29 \equiv 8 \bmod 3$

b. $15 \equiv 4 \bmod 6$

Solution

a. Find $\dfrac{29 - 8}{3} = \dfrac{21}{3} = 7$. Because 7 is an integer, $29 \equiv 8 \bmod 3$ is a true congruence.

b. Find $\dfrac{15 - 4}{6} = \dfrac{11}{6}$. Because $\frac{11}{6}$ is not an integer, $15 \equiv 4 \bmod 6$ is not a true congruence.

▼ **check your progress 2**   Determine whether the congruence is true.

a. $7 \equiv 12 \bmod 5$

b. $15 \equiv 1 \bmod 8$

Solution   *See page S27.*

For $29 \equiv 8 \bmod 3$ given in Example 2, note that $29 \div 3$ (the modulus) $= 9$ remainder 2 and that $11 \div 3$ (the modulus) $= 3$ remainder 2. Both 29 and 8 have the same remainder when divided by the modulus. This leads to an important alternate method to determine a true congruence. If $a \equiv b \bmod n$ and *a* and *b* are whole numbers, then *a* and *b* have the same remainder when divided by *n*.

question   Using the alternate method, is $33 \equiv 49 \bmod 4$ a true congruence?

Now suppose today is Friday. To determine the day of the week 16 days from now, we observe that 14 days from now the day will be Friday, so 16 days from now the day will be Sunday. Note that the remainder when 16 is divided by 7 is 2, or, using modular notation, $16 \equiv 2 \bmod 7$. The 2 signifies 2 days after Friday, which is Sunday.

---

answer   Yes. $33 \div 4 = 8$ remainder 1, and $49 \div 4 = 12$ remainder 1. Both 33 and 49 have the same remainder when divided by 4.

▼ **example 3** a Day of the Week

July 4, 2010, was a Sunday. What day of the week is July 4, 2015?

Solution
There are 5 years between the two dates. Each year has 365 days except 2012, which has one extra day because it is a leap year. So the total number of days between the two dates is $5 \cdot 365 + 1 = 1826$. Because $1826 \div 7 = 260$ remainder 6, $1826 \equiv 6 \bmod 7$. Any multiple of 7 days past a given day will be the same day of the week. So the day of the week 1826 days after July 4, 2010, will be the same as the day 6 days after July 4, 2010. Thus July 4, 2015, will be a Saturday.

▼ **check your progress 3** In 2008, Abraham Lincoln's birthday fell on Tuesday, February 12. On what day of the week does Lincoln's birthday fall in 2017?

Solution *See page S27.*

MATHMATTERS A Leap-Year Formula

The calculation in Example 3 required that we consider whether the intervening years contained a leap year. There is a formula, based on modular arithmetic, that can be used to determine which years are leap years.

The calendar we use today is called the Gregorian calendar. This calendar differs from the Julian calendar (see the Historical Note on page 475) in that leap years do not always occur every fourth year. Here is the rule: Let Y be the year. If $Y \equiv 0 \bmod 4$, then Y is a leap year unless $Y \equiv 0 \bmod 100$. In that case, Y is not a leap year unless $Y \equiv 0 \bmod 400$. Then Y is a leap year.

Using this rule, 2008 is a leap year because $2008 \equiv 0 \bmod 4$, and 2013 is not a leap year because $2013 \not\equiv 0 \bmod 4$. The year 1900 was *not* a leap year because $1900 \equiv 0 \bmod 100$ but $1900 \not\equiv 0 \bmod 400$. The year 2000 was a leap year because $2000 \equiv 0 \bmod 100$ and $2000 \equiv 0 \bmod 400$.

Arithmetic Operations Modulo *n*

Arithmetic modulo *n*, where *n* is a natural number, uses a variation of the standard rules of arithmetic we have used before. Perform the arithmetic operation and then divide by the modulus. The answer is the remainder. Thus the result of an arithmetic operation mod *n* is always a whole number less than *n*.

TAKE NOTE ✓

Recall that the natural numbers are the usual counting numbers: 1, 2, 3, 4, ... The set of whole numbers consists of the natural numbers and zero.

TAKE NOTE ✓

Remember that *m* mod *n* means the remainder when *m* is divided by *n*. For Example 4, we must find the remainder of the sum of $23 + 38$ when divided by 12.

▼ **example 4** Addition Modulo *n*

Evaluate: $(23 + 38) \bmod 12$

Solution
Add $23 + 38$ to produce 61. Then divide by the modulus, 12. The answer is the remainder.

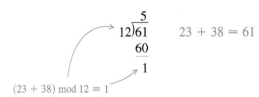

$$23 + 38 = 61$$

$(23 + 38) \bmod 12 \equiv 1$

The answer is 1.

▼ **check your progress** **4**   Evaluate: $(51 + 72) \bmod 3$

Solution   *See page S27.*

In modular arithmetic, adding the modulus to a number does not change the value of the number. For instance,

$13 \equiv 6 \bmod 7$		$10 \equiv 1 \bmod 3$	
$20 \equiv 6 \bmod 7$	• Add 7 to 13.	$13 \equiv 1 \bmod 3$	• Add 3 to 10.
$27 \equiv 6 \bmod 7$	• Add 7 to 20.	$16 \equiv 1 \bmod 3$	• Add 3 to 13.

INSTRUCTOR NOTE

Now is a good time for students to explore the Excursion on page 481, where they can use modular arithmetic to determine the day of the week on which any date far into the future or past falls.

To understand why the value does not change, consider $7 \equiv 0 \bmod 7$ and $3 \equiv 0 \bmod 3$. That is, in mod 7 arithmetic, 7 is equivalent to 0; in mod 3 arithmetic, 3 is equivalent to 0. Just as adding 0 to a number does not change the value of the number in regular arithmetic, in modular arithmetic adding the modulus to a number does not change the value of the number. This property of modular arithmetic is sometimes used in subtraction.

It is possible to use negative numbers modulo n. For instance,

$$-2 \equiv 5 \bmod 7 \text{ because } \frac{-2 - 5}{7} = \frac{-7}{7} = -1, \text{ an integer.}$$

Suppose we want to find x so that $-15 \equiv x \bmod 6$. Using the definition of modulo n, we need to find x so that $\dfrac{-15 - x}{6}$ is an integer. To do this, rewrite the expression and then try various values of x from 0 to the modulus until the value of the expression is an integer.

$$\frac{-15 - x}{6} = \frac{-(15 + x)}{6}$$

When $x = 0$, $\dfrac{-(15 + 0)}{6} = -\dfrac{15}{6}$, not an integer.

When $x = 1$, $\dfrac{-(15 + 1)}{6} = -\dfrac{16}{6} = -\dfrac{8}{3}$, not an integer.

When $x = 2$, $\dfrac{-(15 + 2)}{6} = -\dfrac{17}{6}$, not an integer.

When $x = 3$, $\dfrac{-(15 + 3)}{6} = -\dfrac{18}{6} = -3$, an integer.

$$-15 \equiv 3 \bmod 6$$

It may be necessary to use this idea when subtracting in modular arithmetic.

▼ **example** **5**   **Subtraction Modulo n**

Evaluate each of the following.

a. $(33 - 16) \bmod 6$ **b.** $(14 - 27) \bmod 5$

Solution

a. Subtract $33 - 16 = 17$. The result is positive. Divide the difference by the modulus, 6. The answer is the remainder.

$$\begin{array}{r} 2 \\ 6\overline{)17} \\ \underline{12} \\ 5 \end{array}$$

$(33 - 16) \bmod 6 \equiv 5$

b. Subtract $14 - 27 = -13$. Because the answer is negative, we must find x so that $-13 \equiv x \bmod 5$. Thus we must find x so that the value of $\dfrac{-13 - x}{5} = \dfrac{-(13 + x)}{5}$ is an integer. Trying the whole number values of x less than 5, the modulus, we find that when $x = 2$, $\dfrac{-(13 + 2)}{5} = -\dfrac{15}{5} = -3$.

$(14 - 27) \bmod 5 \equiv 2$

▼ **check your progress 5** Evaluate: $(21 - 43) \bmod 7$

Solution *See page S27.*

The methods of adding and subtracting in modular arithmetic can be used for clock arithmetic and days-of-the-week arithmetic.

▼ **example 6** **Calculating Times**

TAKE NOTE

In Example 6, repeatedly adding the modulus to the difference results in the following.

$-52 + 12 = -40$
$-40 + 12 = -28$
$-28 + 12 = -16$
$-16 + 12 = -4$
$-4 + 12 = 8$

Disregarding A.M. or P.M., if it is 5 o'clock now, what time was it 57 hours ago?

Solution
The time can be determined by calculating $(5 - 57) \bmod 12$. Because $5 - 57 = -52$ is a negative number, find a whole number x less than the modulus 12, so that $-52 \equiv x \bmod 12$. This means to find x so that $\dfrac{-52 - x}{12} = \dfrac{-(52 + x)}{12}$ is an integer.

Evaluating the expression for whole number values of x less than 12, we have, when $x = 8$, $\dfrac{-(52 + 8)}{12} = -\dfrac{60}{12} = -5$, an integer. Thus $(5 - 57) \bmod 12 \equiv 8$. Therefore, if it is 5 o'clock now, 57 hours ago it was 8 o'clock.

▼ **check your progress 6** If today is Tuesday, what day of the week will it be 93 days from now?

Solution *See page S27.*

Problems involving multiplication can also be performed modulo n.

▼ **example 7** **Multiplication Modulo n**

Evaluate: $(15 \cdot 23) \bmod 11$

Solution
Find the product $15 \cdot 23$ and then divide by the modulus, 11. The answer is the remainder.

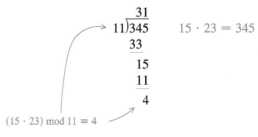

$$\begin{array}{r} 31 \\ 11\overline{)345} \\ 33 \\ \hline 15 \\ 11 \\ \hline 4 \end{array} \qquad 15 \cdot 23 = 345$$

$(15 \cdot 23) \bmod 11 \equiv 4$

The answer is 4.

▼ check your progress **7**   Evaluate: $(33 \cdot 41) \bmod 17$

Solution   *See page S27.*

Solving Congruence Equations

Solving a congruence equation means finding all whole number values of the variable for which the congruence is true.

For example, to solve $3x + 5 \equiv 3 \bmod 4$, we search for whole number values of x for which the congruence is true.

$$3(0) + 5 \not\equiv 3 \bmod 4$$
$$3(1) + 5 \not\equiv 3 \bmod 4$$
$$3(2) + 5 \equiv 3 \bmod 4 \qquad \text{2 is a solution.}$$
$$3(3) + 5 \not\equiv 3 \bmod 4$$
$$3(4) + 5 \not\equiv 3 \bmod 4$$
$$3(5) + 5 \not\equiv 3 \bmod 4$$
$$3(6) + 5 \equiv 3 \bmod 4 \qquad \text{6 is a solution.}$$

If we continued trying values, we would find that 10 and 14 are also solutions. Note that the solutions 6, 10, and 14 are all congruent to 2 modulo 4. In general, once a solution is determined, additional solutions can be found by repeatedly adding the modulus to the original solution. Thus the solutions of $3x + 5 \equiv 3 \bmod 4$ are 2, 6, 10, 14, 18, ...

When solving a congruence equation, it is necessary to check only the whole numbers less than the modulus. For the congruence equation $3x + 5 \equiv 3 \bmod 4$, we needed to check only 0, 1, 2, and 3. Each time a solution is found, additional solutions can be found by repeatedly adding the modulus to it.

A congruence equation can have more than one solution among the whole numbers less than the modulus. The next example illustrates that you must check *all* whole numbers less than the modulus.

▼ example **8**   **Solve a Congruence Equation**

Solve: $2x + 1 \equiv 3 \bmod 10$

Solution
Beginning with 0, substitute each whole number less than 10 into the congruence equation.

$x = 0$	$2(0) + 1 \not\equiv 3 \bmod 10$	Not a solution
$x = 1$	$2(1) + 1 \equiv 3 \bmod 10$	A solution
$x = 2$	$2(2) + 1 \not\equiv 3 \bmod 10$	Not a solution
$x = 3$	$2(3) + 1 \not\equiv 3 \bmod 10$	Not a solution
$x = 4$	$2(4) + 1 \not\equiv 3 \bmod 10$	Not a solution
$x = 5$	$2(5) + 1 \not\equiv 3 \bmod 10$	Not a solution
$x = 6$	$2(6) + 1 \equiv 3 \bmod 10$	A solution
$x = 7$	$2(7) + 1 \not\equiv 3 \bmod 10$	Not a solution
$x = 8$	$2(8) + 1 \not\equiv 3 \bmod 10$	Not a solution
$x = 9$	$2(9) + 1 \not\equiv 3 \bmod 10$	Not a solution

The solutions between 0 and 9 are 1 and 6; the remaining solutions are determined by repeatedly adding the modulus, 10, to these solutions. The solutions are 1, 6, 11, 16, 21, 26, ...

▼ check your progress 8   Solve: $4x + 1 \equiv 5 \bmod 12$

Solution   *See page S27.*

Not all congruence equations have a solution. For instance, $5x + 1 \equiv 3 \bmod 5$ has no solution, as shown below.

$x = 0$	$5(0) + 1 \not\equiv 3 \bmod 5$	Not a solution
$x = 1$	$5(1) + 1 \not\equiv 3 \bmod 5$	Not a solution
$x = 2$	$5(2) + 1 \not\equiv 3 \bmod 5$	Not a solution
$x = 3$	$5(3) + 1 \not\equiv 3 \bmod 5$	Not a solution
$x = 4$	$5(4) + 1 \not\equiv 3 \bmod 5$	Not a solution

Because no whole number value of x less than the modulus is a solution, there is no solution.

Additive and Multiplicative Inverses in Modular Arithmetic

Recall that if the sum of two numbers is 0, then the numbers are *additive inverses* of each other. For instance, $8 + (-8) = 0$, so 8 is the additive inverse of -8, and -8 is the additive inverse of 8.

The same concept applies in modular arithmetic. For example, $(3 + 5) \equiv 0 \bmod 8$. Thus, in mod 8 arithmetic, 3 is the additive inverse of 5, and 5 is the additive inverse of 3. **Here we consider only those whole numbers smaller than the modulus.** Note that $3 + 5 = 8$; that is, the sum of a number and its additive inverse equals the modulus. Using this fact, we can easily find the additive inverse of a number for any modulus. For instance, in mod 11 arithmetic, the additive inverse of 5 is 6 because $5 + 6 = 11$.

▼ example 9   Find the Additive Inverse

Find the additive inverse of 7 in mod 16 arithmetic.

Solution
In mod 16 arithmetic, $7 + 9 = 16$, so the additive inverse of 7 is 9.

▼ check your progress 9   Find the additive inverse of 6 in mod 12 arithmetic.

Solution   *See page S27.*

If the product of two numbers is 1, then the numbers are **multiplicative inverses** of each other. For instance, $2 \cdot \frac{1}{2} = 1$, so 2 is the multiplicative inverse of $\frac{1}{2}$, and $\frac{1}{2}$ is the multiplicative inverse of 2. The same concept applies to modular arithmetic (although the multiplicative inverses will always be natural numbers). For example, in mod 7 arithmetic, 5 is the multiplicative inverse of 3 (and 3 is the multiplicative inverse of 5) because $5 \cdot 3 \equiv 1 \bmod 7$. (Here we will concern ourselves only with natural numbers less than the modulus.) To find the multiplicative inverse of $a \bmod m$, solve the modular equation $ax \equiv 1 \bmod m$ for x.

▼ example 10   Find a Multiplicative Inverse

In mod 7 arithmetic, find the multiplicative inverse of 2.

TAKE NOTE

In mod m arithmetic, every number has an additive inverse but not necessarily a multiplicative inverse. For instance, in mod 12 arithmetic, 3 does not have a multiplicative inverse. You should verify this by trying to solve the equation $3x \equiv 1 \bmod 12$ for x.

A similar situation occurs in standard arithmetic. The number 0 does not have a multiplicative inverse because there is no solution of the equation $0x = 1$.

Solution

To find the multiplicative inverse of 2, solve the equation $2x \equiv 1 \bmod 7$ by trying different natural number values of x less than the modulus.

$$2x \equiv 1 \bmod 7$$
$$2(1) \not\equiv 1 \bmod 7 \qquad \bullet \text{ Try } x = 1.$$
$$2(2) \not\equiv 1 \bmod 7 \qquad \bullet \text{ Try } x = 2.$$
$$2(3) \not\equiv 1 \bmod 7 \qquad \bullet \text{ Try } x = 3.$$
$$2(4) \equiv 1 \bmod 7 \qquad \bullet \text{ Try } x = 4.$$

In mod 7 arithmetic, the multiplicative inverse of 2 is 4.

▼ **check your progress 10** Find the multiplicative inverse of 5 in mod 11 arithmetic.

Solution *See page S28.*

EXCURSION

Computing the Day of the Week

A function that is related to the modulo function is called the *floor function*. In the modulo function, we determine the remainder when one number is divided by another. In the floor function, we determine the quotient (and ignore the remainder) when one number is divided by another. The symbol for the floor function is $\lfloor \ \rfloor$. Here are some examples.

$$\left\lfloor \frac{2}{3} \right\rfloor = 0, \quad \left\lfloor \frac{10}{2} \right\rfloor = 5, \quad \left\lfloor \frac{17}{2} \right\rfloor = 8, \quad \text{and} \quad \left\lfloor \frac{2}{\sqrt{2}} \right\rfloor = 1$$

Using the floor function, we can write a formula that gives the day of the week for any date on the Gregorian calendar. The formula, known as Zeller's congruence, is given by

$$x \equiv \left(\left\lfloor \frac{13m - 1}{5} \right\rfloor + \left\lfloor \frac{y}{4} \right\rfloor + \left\lfloor \frac{c}{4} \right\rfloor + d + y - 2c \right) \bmod 7$$

where

d is the day of the month

m is the month using 1 for March, 2 for April, ..., 10 for December; January and February are assigned the values 11 and 12, respectively

y is the last two digits of the year if the month is March through December; if the month is January or February, y is the last two digits of the year *minus 1*

c is the first two digits of the year

x is the day of the week (using 0 for Sunday, 1 for Monday, ..., 6 for Saturday)

For example, to determine the day of the week on July 4, 1776, we have $c = 17$, $y = 76$, $d = 4$, and $m = 5$. Using these values, we can calculate x.

$$x \equiv \left(\left\lfloor \frac{13(5) - 1}{5} \right\rfloor + \left\lfloor \frac{76}{4} \right\rfloor + \left\lfloor \frac{17}{4} \right\rfloor + 4 + 76 - 2(17) \right) \bmod 7$$

$$\equiv (12 + 19 + 4 + 4 + 76 - 34) \bmod 7$$

$$\equiv 81 \bmod 7$$

Solving $x \equiv 81 \bmod 7$ for x, we get $x = 4$. Therefore, July 4, 1776, was a Thursday.

EXCURSION EXERCISES

1. Determine the day of the week on which you were born.

2. Determine the day of the week on which Abraham Lincoln's birthday (February 12) will fall in 2155.

3. Determine the day of the week on which January 1, 2020, will fall.

4. Determine the day of the week on which Valentine's Day (February 14) 1950 fell

EXERCISE SET 8.1 (Suggested Assignment: The Enhanced WebAssign Exercises)

■ In Exercises 1 to 16, evaluate each expression, where ⊕ and ⊖ indicate addition and subtraction, respectively, using a 12-hour clock.

Robert Brenner/PhotoEdit, Inc.

1. 3 ⊕ 5
2. 6 ⊕ 7
3. 8 ⊕ 4
4. 5 ⊕ 10
5. 11 ⊕ 3
6. 8 ⊕ 8
7. 7 ⊕ 9
8. 11 ⊕ 10
9. 10 ⊖ 6
10. 2 ⊖ 6
11. 3 ⊖ 8
12. 5 ⊖ 7
13. 10 ⊖ 11
14. 5 ⊖ 5
15. 4 ⊖ 9
16. 1 ⊖ 4

■ **Military Time** In Exercises 17 to 24, evaluate each expression, where △⃒ and △ indicate addition and subtraction, respectively, using military time. (Military time uses a 24-hour clock, where 2:00 A.M. is equivalent to 0200 hours and 10 P.M. is equivalent to 2200 hours.)

17. 1800 △⃒ 0900
18. 1600 △⃒ 1200
19. 0800 △⃒ 2000
20. 1300 △⃒ 1300
21. 1000 △ 1400
22. 1800 △ 1900
23. 0200 △ 0500
24. 0600 △ 2200

■ In Exercises 25 to 28, evaluate each expression, where ⊞ and ⊟ indicate addition and subtraction, respectively, using days-of-the-week arithmetic.

25. 6 ⊞ 4
26. 3 ⊞ 5
27. 2 ⊟ 3
28. 3 ⊟ 6

■ In Exercises 29 to 38, determine whether the congruence is true or false.

29. 5 ≡ 8 mod 3
30. 11 ≡ 15 mod 4
31. 5 ≡ 20 mod 4
32. 7 ≡ 21 mod 3
33. 21 ≡ 45 mod 6
34. 18 ≡ 60 mod 7
35. 88 ≡ 5 mod 9
36. 72 ≡ 30 mod 5
37. 100 ≡ 20 mod 8
38. 25 ≡ 85 mod 12

39. List five different natural numbers that are congruent to 8 modulo 6.

40. List five different natural numbers that are congruent to 10 modulo 4.

■ In Exercises 41 to 62, perform the modular arithmetic.

41. $(9 + 15) \bmod 7$
42. $(12 + 8) \bmod 5$
43. $(5 + 22) \bmod 8$
44. $(50 + 1) \bmod 15$
45. $(42 + 35) \bmod 3$
46. $(28 + 31) \bmod 4$
47. $(37 + 45) \bmod 12$
48. $(62 + 21) \bmod 2$
49. $(19 - 6) \bmod 5$
50. $(25 - 10) \bmod 4$
51. $(48 - 21) \bmod 6$
52. $(60 - 32) \bmod 9$
53. $(8 - 15) \bmod 12$
54. $(3 - 12) \bmod 4$
55. $(15 - 32) \bmod 7$
56. $(24 - 41) \bmod 8$
57. $(6 \cdot 8) \bmod 9$
58. $(5 \cdot 12) \bmod 4$
59. $(9 \cdot 15) \bmod 8$
60. $(4 \cdot 22) \bmod 3$
61. $(14 \cdot 18) \bmod 5$
62. $(26 \cdot 11) \bmod 15$

■ **Clocks and Calendars** In Exercises 63 to 70, use modular arithmetic to determine each of the following.

63. Disregarding A.M. or P.M., if it is now 7 o'clock,

   a. what time will it be 59 hours from now?

   b. what time was it 62 hours ago?

64. Disregarding A.M. or P.M., if it is now 2 o'clock,

   a. what time will it be 40 hours from now?

   b. what time was it 34 hours ago?

65. If today is Friday,

    a. what day of the week will it be 25 days from now?

    b. what day of the week was it 32 days ago?

66. If today is Wednesday,

    a. what day of the week will it be 115 days from now?

    b. what day of the week was it 81 days ago?

67. In 2005, Halloween (October 31) fell on a Monday. On what day of the week will Halloween fall in the year 2015?

68. In 2002, April Fool's Day (April 1) fell on a Monday. On what day of the week will April Fool's Day fall in 2013?

69. Valentine's Day (February 14) fell on a Tuesday in 2006. On what day of the week will Valentine's Day fall in 2020?

IN THE NEWS

Cinco de Mayo and the Battle of Puebla

Cinco de Mayo, celebrated on May 5 as its name implies, owes its origins to the Battle of Puebla, which took place on May 5, 1862. The Mexican army overcame seemingly insurmountable odds to defeat invading French forces from conquering the state of Puebla. The victory remains a cause for commemoration nearly 150 years later.

SOURCE: www.mtv.com

---

70. Cinco de Mayo fell on a Wednesday in 2010. On what day of the week will Cinco de Mayo fall in 2020?

Jonathan Nourok/PhotoEdit, Inc.

■ In Exercises 71 to 82, find all whole number solutions of the congruence equation.

71. $x \equiv 10 \bmod 3$

72. $x \equiv 12 \bmod 5$

73. $2x \equiv 12 \bmod 5$

74. $3x \equiv 8 \bmod 11$

75. $(2x + 1) \equiv 5 \bmod 4$

76. $(3x + 1) \equiv 4 \bmod 9$

77. $(2x + 3) \equiv 8 \bmod 12$

78. $(3x + 12) \equiv 7 \bmod 10$

79. $(2x + 2) \equiv 6 \bmod 4$

80. $(5x + 4) \equiv 2 \bmod 8$

81. $(4x + 6) \equiv 5 \bmod 8$

82. $(4x + 3) \equiv 3 \bmod 4$

■ In Exercises 83 to 88, find the additive inverse and the multiplicative inverse, if it exists, of the given number.

83. 4 in modulo 9 arithmetic

84. 4 in modulo 5 arithmetic

85. 7 in modulo 10 arithmetic

86. 11 in modulo 16 arithmetic

87. 3 in modulo 8 arithmetic

88. 6 in modulo 15 arithmetic

■ Modular division can be performed by considering the related multiplication problem. For instance, if $5 \div 7 = x$, then $x \cdot 7 = 5$. Similarly, the quotient $(5 \div 7) \bmod 8$ is the solution to the congruence equation $x \cdot 7 \equiv 5 \bmod 8$, which is 3. In Exercises 89 to 94, find the given quotient.

89. $(2 \div 7) \bmod 8$

90. $(4 \div 5) \bmod 8$

91. $(6 \div 4) \bmod 9$

92. $(2 \div 3) \bmod 5$

93. $(5 \div 6) \bmod 7$

94. $(3 \div 4) \bmod 7$

EXTENSIONS

Critical Thinking

95. Verify that the division $5 \div 8$ has no solution in modulo 8 arithmetic.

96. Verify that the division $4 \div 4$ has more than one solution in modulo 10 arithmetic.

97. Disregarding A.M. or P.M., if it is now 3:00, what time of day will it be in 3500 hours?

98. Using military time, if it is currently 1100 hours, determine what time of day it will be in 4250 hours.

99. There is only one whole number solution between 0 and 11 of the congruence equation

$$(x^2 + 3x + 7) \equiv 2 \bmod 11$$

Find the solution.

100. There are only two whole number solutions between 0 and 27 of the congruence equation

$$(x^2 + 5x + 4) \equiv 1 \bmod 27$$

Find the solutions.

Explorations

101. Rolling Codes Some garage door openers are programmed using modular arithmetic. See page 473. The basic idea is that there is a computer chip in both the transmitter in your car and the receiver in your garage that are programmed exactly the same way. When the transmitter is pressed, two things happen. First, the transmitter sends a number to the receiver; second, the computer chip uses the sent number and modular arithmetic to create a new number that is sent the next time the transmitter is pressed. Here is a simple example using a modulus of 37. A real transmitter and reciever would use a number with at least 40 digits.

$x_0 = 23$ This number is called the seed.

$x_1 \equiv (5x_0 + 17) \bmod 37$ This is the modular formula.

$\equiv (5 \cdot 23 + 17) \bmod 37$ Using this formula, a new number, 21, is calculated when the transmitter is pressed.

$\equiv 132 \bmod 37$

$\equiv 21$

$x_2 \equiv (5x_1 + 17) \bmod 37$ When the transmitter is pressed again, the last number calculated, 21, is used to find the next number.

$\equiv (5 \cdot 21 + 17) \bmod 37$

$\equiv 122 \bmod 37$

$\equiv 11$

This continues each time the transmitter is pressed. A general form for this formula is written as $x_{n+1} \equiv (5x_n + 17) \bmod 37$ and is called a recursive formula. The next number x_{n+1} is calculated using the previous number x_n.

a. Use the recursive formula $x_{n+1} \equiv (2x_n + 13) \bmod 11$ and seed $x_0 = 3$ to find the numbers x_1 to x_9.

b. Find x_{10}. How is this number related to x_0? (Remember: The last number you calculated in part a is x_9.)

c. Why might this recursive formula not be a good one for an actual garage door opener?

102. Many people consider the 13th of the month an unlucky day when it falls on a Friday.

a. Does every year have a Friday the 13th? Explain why or why not.

b. What is the greatest number of Friday the 13ths that can occur in one year? Can you find a recent year in which this number occurred? *Hint:* A non-leap year can have two consecutive months where the 13th is a Friday.

---

section **8.2** Applications of Modular Arithmetic

ISBN and UPC

Every book that is cataloged in the Library of Congress must have an ISBN (International Standard Book Number). This 13-digit number was created to help ensure that orders for books are filled accurately and that books are catalogued correctly.

The first three digits of an ISBN are 978, the next digit indicates the country in which the publisher is incorporated (0, and sometimes 1, for books written in English), the next two to seven digits indicate the publisher, the next group of digits indicates the title of the book, and the last digit (the 13th one) is called a **check digit**.

If we label the first digit of an ISBN d_1, the second digit d_2, and so on to the 13th digit d_{13}, then the check digit is chosen to satisfy the following congruence.

> ▼ **Formula for the ISBN Check Digit**
>
> $d_{13} = 10 - (d_1 + 3d_2 + d_3 + 3d_4 + d_5 + 3d_6 + d_7 + 3d_8 + d_9 + 3d_{10} + d_{11} + 3d_{12}) \bmod 10$
> If $d_{13} = 10$, then the check digit is 0.

It is this check digit that is used to ensure accuracy. For instance, the ISBN for the fourth edition of the American Heritage Dictionary is 978-0-395-82517-4. Suppose, however, that a bookstore clerk sends an order for the *American Heritage Dictionary* and inadvertently enters the number 978-0-395-28517-4, where the clerk transposed the 8 and 2 in the five numbers that identify the book.

Correct ISBN: 978-0-395-82517-4
Incorrect ISBN: 978-0-395-28517-4

The receiving clerk calculates the check digit as follows.

$$d_{13} \equiv 10 - [9 + 3(7) + 8 + 3(0) + 3 + 3(9) + 5 + 3(2) + 8 + 3(5) + 1 + 3(7)] \bmod 10$$
$$\equiv 10 - 124 \bmod 10$$
$$\equiv 10 - 4 = 6$$

Because the check digit is 6 and not 4 as it should be, the receiving clerk knows that an incorrect ISBN has been sent. Transposition errors are among the most frequent errors that occur. The ISBN coding system will catch most of them.

▼ **example 1**   **Determine a Check Digit for an ISBN**

Determine the ISBN check digit for the book *The Equation that Couldn't Be Solved* by Mario Livio. The first 12 digits of the ISBN are 978-0-7432-5820-?

Solution

$$d_{13} \equiv 10 - [9 + 3(7) + 8 + 3(0) + 7 + 3(4) + 3 + 3(2) + 5 + 3(8) + 2 + 3(0)] \bmod 10$$
$$\equiv 10 - 97 \bmod 10$$
$$\equiv 10 - 7 = 3$$

The check digit is 3.

▼ **check your progress 1**   A purchase order for the book *The Mathematical Tourist* by Ivars Peterson includes the ISBN 978-0-760-73261-6. Determine whether this is a valid ISBN.

Solution   *See page S28.*

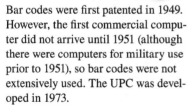

5   00041 10076   4

Another coding scheme that is closely related to the ISBN is the UPC (Universal Product Code). This number is placed on many items and is particularly useful in grocery stores. A check-out clerk passes the product by a scanner, which reads the number from a bar code and records the price on the cash register. If the price of an item changes for a promotional sale, the price is updated in the computer, thereby relieving a clerk of having to reprice each item. In addition to pricing items, the UPC gives the store manager accurate information about inventory and the buying habits of the store's customers.

The UPC is a 12-digit number that satisfies a congruence equation that is similar to the one for ISBNs. The last digit is the check digit. If we label the 12 digits of the UPC as $d_1, d_2, \ldots, d_{12}$, we can write a formula for the UPC check digit d_{12}.

▼ **Formula for the UPC Check Digit**

$d_{12} \equiv 10 - (3d_1 + d_2 + 3d_3 + d_4 + 3d_5 + d_6 + 3d_7 + d_8 + 3d_9 + d_{10} + 3d_{11}) \bmod 10$
If $d_{12} = 10$, then the check digit is 0.

WALT DISNEY PICTURES/THE KOBAL COLLECTION

▼ **example 2** **Determine the Check Digit of a UPC**

Find the check digit for the DVD release of the film *Alice in Wonderland*. The first 11 digits are 7-86936-79798-?

Solution

$d_{12} \equiv 10 - [3(7) + 8 + 3(6) + 9 + 3(3) + 6 + 3(7) + 9 + 3(7) + 9 + 3(8)] \bmod 10$

$\equiv 10 - 155 \bmod 10$

$\equiv 10 - 5 = 5$

The check digit is 5.

▼ **check your progress 2** Is 1-32342-65933-9 a valid UPC?

Solution *See page S28.*

The ISBN and UPC coding systems will normally catch transposition errors. There are instances, however, when they do not.

The UPC for Crisco Puritan Canola Oil with Omega-DHA is

0-51500-24275-9

Suppose, however, that the product code is written 0-51500-24725-9, where the 2 and 7 have been transposed. Calculating the check digit, we have

$d_{12} \equiv 10 - [3(0) + 5 + 3(1) + 5 + 3(0) + 0 + 3(2) + 4 + 3(7) + 2 + 3(5)] \bmod 10$

$\equiv 10 - 61 \bmod 10$

$\equiv 10 - 1 = 9$

The same check digit is calculated, yet the UPC has been entered incorrectly. This was an unfortunate coincidence; if any other two digits were transposed, the result would have given a different check digit and the error would have been caught. It can be shown that the ISBN and UPC coding methods will not catch a transposition error of adjacent digits a and b if $|a - b| = 5$. For the Canola Oil UPC, $|7 - 2| = 5$.

Credit Card Numbers

Companies that issue credit cards also use modular arithmetic to determine whether a credit card number is valid. This is especially important in e-commerce, where credit card information is frequently sent over the Internet. The primary coding method is based on the *Luhn algorithm,* which uses mod 10 arithmetic.

Credit card numbers are normally 13 to 16 digits long. The first one to four digits are used to identify the card issuer. The table below shows the identification prefixes used by four popular card issuers.

Card issuer	Prefix	Number of digits
MasterCard	51 to 55	16
Visa	4	13 or 16
American Express	34 or 37	15
Discover	6011	16

The Luhn algorithm, used to determine whether a credit card number is valid, is calculated as follows: Beginning with the next-to-last digit (the last digit is the check digit)

POINT OF INTEREST

Hans Peter Luhn was born in Germany and later joined IBM in the United States as a senior research engineer. He was awarded a patent in 1960 for the algorithm that bears his name. Credit card companies began using the algorithm shortly after its creation. The algorithm is in the public domain and is still widely used to identify valid card numbers.

and reading from right to left, double every other digit. If a digit becomes a two-digit number after being doubled, treat the number as two individual digits. Now find the sum of the new list of digits; the final sum must equal 0 mod 10. The Luhn algorithm is demonstrated in the next example.

▼ example 3   Determine a Valid Credit Card Number

Determine whether 5234 8213 3410 1298 is a valid credit card number.

Solution
Highlight every other digit, beginning with the next-to-last digit and reading from right to left.

    5 2 3 4 8 2 1 3 3 4 1 0 1 2 9 8

Next double each of the highlighted digits.

    10 2 6 4 16 2 2 3 6 4 2 0 2 2 18 8

Finally, add all digits, treating two-digit numbers as two single digits.

$$(1 + 0) + 2 + 6 + 4 + (1 + 6) + 2 + 2 + 3 + 6 + 4 + 2 + 0 +$$
$$2 + 2 + (1 + 8) + 8 = 60$$

Because $60 \equiv 0$ mod 10, this is a valid credit card number.

▼ check your progress 3   Is 6011 0123 9145 2317 a valid credit card number?

Solution   *See page S28.*   ◄

Cryptology

Related to codes on books and grocery items are secret codes. These codes are used to send messages between people, companies, or nations. It is hoped that by devising a code that is difficult to break, the sender can prevent the communication from being read if it is intercepted by an unauthorized person. **Cryptology** is the study of making and breaking secret codes.

Before we discuss how messages are coded, we need to define a few terms. **Plaintext** is a message before it is coded. The line

SHE WALKS IN BEAUTY LIKE THE NIGHT

from Lord Byron's poem "She Walks in Beauty" is in plaintext. **Ciphertext** is the message after it has been written in code. The line

ODA SWHGO EJ XAWQPU HEGA PDA JECDP

is the same line of the poem in ciphertext.

The method of changing from plaintext to ciphertext is called **encryption**. The line from the poem was encrypted by substituting each letter in plaintext with the letter that is 22 letters after that letter in the alphabet. (Continue from the beginning when the end of the alphabet is reached.) This is called a *cyclical coding scheme* because each letter of the alphabet is shifted the same number of positions. The original alphabet and the substitute alphabet are shown below.

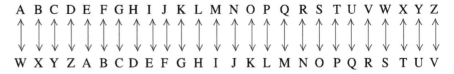

A B C D E F G H I J K L M N O P Q R S T U V W X Y Z

W X Y Z A B C D E F G H I J K L M N O P Q R S T U V

question Using the cyclical coding scheme described on the previous page where each letter is replaced by the one 22 letters after the letter, what is the plaintext word that corresponds to the ciphertext YKZA?

To **decrypt** a message means to take the ciphertext message and write it in plaintext. If a cryptologist thinks a message has been encrypted using a cyclical substitution code like the one shown above, the key to the code can be found by taking a word from the message (usually one of the longer words) and continuing the alphabet for each letter of the word. When a recognizable word appears, the key can be determined. This method is shown below using the ciphertext word XAWQPU.

X	A	W	Q	P	U	
Y	B	X	R	Q	V	Shift
Z	C	Y	S	R	W	four
A	D	Z	T	S	X	positions
B	E	A	U	T	Y	

Once a recognizable word has been found (BEAUTY), count the number of positions that the letters have been shifted (four, in this case). To decode the message, substitute the letter of the normal alphabet that comes four positions after the letter in the ciphertext.

O D A   S H E

Cyclical encrypting using the alphabet is related to modular arithmetic. We begin with the normal alphabet and associate each letter with a number as shown in Table 8.1.

TABLE 8.1 Numerical Equivalents for the Letters of the Alphabet

A	B	C	D	E	F	G	H	I	J	K	L	M	N	O	P	Q	R	S	T	U	V	W	X	Y	Z
1	2	3	4	5	6	7	8	9	10	11	12	13	14	15	16	17	18	19	20	21	22	23	24	25	0

If the encrypting code is to shift each letter of the plaintext message m positions, then the corresponding letter in the ciphertext message is given by $c \equiv (p + m) \bmod 26$, where p is the numerical equivalent of the plaintext letter and c is the numerical equivalent of the ciphertext letter. The letter Z is coded as 0 because $26 \equiv 0 \bmod 26$.

Each letter in Lord Byron's poem was shifted 22 positions ($m = 22$) to the right. To code the plaintext letter S in the word SHE, we use the congruence $c \equiv (p + m) \bmod 26$.

$c \equiv (p + m) \bmod 26$

$c \equiv (19 + 22) \bmod 26$ • $p = 19$ (S is the 19th letter.)
                                   $m = 22$, the number of
$c \equiv 41 \bmod 26$ positions the letter is shifted

$c = 15$

The 15th letter is O. Thus S is coded as O.

answer CODE

Once plaintext has been converted to ciphertext, there must be a method by which the person receiving the message can return the message to plaintext. For the cyclical code, the congruence is $p \equiv (c + n) \bmod 26$, where p and c are defined as before and $n = 26 - m$. The letter O in ciphertext is decoded below using the congruence $p \equiv (c + n) \bmod 26$.

$p \equiv (c + n) \bmod 26$

$p \equiv (15 + 4) \bmod 26$ • $c = 15$ (O is the 15th letter.)
                        $n = 26 - m = 26 - 22 = 4$

$p \equiv 19 \bmod 26$

$p = 19$

The 19th letter is S. Thus O is decoded as S.

MATHMATTERS   A Cipher of Caesar

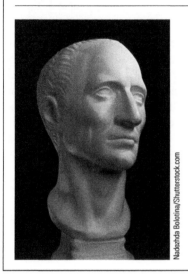

Julius Caesar supposedly used a cyclical alphabetic encrypting code to communicate with his generals. The messages were encrypted using the congruence $c \equiv (p + 3) \bmod 26$ (in modern notation). Using this coding scheme, the message BEWARE THE IDES OF MARCH would be coded as EHZDUH WKH LGHV RI PDUFK.

▼ example 4   Write Messages Using Cyclical Coding

Use the cyclical alphabetic encrypting code that shifts each letter 11 positions to
a. code CATHERINE THE GREAT.   **b.** decode TGLY ESP EPCCTMWP.

Solution

a. The encrypting congruence is $c \equiv (p + 11) \bmod 26$. Replace p by the numerical equivalent of each letter of plaintext and determine c. The results for CATHERINE are shown below.

C	$c \equiv (3 + 11) \bmod 26 \equiv 14 \bmod 26 \equiv 14$	Code C as N.
A	$c \equiv (1 + 11) \bmod 26 \equiv 12 \bmod 26 \equiv 12$	Code A as L.
T	$c \equiv (20 + 11) \bmod 26 \equiv 31 \bmod 26 \equiv 5$	Code T as E.
H	$c \equiv (8 + 11) \bmod 26 \equiv 19 \bmod 26 \equiv 19$	Code H as S.
E	$c \equiv (5 + 11) \bmod 26 \equiv 16 \bmod 26 \equiv 16$	Code E as P.
R	$c \equiv (18 + 11) \bmod 26 \equiv 29 \bmod 26 \equiv 3$	Code R as C.
I	$c \equiv (9 + 11) \bmod 26 \equiv 20 \bmod 26 \equiv 20$	Code I as T.
N	$c \equiv (14 + 11) \bmod 26 \equiv 25 \bmod 26 \equiv 25$	Code N as Y.
E	$c \equiv (5 + 11) \bmod 26 \equiv 16 \bmod 26 \equiv 16$	Code E as P.

Continuing, the plaintext would be coded as NLESPCTYP ESP RCPLE.

b. Because $m = 11, n = 26 - 11 = 15$. The ciphertext is decoded by using the congruence $p \equiv (c + 15) \bmod 26$. The results for TGLY are shown below.

T	$c \equiv (20 + 15) \bmod 26 \equiv 35 \bmod 26 \equiv 9$	Decode T as I.
G	$c \equiv (7 + 15) \bmod 26 \equiv 22 \bmod 26 \equiv 22$	Decode G as V.
L	$c \equiv (12 + 15) \bmod 26 \equiv 27 \bmod 26 \equiv 1$	Decode L as A.
Y	$c \equiv (25 + 15) \bmod 26 \equiv 40 \bmod 26 \equiv 14$	Decode Y as N.

Continuing, the ciphertext would be decoded as IVAN THE TERRIBLE.

▼ **check your progress 4** Use the cyclical alphabetic encrypting code that shifts each letter 17 positions to

a. encode ALPINE SKIING. **b.** decode TIFJJ TFLEKIP JBZZEX.

Solution *See page S28.* ◀

The practicality of a cyclical alphabetic coding scheme is limited because it is relatively easy for a cryptologist to determine the coding scheme. (Recall the method used for the line from Lord Byron's poem.) A coding scheme that is a little more difficult to break is based on the congruence $c \equiv (ap + m) \bmod 26$, where a and 26 do not have a common factor. (For instance, a cannot be 14 because 14 and 26 have a common factor of 2.) The reason why a and 26 cannot have a common factor is related to the procedure for determining the decoding congruence. However, we will leave that discussion to other math courses.

▼ **example 5** Encode a Message

Use the congruence $c \equiv (5p + 2) \bmod 26$ to encode the message LASER PRINTER.

Solution
The encrypting congruence is $c \equiv (5p + 2) \bmod 26$. Replace p by the numerical equivalent of each letter from Table 8.1 and determine c. The results for LASER are shown below.

L	$c \equiv (5 \cdot 12 + 2) \bmod 26 \equiv 62 \bmod 26 = 10$	Code L as J.
A	$c \equiv (5 \cdot 1 + 2) \bmod 26 \equiv 7 \bmod 26 = 7$	Code A as G.
S	$c \equiv (5 \cdot 19 + 2) \bmod 26 \equiv 97 \bmod 26 = 19$	Code S as S.
E	$c \equiv (5 \cdot 5 + 2) \bmod 26 \equiv 27 \bmod 26 = 1$	Code E as A.
R	$c \equiv (5 \cdot 18 + 2) \bmod 26 \equiv 92 \bmod 26 = 14$	Code R as N.

Continuing, the plaintext is coded in ciphertext as JGSAN DNUTXAN.

▼ **check your progress 5** Use the congruence $c \equiv (3p + 1) \bmod 26$ to encode the message COLOR MONITOR.

Solution *See page S29.* ◀

Decoding a message that was encrypted using the congruence $c \equiv (ap + m) \bmod n$ requires solving the congruence for p. The method relies on multiplicative inverses, which were discussed in Section 8.1.

Here we solve the congruence used in Example 5 for p.

$$c = 5p + 2$$

$$c - 2 = 5p \qquad \text{• Subtract 2 from each side of the equation.}$$

$$21(c - 2) = 21(5p) \qquad \text{• Multiply each side of the equation by the multiplicative inverse of 5. Because } 21 \cdot 5 \equiv 1 \bmod 26, \text{ multiply each side by 21.}$$

$$[21(c - 2)] \bmod 26 \equiv p$$

TAKE NOTE ✓

$21(5p) \equiv p \bmod 26$ because $21 \cdot 5 = 105$ and $105 \div 26 = 4$ remainder 1.

Using this congruence equation, we can decode the ciphertext message JGSAN DNUTXAN. The method for decoding JGSAN is shown below.

J	$[21(10 - 2)] \bmod 26 \equiv 168 \bmod 26 \equiv 12$	Decode J as L.
G	$[21(7 - 2)] \bmod 26 \equiv 105 \bmod 26 \equiv 1$	Decode G as A.
S	$[21(19 - 2)] \bmod 26 \equiv 357 \bmod 26 \equiv 19$	Decode S as S.
A	$[21(1 - 2)] \bmod 26 \equiv (-21) \bmod 26 \equiv 5$	Decode A as E.
N	$[21(14 - 2)] \bmod 26 \equiv 252 \bmod 26 \equiv 18$	Decode N as R.

Note that to decode A it was necessary to determine $(-21) \bmod 26$. Recall that this requires adding the modulus until a whole number less than 26 results. Because $(-21) + 26 = 5$, we have $(-21) \bmod 26 \equiv 5$.

▼ **example 6**   **Decode a Message**

Decode the message ACXUT CXRT, which was encrypted using the congruence $c \equiv (3p + 5) \bmod 26$.

Solution
Solve the congruence equation for p.

$$c = 3p + 5$$

$$c - 5 = 3p$$

$$9(c - 5) = 9(3p) \qquad \text{• } 9(3) = 27 \text{ and } 27 \equiv 1 \bmod 26$$

$$[9(c - 5)] \bmod 26 \equiv p$$

The decoding congruence is $p \equiv [9(c - 5)] \bmod 26$.

Using this congruence, we will show the details for decoding ACXUT.

A	$[9(1 - 5)] \bmod 26 \equiv (-36) \bmod 26 \equiv 16$	Decode A as P.
C	$[9(3 - 5)] \bmod 26 \equiv (-18) \bmod 26 \equiv 8$	Decode C as H.
X	$[9(24 - 5)] \bmod 26 \equiv 171 \bmod 26 \equiv 15$	Decode X as O.
U	$[9(21 - 5)] \bmod 26 \equiv 144 \bmod 26 \equiv 14$	Decode U as N.
T	$[9(20 - 5)] \bmod 26 \equiv 135 \bmod 26 \equiv 5$	Decode T as E.

Continuing, we would decode the message as PHONE HOME.

▼ **check your progress 6**   Decode the message IGHT OHGG, which was encrypted using the congruence $c \equiv (7p + 1) \bmod 26$.

Solution   *See page S29.*   ◄

MATHMATTERS   The Enigma Machine

markrhiggins/Shutterstock.com

During World War II, Nazi Germany encoded its transmitted messages using an Enigma machine, first patented in 1919. Each letter of the plaintext message was substituted with a different letter, but the machine changed the substitutions throughout the message so that the ciphertext appeared more random and would be harder to decipher if the message were intercepted. The Enigma machine accomplished its task by using three wheels of letters (chosen from a box of five) that rotated as letters were typed. In addition, electrical sockets corresponding to letters were connected in pairs with wires. The choice and order of the three wheels, their starting positions, and the arrangement of the electrical wires determined how the machine would encode a message. The receiver of the message needed to know the setup of the wheels and wires; with the receiver's Enigma machine configured identically, the ciphertext could be decoded. There are a staggering 150 quintillion different ways to configure the Enigma machine, making it very difficult to decode messages without knowing the setup that was used. A team of mathematicians and other code breakers, led by Alan Turing, was assembled by the British government in an attempt to decode the German messages. The team was eventually successful, and their efforts helped change the course of World War II. One aspect of the Enigma machine that aided the code breakers was the fact that the machine would never substitute a letter with itself.

As part of its *Nova* series, PBS aired an excellent account of the breaking of the Enigma codes during World War II entitled *Decoding Nazi Secrets*. Information is available at http://www.pbs.org/wgbh/nova/decoding/.

EXCURSION

Photolibrary

Sophie Germain (zhĕ r-män) (1776–1831) was born in Paris, France. Because enrollment in the university she wanted to attend was available only to men, Germain attended under the name Antoine-August Le Blanc. Eventually her ruse was discovered, but not before she came to the attention of Pierre Lagrange, one of the best mathematicians of the time. He encouraged her work and became a mentor to her.

A certain type of prime number named after Germain is called a Germain prime number. It is a prime p such that p and $2p + 1$ are both prime. For instance, 11 is a Germain prime because 11 and $2(11) + 1 = 23$ are both prime numbers. Germain primes are used in public key cryptography.

Public Key Cryptography

In **public key cryptography**, there are two keys created, one for encoding a message (the public key) and one for decoding the message (the private key). One form of this scheme is known as RSA, from the first letters of the last names of Ron Rivest, Adi Shamir, and Leonard Adleman, who published the method in 1977 while professors at the Massachusetts Institute of Technology. This method uses modular arithmetic in a very unique way.

To create a coding and decoding scheme in RSA cryptography, we begin by choosing two large, distinct prime numbers p and q, say, $p = 59$ and $q = 83$. In practice, these would be prime numbers that are 200 or more digits long.

Find the product of p and q, $n = p \cdot q = 59 \cdot 83 = 4897$. The value of n is the modulus. Now find the product $z = (p - 1)(q - 1) = 58 \cdot 82 = 4756$ and randomly choose a number e, where $1 < e < z$ and e and z have no common factors. We will choose $e = 129$. Solve the modular equation $ed = 1 \bmod z$ for d. For the numbers we are using, we must solve $129d \equiv 1 \bmod 4756$. We won't go into the details but the solution is $d = 1401$. Recall that because $129 \cdot 1401 \equiv 1 \bmod 4756$, 1401 is the multiplicative inverse of 129 mod 4756.

To receive encrypted messages, Olivia posts these values of n and e to a public key encryption service, called a certificate authority, which guarantees the integrity of her public key. If Henry wants to send Olivia a message, he codes his message as a number using a number for each letter of the alphabet. For instance, he might use $A = 11, B = 12, C = 13, \ldots$, and $Z = 36$. For Henry to send the message MATH, he would use the numbers 23 11 30 18 and code those numbers using Olivia's public key,

N^{129} mod 4897 $\equiv M$, where N is the plaintext (23 11 30 18), and M is the ciphertext, the result of using the modular equation. This is shown below.

$$23^{129} \text{ mod } 4897 \equiv 3065 \qquad 11^{129} \text{ mod } 4897 \equiv 2001$$

$$30^{129} \text{ mod } 4897 \equiv 957 \qquad 18^{129} \text{ mod } 4897 \equiv 2753$$

Thus Henry sends Olivia the ciphertext numbers 3065, 2001, 957, and 2753.

When Olivia receives the message, she uses her private key, M^{1401} mod 4897 $\equiv N$, where M is the ciphertext she received from Henry and N is the original plaintext, to decode the message. Olivia calculates

$$3065^{1401} \text{ mod } 4897 \equiv 23 \qquad 2001^{1401} \text{ mod } 4897 \equiv 11$$

$$957^{1401} \text{ mod } 4897 \equiv 30 \qquad 2753^{1401} \text{ mod } 4897 \equiv 18$$

She decodes the message as 23 11 30 18 or MATH.

TAKE NOTE

✓

The calculations at the right require a calculator that has been programmed to compute these values. These TI-83/84 programs are available on our website at http://www.cengage.com/math/aufmann.

EXCURSION EXERCISES

You will need the RSA program for the TI-83/84 calculators that can be downloaded from our website at http://www.cengage.com/math/aufmann to do these exercises. For exercises 1 to 4, use the values for n, e, and d that we used above. Code a word using $A = 11, B = 12, C = 13, \dots,$ and $Z = 36$.

1. What is the ciphertext for the word CODE?

2. What is the ciphertext for the word HEART?

3. What is the word for a message that was received as 170, 607, 1268, 3408, 141?

4. What is the word for the message that was received as 1553, 3532, 2001, 3688, 957, 1941?

5. You can try this with your own numbers.

    a. Choose two prime numbers larger than 50. These are p and q.

    b. Find $n = p \cdot q$.

    c. Find $z = (p - 1)(q - 1)$.

    d. Choose a number e such that $1 < e < z$ and e and z have no common factors.

    e. Solve $de \equiv 1 \text{ mod } z$. You will need the program MODINV from our website at http://www.cengage.com/math/aufmann.

Your public encryption key is N^e mod $n \equiv M$. You can give this to a friend who could then send you a message. You would decode using your private decryption key M^d mod $n \equiv N$. To perform these calculations, use the RSA program above.

EXERCISE SET **8.2**   (Suggested Assignment: The Enhanced WebAssign Exercises and Exercises 63 and 65)

■ **ISBN Numbers** In Exercises 1 to 6, determine whether the given number is a valid ISBN.

1. 978-0-281-44268-5

2. 978-1-55690-182-9

3. 978-0-671-51983-4

4. 978-0-614-35945-2

5. 978-0-143-03943-3

6. 978-0-231-10324-1

■ **ISBN Numbers** In Exercises 7 to 14, determine the correct check digit for each ISBN.

7. *The King's Speech: How One Man Saved the British Monarchy* by Mark Logue and Peter Conradi; 978-1-4027-8676-?

8. *The Hobbit* by J.R.R. Tolkien; 978-0-345-27257-?

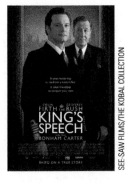

9. *Facebook for Dummies* by Leah Pearlman and Carolyn Abram; 978-0-470-87804-?

10. *Inside of a Dog: What Dogs See, Smell, and Know* by Alexandra Horowitz; 978-1-4165-8343-?

11. *Freakonomics: A Rogue Economist Explores the Hidden Side of Everything* by Steven D. Levitt and Stephen J. Dubner; 978-0-06-073133-?

12. *The Girl with the Dragon Tattoo* by Stieg Larsson; 978-0-307-47347-?

13. *Relativity: The Special and General Theory* by Albert Einstein; 978-0-517-88441-?

14. *Cleopatra: A Life* by Stacy Schiff; 978-0-316-00192-?

■ **UPC Codes** In Exercises 15 to 22, determine the correct check digit for the UPC.

15. 0-79893-46500-? (Organics Honey)

16. 6-53569-39973-? (Scrabble)

17. 7-14043-01126-? (Monopoly)

18. 0-32031-13439-? (Beethoven's 9th Symphony, DVD)

19. 8-85909-19432-? (16 GB iPad with WiFi)

20. 0-33317-20083-? (TI-84 Silver Edition calculator)

21. 0-41790-22106-? (Bertolli Classico olive oil)

22. 0-71818-02100-? (Guittard real semisweet chocolate chips)

■ **Money Orders** Some money orders have serial numbers that consist of a 10-digit number followed by a check digit. The check digit is chosen to be the sum of the first 10 digits mod 9. In Exercises 23 to 26, determine the check digit for the given money order serial number.

23. 0316615498-?

24. 5492877463-?

25. 1331497533-?

26. 3414793288-?

■ **Air Travel** Many printed airline tickets contain a 10-digit document number followed by a check digit. The check digit is chosen to be the sum of the first 10 digits mod 7. In Exercises 27 to 30, determine whether the given number is a valid document number.

27. 1182649758 2

28. 1260429984 4

29. 2026178914 5

30. 2373453867 6

■ **Credit Card Numbers** In Exercises 31 to 38, determine whether the given credit card number is a valid number.

31. 4417-5486-1785-6411

32. 5164-8295-1229-3674

33. 5591-4912-7644-1105

34. 6011-4988-1002-6487

35. 6011-0408-4977-3158

36. 4896-4198-8760-1970

37. 3715-548731-84466

38. 3401-714339-12041

■ **Encryption** In Exercises 39 to 44, encode the message by using a cyclical alphabetic encrypting code that shifts the message the stated number of positions.

39. 8 positions: THREE MUSKETEERS

40. 6 positions: FLY TONIGHT

41. 12 positions: IT'S A GIRL

42. 9 positions: MEET AT NOON

43. 3 positions: STICKS AND STONES

44. 15 positions: A STITCH IN TIME

■ **Decoding** In Exercises 45 to 48, use a cyclical alphabetic encrypting code that shifts the letters the stated number of positions to decode the encrypted message.

45. 18 positions: SYW GX WFDAYZLWFEWFL

46. 20 positions: CGUACHUNCIH LOFYM NBY QILFX

47. 15 positions: UGXTCS XC CTTS

48. 8 positions: VWJWLG QA XMZNMKB

■ **Decoding** In Exercises 49 to 52, use a cyclical alphabetic encrypting code to decode the encrypted message.

49. YVIBZM RDGG MJWDINJI

50. YBZAM HK YEBZAM

51. UDGIJCT RDDZXT

52. AOB HVS HCFDSRCSG

53. **Encryption** Julius Caesar supposedly used an encrypting code equivalent to the congruence $c \equiv (p + 3) \bmod 26$. Use the congruence to encrypt the message "men willingly believe what they wish."

54. **Decoding** Julius Caesar supposedly used an encrypting code equivalent to the congruence $c \equiv (p + 3) \bmod 26$. Use the congruence to decrypt the message WKHUH DUH QR DFFLGHQWV.

55. **Encryption** Use the encrypting congruence $c \equiv (3p + 2) \bmod 26$ to code the message TOWER OF LONDON.

56. **Encryption** Use the encrypting congruence $c \equiv (5p + 3) \bmod 26$ to code the message DAYLIGHT SAVINGS TIME.

57. **Encryption** Use the encrypting congruence $c \equiv (7p + 8) \bmod 26$ to code the message PARALLEL LINES.

58. **Encryption** Use the encrypting congruence $c \equiv (3p + 11) \bmod 26$ to code the message NONE SHALL PASS.

59. **Decoding** Decode the message LOFT JGMK LBS MNWMK that was encrypted using the congruence $c \equiv (3p + 4) \bmod 26$.

60. Decoding   Decode the message BTYW SCRBKN UCYN that was encrypted using the congruence $c \equiv (5p + 6) \bmod 26$.

61. Decoding   Decode the message SNUUHQ FM VFALHDZ that was encrypted using the congruence $c \equiv (5p + 9) \bmod 26$.

62. Decoding   Decode the message GBBZ OJQBWJ TBR GJHI that was encrypted using the congruence $c \equiv (7p + 1) \bmod 26$.

EXTENSIONS

Critical Thinking

63. **Money Orders**   Explain why the method used to determine the check digit of the money order serial numbers in Exercises 23 to 26 will not detect a transposition of digits among the first 10 digits.

64. **UPC Codes**   Explain why the transposition of adjacent digits a and b in a UPC will go undetected only when $|a - b| = 5$.

Explorations

65. Banking   When banks process an electronic funds transfer (EFT), they assign a nine-digit routing number, where the ninth digit is a check digit. The check digit is chosen such that when the nine digits are multiplied in turn by the numbers 3, 7, 1, 3, 7, 1, 3, 7, and 1, the sum of the resulting products is 0 mod 10. For instance, 123456780 is a valid routing number because

$$1(3) + 2(7) + 3(1) + 4(3) + 5(7) + 6(1)$$
$$+ 7(3) + 8(7) + 0(1) = 150$$

and $150 \equiv 0 \bmod 10$.

a. Compute the check digit for the routing number 72859372-?.

b. Verify that 584926105 is a valid routing number.

c. If a bank clerk inadvertantly types the routing number in part b as 584962105, will the computer be able to detect the error?

d. A bank employee accidentally transposed the 6 and the 1 in the routing number from part b and entered the number 584921605 into the computer system, but the computer did not detect an error. Explain why.

66. Bar Codes   The Codabar system is a bar code method developed in 1972 that is commonly used in libraries.

Each bar code contains a 14-digit number, where the 14th digit is a check digit. The various check digit schemes discussed in this section are designed to detect errors, but they cannot catch every possible error. The Codabar method is more thorough and will detect any single mistyped digit as well as any adjacent digits that have been transposed, with the exception of 9–0 and 0–9. To determine the check digit, label the first 13 digits a_1 through a_{13}. The check digit is given by

$$(-2a_1 - a_2 - 2a_3 - a_4 - 2a_5 - a_6 - 2a_7 - a_8$$
$$- 2a_9 - a_{10} - 2a_{11} - a_{12} - 2a_{13} - r) \bmod 10$$

where r is the number of digits from the digits a_1, a_3, a_5, a_7, a_9, a_{11}, and a_{13} that are greater than or equal to 5. For example, the bar code with the number 3194870254381-? has digits a_3, a_5, and a_9 that are greater than or equal to 5. Thus $r = 3$ and

$$-2(3) - 1 - 2(9) - 4 - 2(8) - 7 - 2(0)$$
$$- 2 - 2(5) - 4 - 2(3) - 8 - 2(1) - 3 = -87$$
$$\equiv 3 \bmod 10$$

The check digit is 3.

a. Find the check digit for the bar code number 1982273895263-?.

b. Find the check digit for the bar code number 6328293856384-?.

section **8.3**   **Introduction to Group Theory**

Introduction to Groups

An **algebraic system** is a set of elements along with one or more operations for combining the elements. The real numbers with the operations of addition and multiplication are an example of an algebraic system. Mathematicians classify this particular algebraic system as a *field*.

In the previous two sections of this chapter we discussed operations modulo *n*. Consider the set {0, 1, 2, 3, 4, 5} and addition modulo 6. The set of elements is {0, 1, 2, 3, 4, 5} and the operation is addition modulo 6. In this case there is only one operation. This is an example of an algebraic system called a *group*.

In addition to fields and groups, there are other types of algebraic systems. It is the properties of an algebraic system that distinguish it from another algebraic system. In this section, we will focus on groups.

HISTORICAL NOTE

Emmy Noether
(nōh'ŭh thǐr)
(1882–1935) made many contributions to mathematics and theoretical physics. One theorem she proved, sometimes referred to as Noether's theorem, relates to symmetries in physics. We will discuss symmetries of geometric figures later in this section.

Some of Noether's work led to several of the concepts of Einstein's general theory of relativity. In a letter to a colleague, Einstein described Noether's work as "penetrating, mathematical thinking."

▼ A Group

A **group** is a set of elements, with one operation, that satisfies the following four properties.

1. The set is closed with respect to the operation.

2. The operation satisfies the associative property.

3. There is an identity element.

4. Each element has an inverse.

Note from the above definition that a group is an algebraic system with *one* operation and that the operation must have certain characteristics. The first of these characteristics is that the set is *closed* with respect to the operation. **Closure** means that if any two elements are combined using the operation, the result must be an element of the set.

For example, the set {0, 1, 2, 3, 4, 5} with addition modulo 6 as the operation is closed. If we add two numbers of this set, modulo 6, the result is always a number of the set. For instance, $(3 + 5) \bmod 6 \equiv 2$ and $(1 + 3) \bmod 6 \equiv 4$.

As another example, consider the whole numbers {0, 1, 2, 3, 4, ...} with multiplication as the operation. If we multiply two whole numbers, the result is a whole number. For instance, $5 \cdot 9 = 45$ and $12 \cdot 15 = 180$. Thus the set of whole numbers is closed using multiplication as the operation. However, the set of whole numbers is not closed with division as the operation. For example, even though $36 \div 9 = 4$ (a whole number), $6 \div 4 = 1.5$, which is not a whole number. Therefore, the set of whole numbers is not closed with respect to division.

question **a.** Is the set of whole numbers closed with respect to addition?

b. Is the set of whole numbers closed with respect to subtraction?

The second requirement of a group is that the operation must satisfy the associative property. Recall that the associative property of addition states that $a + (b + c) = (a + b) + c$. Addition modulo 6 is an associative operation. For instance, if we use the symbol $\triangledown$ to represent addition modulo 6, then

$$2 \triangledown (5 \triangledown 3) = 2 \triangledown (8 \bmod 6) = 2 \triangledown 2 = 4$$

and

$$(2 \triangledown 5) \triangledown 3 = (7 \bmod 6) \triangledown 3 = 1 \triangledown 3 = 4$$

Thus $2 \triangledown (5 \triangledown 3) = (2 \triangledown 5) \triangledown 3$.

Although there are some operations that do not satisfy the associative property, we will assume that an operation used in this section is associative.

---

answer **a.** Yes. The sum of any two whole numbers is a whole number. **b.** No. The difference between any two whole numbers is not always a whole number. For instance, $5 - 8 = -3$, but -3 is not a whole number.

The third requirement of a group is that the set must contain an *identity element*. An **identity element** is an element that, when combined with a second element using the group's operation, always returns the second element. As an illustration, if zero is added to a number, there is no change: $6 + 0 = 6$ and $0 + 1.5 = 1.5$. The number 0 is called an *additive identity*. Similarly, if we multiply any number by 1, there is no change: $\frac{1}{2} \cdot 1 = \frac{1}{2}$ and $1 \cdot 5 = 5$. The number 1 is called a *multiplicative identity*. For the set $\{0, 1, 2, 3, 4, 5\}$ with addition modulo 6 as the operation, the identity element is 0. As we will soon see, an identity element does not always have to be zero or one.

The numbers 3 and -3 are called *additive inverses*. Adding these two numbers results in the additive identity: $3 + (-3) = 0$. The numbers $\frac{2}{3}$ and $\frac{3}{2}$ are called *reciprocals* or *multiplicative inverses*. Multiplying these two numbers results in the multiplicative identity: $\frac{2}{3} \cdot \frac{3}{2} = 1$. The last requirement of a group is that each element must have an *inverse*. This is a little more difficult to see for the set $\{0, 1, 2, 3, 4, 5\}$ with addition modulo 6 as the operation. However, using addition modulo 6, we have

$$(0 + 0) \bmod 6 \equiv 0 \qquad (2 + 4) \bmod 6 \equiv 0$$

$$(1 + 5) \bmod 6 \equiv 0 \qquad (3 + 3) \bmod 6 \equiv 0$$

The above equations show that 0 is its own inverse, 1 and 5 are inverses, 2 and 4 are inverses, and 3 is its own inverse. Therefore, every element of this set has an inverse.

We have shown that the set $\{0, 1, 2, 3, 4, 5\}$ with addition modulo 6 as the operation is a group because it satisfies the four conditions stated in the definition of a group (see page 496).

▼ example 1 Verify the Properties of a Group

Show that the integers with addition as the operation form a group.

Solution
We must show that the four properties of a group are satisfied.

1. The sum of two integers is always an integer. For instance,

$$-12 + 5 = -7 \qquad \text{and} \qquad -14 + (-21) = -35$$

Therefore, the integers are closed with respect to addition.

2. The associative property of addition holds true for the integers.

3. The identity element is 0, which is an integer. Therefore, the integers have an identity element for addition.

4. Each element has an inverse. If a is an integer, then $-a$ is the inverse of a.

Because each of the four conditions of a group is satisfied, the integers with addition as the operation form a group.

▼ check your progress 1 Does the set $\{1, 2, 3\}$ with operation multiplication modulo 4 form a group?

Solution *See page S29.*

Recall that the **commutative property** for an operation states that the order in which two elements are combined does not affect the result. For each of the groups discussed thus far, the operation has satisfied the commutative property. For example, the group $\{0, 1, 2, 3, 4, 5\}$ with addition modulo 6 satisfies the commutative property, since, for instance, $2 + 5 = 5 + 2$. Groups in which the operation satisfies the commutative property are called **commutative groups** or **abelian groups**, after Niels Abel. The type of group we will look at next is an example of a *nonabelian group*. A **nonabelian group** is a group whose operation does not satisfy the commutative property.

Symmetry Groups

The concept of a group is very general. The elements that make up a group do not have to be numbers, and the operation does not have to be addition or multiplication. Also, a group does not have to satisfy the commutative property. We will now look at another group, called a *symmetry group,* an extension of which plays an important role in the study of atomic reactions. Symmetry groups are based on *regular polygons*. A **regular polygon** is a polygon all of whose sides have the same length and all of whose angles have the same measure.

Consider two equilateral triangles, one placed inside the other, with their vertices numbered clockwise from 1 to 3. The larger triangle is the *reference triangle*. If we pick up the smaller triangle, there are several different ways in which we can set it back in its place. Each possible positioning of the inner triangle will be an element of a group. For instance, we can pick up the triangle and replace it exactly as we found it. We will call this position *I*; it will represent no change in position.

Now pick up the smaller triangle, rotate it 120° clockwise, and set it down again on top of the reference triangle. The result is shown below.

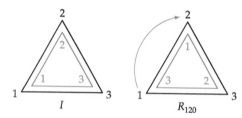

Note that the vertex originally at vertex 1 of the reference triangle is now at vertex 2, 2 is now at 3, and 3 is now at 1. We call the rotation of the triangle 120° clockwise R_{120}.

Now return the smaller triangle to its original position, where the numbers on the vertices of the triangles coincide. Consider a 240° clockwise rotation of the smaller triangle. The result of this rotation is shown below.

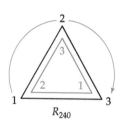

Note that the vertex originally at vertex 1 of the reference triangle is now at vertex 3, 2 is now at 1, and 3 is now at 2. We call the rotation of the triangle 240° clockwise R_{240}.

If the original triangle were rotated 360°, there would be no apparent change. The vertex at 1 would return to 1, the vertex at 2 would return to 2, and the vertex at 3 would return to 3. Because this rotation does not produce a new arrangement of the vertices, we consider it the same as the element we named *I*.

If we rotate the inner triangle *counterclockwise* 120°, the effect is the same as rotating it 240° clockwise. This rotation does not produce a different arrangement of the vertices. Similarly, a counterclockwise rotation of 240° is the same as a clockwise rotation of 120°.

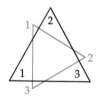

A line of symmetry for a figure is one such that if the figure were folded along that line, the two halves of the figure would match up. See the diagram below.

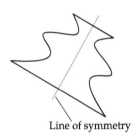

Line of symmetry

In addition to rotating the triangle clockwise 120° and 240° as we did above, we could rotate the triangle about a line of symmetry that goes through a vertex before setting the triangle back down. Because there are three vertices, there are three possible results. These are shown below.

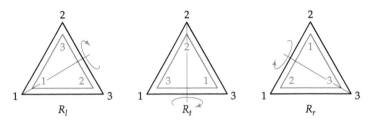

For rotation R_l, the bottom *left* vertex does not change, but vertices 2 and 3 are interchanged. If we rotate the triangle about the line of symmetry through the *top* vertex, rotation R_t, vertex 2 does not change, but vertices 1 and 3 are interchanged. For rotation R_r, the bottom *right* vertex does not change, but vertices 1 and 2 are interchanged.

The six positions of the vertices we have seen thus far are the only possibilities: I (the triangle without any rotation), R_{120}, R_{240}, R_l, R_t, and R_r. These positions are shown below (without the reference triangle).

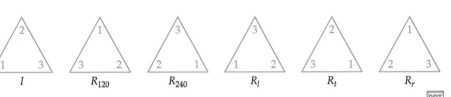

As we will see, these rotations are elements that form a group.

A group must have an operation, a method by which two elements of the group can be combined to form a third element that must also be a member of the group. (Recall that a group operation must be closed.) The operation we will use is called "followed by" and is symbolized by Δ. Next we show an example of how this operation works.

Consider $R_l \Delta R_{240}$. This means we rotate the original triangle, labeled as I, about the line of symmetry through vertex 1 "followed by" (without returning to the original position) a clockwise rotation of 240°. The result is one of the elements of the group, R_t. This operation is shown below.

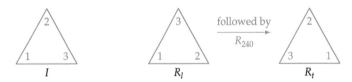

Now try reversing the operation, and consider $R_{240} \Delta R_l$. This means we rotate the original triangle, I, clockwise 240° "followed by" (without returning to the original position) a rotation about the axis of symmetry through the bottom left vertex. Note that the result is an element of the group, namely R_r, as shown below.

From these two examples, $R_l \Delta R_{240} = R_t$ and $R_{240} \Delta R_l = R_r$. Therefore, $R_l \Delta R_{240} \neq R_{240} \Delta_l$, which means the operation "followed by" is not commutative.

▼ **example** **2**   **Perform an Operation of a Symmetry Group**

Find $R_l \Delta R_r$.

Solution
Rotate the original triangle, I, about the line of symmetry through the bottom left vertex, followed by a rotation about the line of symmetry through the bottom right vertex.

I

R_l

followed by
R_r

R_{120}

From the diagram, $R_l \Delta R_r = R_{120}$.

▼ **check your progress** **2**   Find $R_r \Delta R_{240}$.

Solution   *See page S29.*

INSTRUCTOR NOTE

At this point, students would be prepared to try the Excursion on page 503, where they can extend the ideas of a symmetry group to an infinitely large pattern in what are known as *wallpaper groups*.

We have stated that the elements R_{120}, R_{240}, R_l, R_t, R_r, and I, with the operation "followed by," form a group. However, we have not demonstrated this fact, so we will do so now.

As we have seen, the set is closed with respect to the operation Δ. To show that the operation Δ is associative, think about the meaning of $x\Delta(y\Delta z)$ and $(x\Delta y)\Delta z$, where x, y, and z are elements of the group. $x\Delta(y\Delta z)$ means x, followed by the result of y followed by z; $(x\Delta y)\Delta z$ means x followed by y, followed by z. For instance, $R_t\Delta(R_l\Delta R_r) = R_t\Delta R_{120} = R_l$ and $(R_t\Delta R_l)\Delta R_r = R_{120}\Delta R_r = R_l$. All the remaining combinations of elements can be verified similarly.

The identity element is I and can be thought of as "no rotation." To see that each element has an inverse, verify the following:

$$R_{120}\Delta R_{240} = I \qquad R_l\Delta R_l = I \qquad R_t\Delta R_t = I \qquad R_r\Delta R_r = I$$

Your work should show that every element has an inverse. (The inverse of I is I.) Thus the four conditions of a group are satisfied.

To determine the outcome of the "followed by" operation on two elements of the group, we drew triangles and then rotated them as directed by the type of rotation. From a mathematical point of view, it would be nice to have a symbolic (rather than geometric) way of determining the outcome. We can do this by creating a mathematical object that describes the geometric object.

Note that any rotation of the triangle changes the position of the vertices. For R_{120}, the vertex originally at position 1 moved to position 2, the vertex at 2 moved to 3, and the vertex at 3 moved to 1. We can represent this as

$$R_{120} = \begin{pmatrix} 1 & 2 & 3 \\ 2 & 3 & 1 \end{pmatrix}$$

Similarly, for R_t, the vertex originally at position 1 moved to position 3, the vertex at 2 remained at 2, and the vertex at 3 moved to 1. This can be represented as

$$R_t = \begin{pmatrix} 1 & 2 & 3 \\ 3 & 2 & 1 \end{pmatrix}$$

The remaining four elements can be represented as

$$I = \begin{pmatrix} 1 & 2 & 3 \\ 1 & 2 & 3 \end{pmatrix}, R_{240} = \begin{pmatrix} 1 & 2 & 3 \\ 3 & 1 & 2 \end{pmatrix}, R_l = \begin{pmatrix} 1 & 2 & 3 \\ 1 & 3 & 2 \end{pmatrix}, R_r = \begin{pmatrix} 1 & 2 & 3 \\ 2 & 1 & 3 \end{pmatrix}$$

▼ **example** **3**   **Symbolic Notation**

Use symbolic notation to find $R_{120}\Delta R_t$.

Solution

To find the result of $R_{120}\Delta R_t$, we follow the movement of each vertex. The path that vertex 2 follows is highlighted.

$$R_{120}\Delta R_t = \begin{pmatrix} 1 & 2 & 3 \\ 2 & 3 & 1 \end{pmatrix} \Delta \begin{pmatrix} 1 & 2 & 3 \\ 3 & 2 & 1 \end{pmatrix}$$

- $1 \to 2 \to 2$. Thus $1 \to 2$.
- $2 \to 3 \to 1$. Thus $2 \to 1$.
- $3 \to 1 \to 3$. Thus $3 \to 3$.

$$= \begin{pmatrix} 1 & 2 & 3 \\ 2 & 1 & 3 \end{pmatrix} = R_r$$

$R_{120}\Delta R_t = R_r$

▼ **check your progress** **3**   Use symbolic notation to find $R_r \Delta R_{240}$.

Solution   *See page S30.*

MATHMATTERS   A Group of Subatomic Particles

In 1961, Murray Gell-Mann introduced what he called the *eightfold way* (after the Buddha's Eightfold Path to enlightenment and bliss) to categorize subatomic particles into eight different families, where each family was based on a certain symmetry group. Because each family was a group, each family had to satisfy the closure requirement. For that to happen, Gell-Mann's theory indicated that certain particles that had not yet been discovered must exist. Scientists began looking for these particles. One such particle, discovered in 1964, is called omega-minus. Its existence was predicted purely on the basis of the known properties of groups.

Permutation Groups

The triangular symmetry group discussed previously is an example of a special kind of group called a *permutation group*. A **permutation** is a rearrangement of objects. For instance, if we start with the arrangement of objects [♦ ♥ ♣], then one permutation of these objects is the rearrangement [♥ ♣ ♦]. If we consider each permutation of these objects as an element of a set, then the set of all possible permutations forms a group. The elements of the group are not numbers or the objects themselves, but rather the different permutations of the objects that are possible.

If we start with the numbers 1 2 3, we can represent permutations of the numbers using the same symbolic method that we used for the triangular symmetry group. For example, the permutation that rearranges 1 2 3 to 2 3 1 can be written

$$\begin{pmatrix} 1 & 2 & 3 \\ 2 & 3 & 1 \end{pmatrix}$$

which means that 1 is replaced by 2, 2 is replaced by 3, and 3 is replaced by 1.

You may have played with a "15 puzzle" like the one pictured above. There is one empty slot that allows the numbered tiles to move around the board. The goal is to arrange the tiles in numerical order. The puzzle can be analyzed in the context of group theory, where each element of a group is a permutation of the tiles. If one were to remove two adjacent tiles and replace them with their positions switched, group theory could be used to prove that the puzzle becomes impossible to solve.

There are only six distinct permutations of the numbers 1 2 3. We list and label them below. The identity element is named I.

$$I = \begin{pmatrix} 1 & 2 & 3 \\ 1 & 2 & 3 \end{pmatrix}, A = \begin{pmatrix} 1 & 2 & 3 \\ 2 & 3 & 1 \end{pmatrix}, B = \begin{pmatrix} 1 & 2 & 3 \\ 3 & 1 & 2 \end{pmatrix}$$

$$C = \begin{pmatrix} 1 & 2 & 3 \\ 1 & 3 & 2 \end{pmatrix}, D = \begin{pmatrix} 1 & 2 & 3 \\ 3 & 2 & 1 \end{pmatrix}, E = \begin{pmatrix} 1 & 2 & 3 \\ 2 & 1 & 3 \end{pmatrix}$$

The operation for the group is "followed by," which we will again denote by the symbol Δ. One can verify that the six elements along with this operation do indeed form a group.

▼ **example 4**   **Perform an Operation in a Permutation Group**

Find $B\Delta C$, where B and C are elements of the permutation group defined above.

Solution

$B\Delta C = \begin{pmatrix} 1 & 2 & 3 \\ 3 & 1 & 2 \end{pmatrix}\Delta\begin{pmatrix} 1 & 2 & 3 \\ 1 & 3 & 2 \end{pmatrix}$. In the first permutation, 1 is replaced by 3. In

the second permutation, 3 is replaced by 2. When we combine these two actions, 1 is ultimately replaced by 2. Similarly, in the first permutation 2 is replaced by 1, which remains as 1 in the second permutation. 3 is replaced by 2, which is in turn replaced by 3. When these actions are combined, 3 remains as 3. The result of the operation is

$\begin{pmatrix} 1 & 2 & 3 \\ 2 & 1 & 3 \end{pmatrix}$. Thus we see that $B\Delta C = E$.

▼ **check your progress 4**   Compute $E\Delta B$, where E and B are elements of the permutation group defined previously.

Solution   *See page S30.*

▼ **example 5**   **Inverse Element of a Permutation Group**

One of the requirements of a group is that each element must have an inverse. Find the inverse of the element B of the permutation group defined previously.

Solution

Because $B = \begin{pmatrix} 1 & 2 & 3 \\ 3 & 1 & 2 \end{pmatrix}$ replaces 1 with 3, 2 with 1, and 3 with 2, its inverse must

reverse these replacements. Thus we need to replace 3 with 1, 1 with 2, and 2 with 3.

In a nonabelian group, the order in which the two elements are combined with the group operation can affect the result. For an element to be the inverse of another element, combining the elements must give the identity element regardless of the order used.

The inverse is the element $\begin{pmatrix} 1 & 2 & 3 \\ 2 & 3 & 1 \end{pmatrix} = A$. To verify,

$$B\Delta A = \begin{pmatrix} 1 & 2 & 3 \\ 3 & 1 & 2 \end{pmatrix}\Delta\begin{pmatrix} 1 & 2 & 3 \\ 2 & 3 & 1 \end{pmatrix} = \begin{pmatrix} 1 & 2 & 3 \\ 1 & 2 & 3 \end{pmatrix} \text{ and }$$

$$A\Delta B = \begin{pmatrix} 1 & 2 & 3 \\ 2 & 3 & 1 \end{pmatrix}\Delta\begin{pmatrix} 1 & 2 & 3 \\ 3 & 1 & 2 \end{pmatrix} = \begin{pmatrix} 1 & 2 & 3 \\ 1 & 2 & 3 \end{pmatrix}.$$

▼ **check your progress 5**   Find the inverse of the element D of the permutation group defined previously.

Solution   *See page S30.*

If we consider a group with more than three numbers or objects, the number of different permutations increases dramatically. The permutation group of the five numbers 1 2 3 4 5 has 120 elements, and there are over 40,000 permutations of the numbers 1 2 3 4 5 6 7 8. In Exercise Set 8.3, you are asked to examine the permutations of 1 2 3 4.

EXCURSION

Photolibrary

Wallpaper Groups

The symmetry group we studied based on an equilateral triangle involved rotations of the triangle that retained the position of the triangle within a reference triangle. We will now look at symmetries of an infinitely large object.

Imagine that the pattern shown in Figure 8.3 below continues forever in every direction. You can think of it as an infinitely large sheet of wallpaper. Overlaying this sheet is a transparent sheet with the same pattern.

Visualize shifting the transparent sheet of wallpaper upward until the printed cats realign in what appears to be the same position. This is called a **translation**. Any such translation, in which we can shift the paper in one direction until the pattern aligns with its original appearance, is an element of a group. There are several directions in which we can shift the wallpaper (up, down, and diagonally) and an infinite number of distances we can shift the wallpaper. So the group has an infinite number of elements. Several elements are shown in Figure 8.4. Not shifting the paper at all is also considered an element; we will call it *I*, because it is the identity element for the group. As with the symmetry groups, the operation is "followed by," denoted by Δ.

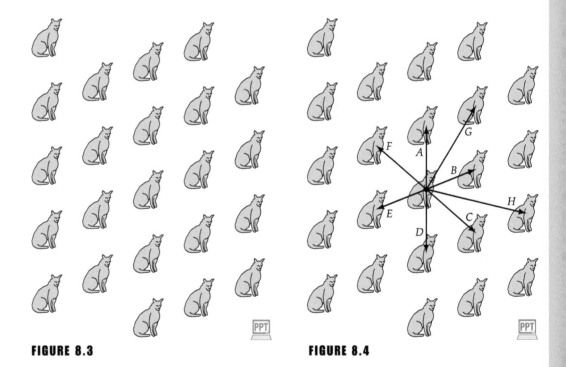

FIGURE 8.3   **FIGURE 8.4**

Some wallpaper patterns have rotational symmetries in addition to translation symmetries. The pattern in Figure 8.5 on the following page can be rotated 180° about the "×" and the wallpaper will appear to be in the exact same position. Thus the wallpaper group derived from this pattern will include this rotation as an element. Excursion Exercise 4 asks you to identify other rotational elements. Note that the group also includes translational elements.

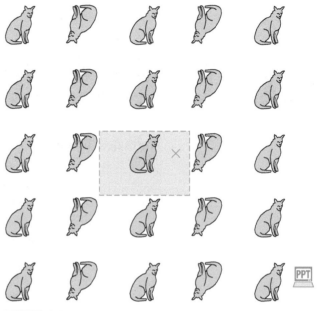

FIGURE 8.5

We can also consider mirror reflections. The wallpaper shown below can be reflected across the dashed line and its appearance will be unchanged. (There are many other vertical axes that could also be used.) So if we consider the group derived from symmetries of this wallpaper, one element will be the reflection of the entire sheet across this axis.

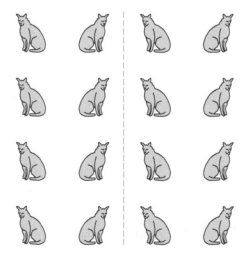

There is a fourth type of symmetry, called a *glide reflection*, that is a combination of a reflection and a shift. Mathematicians have determined that these are the only four types of symmetries possible and that there are only 17 distinct wallpaper groups. Of course, the visual patterns you see on actual wallpaper can appear quite varied, but if we formed groups from their symmetries, each would have to be one of the 17 known groups. (If you wish to see all 17 patterns, an excellent resource is http://www.clarku.edu/~djoyce/wallpaper/.)

EXCURSION EXERCISES

1. Explain why the properties of a group are satisfied by the symmetries of the wallpaper pattern shown in Figure 8.3. (You may assume that the operation is associative.)

2. Verify that in Figure 8.4, $A\Delta B = G$.

3. In Figure 8.6, determine where the black cat will be after applying the following translations illustrated in Figure 8.4.

   a. $F\Delta(D\Delta C)$ **b.** $(C\Delta D)\Delta G$ **c.** $(A\Delta H)\Delta(E\Delta D)$

4. In Figure 8.5, there are three other locations within the dashed box about which the wallpaper can be rotated and align to its original appearance. Find the rotational center points.

5. Consider the wallpaper group formed from symmetries of the pattern shown below in Figure 8.7.

   a. Identify three different elements of the group derived from translations.

   b. Find two different center points about which a 180° rotation is an element of the group.

   c. Find two different center points about which a 120° rotation is an element of the group.

   d. Identify three distinct lines that could serve as axes of reflection for elements of the group.

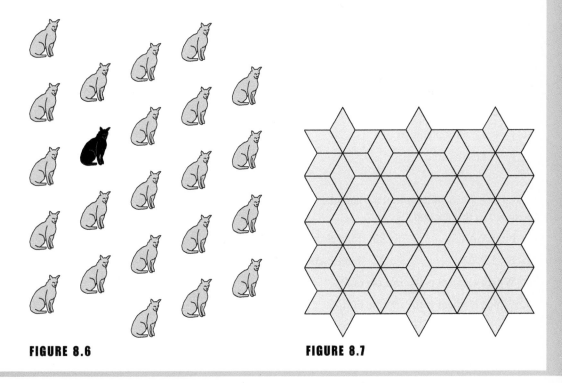

FIGURE 8.6   **FIGURE 8.7**

EXERCISE SET **8.3**   (Suggested Assignment: The Enhanced WebAssign Exercises and Exercise 59)

1. Is the set $\{-1, 1\}$ closed with respect to the operation

   a. multiplication?

   b. addition?

2. Is the set of all even integers closed with respect to

   a. multiplication?

   b. addition?

3. Is the set of all odd integers closed with respect to
   a. multiplication?
   b. addition?

4. Is the set of all negative real numbers closed with respect to
   a. addition?   **b.** subtraction?
   c. multiplication?   **d.** division?

■ In Exercises 5 to 18, determine whether the set forms a group with respect to the given operation. (You may assume the operation is associative.) If the set does not form a group, determine which properties fail.

5. The even integers; addition

6. The even integers; multiplication

7. The real numbers; addition

8. The real numbers; division

9. The real numbers; multiplication

10. The real numbers *except* 0; multiplication

11. The rational numbers; addition
    (Recall that a rational number is any number that can be expressed in the form p/q, where p and q are integers, $q \neq 0$.)

12. The positive rational numbers; multiplication

13. {0, 1, 2, 3}; addition modulo 4

14. {0, 1, 2, 3, 4}; addition modulo 6

15. {0, 1, 2, 3}; multiplication modulo 4

16. {1, 2, 3, 4}; multiplication modulo 5

17. {−1, 1}; multiplication

18. {−1, 1}; division

■ Exercises 19 to 24 refer to the triangular symmetry group discussed in this section.

19. Find $R_t \Delta R_{120}$.   **20.** Find $R_{240} \Delta R_r$.

21. Find $R_r \Delta R_t$.   **22.** Find $R_t \Delta R_l$.

23. Which element is the inverse of R_{240}?

24. Which element is the inverse of R_l?

■ We can form another symmetry group by using rotations of a square rather than of a triangle. Start with a square with corners labeled 1 through 4, placed in a reference square.

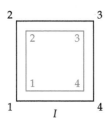

The current position of the inner square is labeled I. We can rotate the inner square clockwise 90° to give an element of the group, R_{90}. Similarly, we can rotate the square 180° or 270° to obtain the elements R_{180} and R_{270}. In addition, we can rotate the square about any of the lines of symmetry shown below.

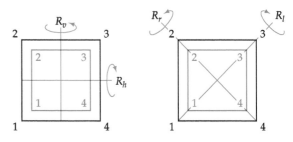

As before, the operation is "followed by," denoted by Δ. Exercises 25 to 34 refer to this symmetry group of the square.

25. Find $R_{90} \Delta R_v$.   **26.** Find $R_r \Delta R_{180}$.

27. Find $R_h \Delta R_{270}$.   **28.** Find $R_l \Delta R_h$.

29. List each of the eight elements of the group in symbolic notation. For instance, $I = \begin{pmatrix} 1 & 2 & 3 & 4 \\ 1 & 2 & 3 & 4 \end{pmatrix}$.

30. Use symbolic notation to find $R_v \Delta R_l$.

31. Use symbolic notation to find $R_h \Delta R_{90}$.

32. Use symbolic notation to find $R_{180} \Delta R_r$.

33. Find the inverse of the element R_{270}.

34. Find the inverse of the element R_v.

■ Exercises 35 to 42 refer to the group of permutations of the numbers 1 2 3 discussed in this section.

35. Find $A \Delta B$.   **36.** Find $B \Delta A$.

37. Find $B \Delta E$.   **38.** Find $D \Delta C$.

39. Find $C \Delta A$.   **40.** Find $B \Delta B$.

41. Find the inverse of the element A.

42. Find the inverse of the element E.

■ We can form a new permutation group by considering all permutations of the numbers 1 2 3 4. Exercises 43 to 50 refer to this group.

43. List all 24 elements of the group. Use symbolic notation, such as $I = \begin{pmatrix} 1 & 2 & 3 & 4 \\ 1 & 2 & 3 & 4 \end{pmatrix}$.

44. Find $\begin{pmatrix} 1 & 2 & 3 & 4 \\ 4 & 2 & 1 & 3 \end{pmatrix} \Delta \begin{pmatrix} 1 & 2 & 3 & 4 \\ 2 & 1 & 4 & 3 \end{pmatrix}$.

45. Find $\begin{pmatrix} 1 & 2 & 3 & 4 \\ 1 & 4 & 2 & 3 \end{pmatrix} \Delta \begin{pmatrix} 1 & 2 & 3 & 4 \\ 2 & 3 & 4 & 1 \end{pmatrix}$.

46. Find $\begin{pmatrix} 1 & 2 & 3 & 4 \\ 3 & 2 & 1 & 4 \end{pmatrix} \Delta \begin{pmatrix} 1 & 2 & 3 & 4 \\ 3 & 4 & 2 & 1 \end{pmatrix}$.

47. Find the inverse of $\begin{pmatrix} 1 & 2 & 3 & 4 \\ 2 & 4 & 1 & 3 \end{pmatrix}$.

48. Find the inverse of $\begin{pmatrix} 1 & 2 & 3 & 4 \\ 3 & 4 & 1 & 2 \end{pmatrix}$.

49. Find the inverse of $\begin{pmatrix} 1 & 2 & 3 & 4 \\ 4 & 3 & 2 & 1 \end{pmatrix}$.

50. Find the inverse of $\begin{pmatrix} 1 & 2 & 3 & 4 \\ 1 & 3 & 4 & 2 \end{pmatrix}$.

■ In Exercises 51 to 58, consider the set of elements $\{a, b, c, d\}$ with the operation ∇ as given in the table below.

∇	a	b	c	d
a	d	a	b	c
b	a	b	c	d
c	b	c	d	a
d	c	d	a	b

For example, to compute $c\nabla d$, find c in the left column and align it with d in the top row. The result is a.

51. Find $a\nabla a$. **52.** Find $d\nabla c$.

53. Find $c\nabla b$. **54.** Find $a\nabla d$.

55. Verify that the set with the operation ∇ is a group. (You may assume that the operation is associative.)

56. Which element of the group is the identity element?

57. Is the group commutative?

58. Identify the inverse of each element.

EXTENSIONS

Critical Thinking

59. a. Verify that the integers $\{1, 2, 3, 4, 5, 6\}$ with the operation multiplication modulo 7 form a group.

b. Verify that the integers $\{1, 2, 3, 4, 5\}$ with the operation multiplication modulo 6 do not form a group.

c. For which values of n are the integers $\{1, 2, ..., n - 1\}$ a group under the operation multiplication modulo n?

60. The set of numbers $\{2^k\}$, where k is any integer, form a group with multiplication as the operation.

a. Which element of the group is the identity element?

b. Verify that the set is closed with respect to multiplication.

c. Which element is the inverse of 2^8?

d. Which element is the inverse of $\frac{1}{32}$?

e. In general, what is the inverse of 2^k?

61. Consider the set $G = \{1, 2, 3, 4, 5\}$ and the operation $\oplus$ defined by $a \oplus b = a + b - 1 - r$, where $r = 5$ if $a + b \geq 7$ and $r = 0$ if $a + b \leq 6$. It can be shown that G forms a group with this operation.

a. Make an operation table (similar to the one given for Exercises 51 to 58) for the group G and the operation $\oplus$.

b. Is the group commutative?

c. Which element is the identity?

d. Find the inverse of each element of the group.

Explorations

62. The **quaternion group** is a famous noncommutative group; we will define the group as the set $Q = \{e, r, s, t, u, v, w, x\}$ and the operation ∇ that is defined by the table below.

∇	e	r	s	t	u	v	w	x
e	e	r	s	t	u	v	w	x
r	r	u	t	w	v	e	x	s
s	s	x	u	r	w	t	e	v
t	t	s	v	u	x	w	r	e
u	u	v	w	x	e	r	s	t
v	v	e	x	s	r	u	t	w
w	w	t	e	v	s	x	u	r
x	x	w	r	e	t	s	v	u

a. Show that the group is noncommutative.

b. Which element is the identity element?

c. Find the inverse of every element in the group.

d. Show that $r^4 = e$, where $r^4 = r\nabla r\nabla r\nabla r$.

e. Show that the set $\{e, r, u, v\}$, which is a subset of Q, with the operation ∇ satisfies the properties of a group on its own. This is called a **subgroup** of the group.

f. Find a subgroup that has only two elements.

CHAPTER 8 | SUMMARY

The following table summarizes essential concepts in this chapter. The references given in the right-hand column list Examples and Exercises that can be used to test your understanding of a concept.

8.1 Modular Arithmetic

Modulo n Two integers a and b are said to be congruent modulo n, where n is a natural number, if $\dfrac{a-b}{n}$ is an integer. In this case, we write $a \equiv b \bmod n$. The number n is called the modulus. The statement $a \equiv b \bmod n$ is called a congruence.	See **Example 2** on page 475, and then try Exercises 11 and 12 on page 510.
Arithmetic Modulo n After performing the operation as usual, divide the result by n; the answer is the remainder.	See **Examples 4, 5, and 7** on pages 476 to 478, and then try Exercises 15, 18, and 19 on page 510.
Solving Congruence Equations Individually check each whole number less than the modulus to see if it satisfies the congruence. Congruence equations may have more than one solution, or none at all.	See **Example 8** on page 479, and then try Exercises 26 and 27 on page 510.
Finding Inverses in Modular Arithmetic To find the additive inverse of a number modulo n, subtract the number from n. To find the multiplicative inverse (if it exists) of a number a, solve the congruence equation $ax \equiv 1 \bmod n$ for x.	See **Example 9 and 10** on page 480, and then try Exercise 29 on page 510.

8.2 Applications of Modular Arithmetic

ISBN Check Digit An ISBN is a 13-digit number used to identify a book. The 13th digit is a check digit. If we label the first digit of an ISBN d_1, the second digit d_2, and so on to the 13th digit as d_{13}, then the check digit is chosen to satisfy the following congruence. $d_{13} = 10 - (d_1 + 3d_2 + d_3 + 3d_4 + d_5 + 3d_6 + d_7 +$ $\qquad\qquad\qquad 3d_8 + d_9 + 3d_{10} + d_{11} + 3d_{12}) \bmod 10$	See **Example 1** on page 485, and then try Exercise 33 on page 510.
UPC Check Digit A UPC is a 12-digit number that used to identify a product such as a DVD, game, or grocery item. If we label the twelve digits of the UPC as $d_1, d_2, \ldots, d_{12}$, then the UPC check digit d_{12} is given by $d_{12} \equiv 10 - (3d_1 + d_2 + 3d_3 + d_4 + 3d_5 + d_6 +$ $\qquad\qquad\qquad 3d_7 + d_8 + 3d_9 + d_{10} + 3d_{11}) \bmod 10$	See **Example 2** on page 486, and then try Exercise 35 on page 510.
Luhn Algorithm for Valid Credit Card Numbers The last digit of the credit card number is a check digit. Beginning with the next-to-last digit and reading from right to left, double every other digit. Treat any resulting two-digit number as two individual digits. Find the sum of the revised set of digits; the check digit is chosen such that the sum is congruent to 0 mod 10.	See **Example 3** on page 487, and then try Exercises 38 and 39 on page 510.
Cyclical Alphabetic Encryption A message can be encrypted by shifting each letter p of the plaintext message m positions through the alphabet. The encoded letter c satisfies the congruence $c \equiv (p + m) \bmod 26$, where p and c are the numerical equivalents of the plaintext and ciphertext letters, respectively. To decode the message, use the congruence $p \equiv (c + n) \bmod 26$, where $n = 26 - m$.	See **Example 4** on page 489, and then try Exercise 43 on page 511.

The $c \equiv (ap + n)$ mod 26 Encryption Scheme Choose a natural number a that does not have any common factors with 26 and a natural number m. A plaintext letter p is encoded to a ciphertext letter c using the congruence $c \equiv (ap + n)$ mod 26. To decode the message, solve the congruence equation $c \equiv (ap + n)$ mod 26 for p.	See **Example 5** on page 490, and then try Exercise 46 on page 511.

8.3 Introduction to Group Theory

Properties of a Group A group is a set of objects with one operation that satisfies the following four properties. **1.** The set is closed with respect to the operation. **2.** The operation satisfies the associative property. **3.** There is an identity element. **4.** Each element has an inverse.	See **Example 1** on page 497, and then try Exercises 47, 49, and 50 on page 511.
Symmetry Group of an Equilateral Triangle The group has six elements, which consist of the different possible rotations that return the triangle to its reference triangle: I R_{120} R_{240} R_l R_t R_r The group operation is "followed by," denoted by Δ. The operation is noncommutative.	See **Example 2** on page 500, and then try Exercise 51 on page 511.
Symbolic Notation for the Symmetry Group of an Equilateral Triangle The six elements of the group are: $I = \begin{pmatrix} 1 & 2 & 3 \\ 1 & 2 & 3 \end{pmatrix}$ $\qquad R_{120} = \begin{pmatrix} 1 & 2 & 3 \\ 2 & 3 & 1 \end{pmatrix}$ $R_{240} = \begin{pmatrix} 1 & 2 & 3 \\ 3 & 1 & 2 \end{pmatrix}$ $\qquad R_l = \begin{pmatrix} 1 & 2 & 3 \\ 1 & 3 & 2 \end{pmatrix}$ $R_t = \begin{pmatrix} 1 & 2 & 3 \\ 3 & 2 & 1 \end{pmatrix}$ $\qquad R_r = \begin{pmatrix} 1 & 2 & 3 \\ 2 & 1 & 3 \end{pmatrix}$ where, for instance, $\begin{pmatrix} 1 & 2 & 3 \\ 2 & 3 & 1 \end{pmatrix}$ signifies that the vertex of the equilateral triangle originally at vertex 1 of the reference triangle moves to vertex 2, the vertex at 2 moves to 3, and the vertex at 3 moves to 1.	See **Example 3** on page 501, and then try Exercises 53 and 54 on page 511.

Permutation Group of the Numbers 1 2 3 The group consists of the six permutations of the numbers 1 2 3:

$$I = \begin{pmatrix} 1 & 2 & 3 \\ 1 & 2 & 3 \end{pmatrix} \qquad A = \begin{pmatrix} 1 & 2 & 3 \\ 2 & 3 & 1 \end{pmatrix}$$

$$B = \begin{pmatrix} 1 & 2 & 3 \\ 3 & 1 & 2 \end{pmatrix} \qquad C = \begin{pmatrix} 1 & 2 & 3 \\ 1 & 3 & 2 \end{pmatrix}$$

$$D = \begin{pmatrix} 1 & 2 & 3 \\ 3 & 2 & 1 \end{pmatrix} \qquad E = \begin{pmatrix} 1 & 2 & 3 \\ 2 & 1 & 3 \end{pmatrix}$$

The group operations is "followed by," denoted by Δ.

See **Examples 4 and 5** on page 502, and then try Exercises 55, 56, and 57 on page 511.

CHAPTER 8 REVIEW EXERCISES

■ In Exercises 1 to 8, evaluate each expression, where $\oplus$ and $\ominus$ indicate addition and subtraction, respectively, using a 12-hour clock.

1. $9 \oplus 5$ **2.** $8 \oplus 6$

3. $10 \oplus 7$ **4.** $7 \oplus 11$

5. $6 \ominus 9$ **6.** $2 \ominus 10$

7. $3 \ominus 4$ **8.** $7 \ominus 12$

■ In Exercises 9 and 10, evaluate each expression, where $\boxplus$ and $\boxminus$ indicate addition and subtraction, respectively, using days-of-the-week arithmetic.

9. $5 \boxplus 5$ **10.** $2 \boxminus 5$

■ In Exercises 11 to 14, determine whether the congruence is true or false.

11. $17 \equiv 2 \bmod 5$ **12.** $14 \equiv 24 \bmod 4$

13. $35 \equiv 53 \bmod 10$ **14.** $12 \equiv 36 \bmod 8$

■ In Exercises 15 to 22, perform the modular arithmetic.

15. $(8 + 12) \bmod 3$ **16.** $(15 + 7) \bmod 6$

17. $(42 - 10) \bmod 8$ **18.** $(19 - 8) \bmod 4$

19. $(7 \cdot 5) \bmod 9$ **20.** $(12 \cdot 9) \bmod 5$

21. $(15 \cdot 10) \bmod 11$ **22.** $(41 \cdot 13) \bmod 8$

■ **Clocks and Calendars** In Exercises 23 and 24, use modular arithmetic to determine each of the following.

23. Disregarding A.M. or P.M., if it is now 5 o'clock,

    a. what time will it be 45 hours from now?

    b. what time was it 71 hours ago?

24. In 2005, April 15 (the day taxes are due in the United States) fell on a Friday. On what day of the week will April 15 fall in 2013?

■ In Exercises 25 to 28, find all whole number solutions of the congruence equation.

25. $x \equiv 7 \bmod 4$ **26.** $2x \equiv 5 \bmod 9$

27. $(2x + 1) \equiv 6 \bmod 5$ **28.** $(3x + 4) \equiv 5 \bmod 11$

■ In Exercises 29 and 30, find the additive inverse and the multiplicative inverse, if it exists, of the given number.

29. 5 in mod 7 arithmetic

30. 7 in mod 12 arithmetic

■ In Exercises 31 and 32, perform the modular division by solving a related multiplication problem.

31. $(2 \div 5) \bmod 7$ **32.** $(3 \div 4) \bmod 5$

■ **ISBN Numbers** In Exercises 33 and 34, determine the correct check digit for each ISBN.

33. *Oliver Twist* by Charles Dickens, 978-1-402-75425-?

34. *Interview with the Vampire* by Anne Rice, 978-0-345-40964-?

■ **UPC Codes** In Exercises 35 and 36, determine the correct check digit for the UPC.

35. 0-29000-07004-? (Planters Almonds)

36. 0-85391-89512-? (*Best in Show* DVD)

■ **Credit Card Numbers** In Exercises 37 to 40, determine whether the given credit card number is a valid number.

37. 5126-6993-4231-2956

38. 5383-0118-3416-5931

39. 3412-408439-82594

40. 6011-5185-8295-8328

■ **Encryption**   In Exercises 41 and 42, encode the message using a cyclical alphabetic encrypting code that shifts the message the stated number of positions.

41. 7 positions: MAY THE FORCE BE WITH YOU

42. 11 positions: CANCEL ALL PLANS

■ **Encryption**   In Exercises 43 and 44, use a cyclical alphabetic encrypting code to decode the encrypted message.

43. PXXM UDLT CXVXAAXF

44. HVS ROM VOG OFFWJSR

45. Encryption   Use the encrypting congruence $c \equiv (3p + 6)$ mod 26 to encrypt the message END OF THE LINE.

46. Encryption   Decode the message WEU LKGGMF NHM NMGN, which was encrypted using the congruence $c \equiv (7p + 4)$ mod 26.

■ In Exercises 47 to 50, determine whether the set is a group with respect to the given operation. (You may assume the operation is associative.) If the set is not a group, determine which properties fail.

47. All rational numbers *except* 0; multiplication

48. All multiples of 3; addition

49. All negative integers; multiplication

50. $\{1, 2, 3, 4, 5, 6, 7, 8, 9, 10\}$; multiplication modulo 11

■ Exercises 51 and 52 refer to the triangular symmetry group discussed in Section 8.3.

51. Find $R_t \Delta R_r$.

52. Find $R_{240} \Delta R_t$.

■ For Exercises 53 and 54, use symbolic notation, shown below, for the operations on an equilateral triangle.

$$I = \begin{pmatrix} 1 & 2 & 3 \\ 1 & 2 & 3 \end{pmatrix}, R_{120} = \begin{pmatrix} 1 & 2 & 3 \\ 2 & 3 & 1 \end{pmatrix}, R_{240} = \begin{pmatrix} 1 & 2 & 3 \\ 3 & 1 & 2 \end{pmatrix}$$

$$R_l = \begin{pmatrix} 1 & 2 & 3 \\ 1 & 3 & 2 \end{pmatrix}, R_t = \begin{pmatrix} 1 & 2 & 3 \\ 3 & 2 & 1 \end{pmatrix}, R_r = \begin{pmatrix} 1 & 2 & 3 \\ 2 & 1 & 3 \end{pmatrix}$$

53. $R_{240} \Delta R_l$

54. $R_t \Delta R_r$

■ For Exercises 55 and 56, use the elements of the permutation group on the numbers 1 2 3 shown below.

$$I = \begin{pmatrix} 1 & 2 & 3 \\ 1 & 2 & 3 \end{pmatrix}, A = \begin{pmatrix} 1 & 2 & 3 \\ 2 & 3 & 1 \end{pmatrix}, B = \begin{pmatrix} 1 & 2 & 3 \\ 3 & 1 & 2 \end{pmatrix}$$

$$C = \begin{pmatrix} 1 & 2 & 3 \\ 1 & 3 & 2 \end{pmatrix}, D = \begin{pmatrix} 1 & 2 & 3 \\ 3 & 2 & 1 \end{pmatrix}, E = \begin{pmatrix} 1 & 2 & 3 \\ 2 & 1 & 3 \end{pmatrix}$$

55. $C \Delta E$

56. $B \Delta A$

57. Find the inverse of D.

58. Find the inverse of A.

CHAPTER 8   TEST

1. Evaluate each expression, where $\oplus$ and $\ominus$ indicate addition and subtraction, respectively, using a 12-hour clock.

  a. $8 \oplus 7$ **b.** $2 \ominus 9$

2. January 1, 2011, was a Saturday. What day of the week is January 1, 2018?

3. Determine whether the congruence is true or false.

  a. $8 \equiv 20$ mod 3   **b.** $61 \equiv 38$ mod 7

■ In Exercises 4 to 6, perform the modular arithmetic.

4. $(25 + 9)$ mod 6

5. $(31 - 11)$ mod 7

6. $(5 \cdot 16)$ mod 12

7. Scheduling   Disregarding A.M. or P.M., if it is now 3 o'clock, use modular arithmetic to determine

  a. what time it will be 27 hours from now.

  b. what time it was 58 hours ago.

■ In Exercises 8 and 9, find all whole number solutions of the congruence equation.

8. $x \equiv 5$ mod 9

9. $(2x + 3) \equiv 1$ mod 4

10. Find the additive inverse and the multiplicative inverse, if it exists, of 5 in modulo 9 arithmetic.

11. ISBN Number   Determine the correct check digit for the ISBN 978-0-739-49424-? (*Dictionary of American Slang, 4th Edition*, Barbara Ann Kipfer, Editor)

12. UPC Codes   Determine the correct check digit for the UPC 0-72878-27533-? (Herdez Salsa Verde).

13. Credit Card Numbers   Determine whether the credit card number 4232-8180-5736-4876 is a valid number.

14. Encryption   Encrypt the plaintext message REPORT BACK using the cyclical alphabetic encrypting code that shifts letters 10 positions.

15. Encryption   Decode the message UTSTG DPFM that was encrypted using the congruence $c \equiv (3p + 5)$ mod 26.

16. Determine whether the set {1, 2, 3, 4} with the operation multiplication modulo 5 forms a group.

17. Determine whether the set of all odd integers with multiplication as the operation is a group. (You may assume the operation is associative.) If the set is not a group, explain which properties fail.

18. In the symmetry group of an equilateral triangle, determine the result of the operation.

    a. $R_{120} \Delta R_l$ **b.** $R_t \Delta R_{240}$

19. In the permutation group of the numbers 1 2 3, determine the result of the operation.

$$\begin{pmatrix} 1 & 2 & 3 \\ 3 & 1 & 2 \end{pmatrix} \Delta \begin{pmatrix} 1 & 2 & 3 \\ 3 & 2 & 1 \end{pmatrix}$$

20. In the permutation group of the numbers 1 2 3 with operation Δ, find the inverse of $\begin{pmatrix} 1 & 2 & 3 \\ 3 & 1 & 2 \end{pmatrix}$.

Photolibrary

12

Combinatorics and Probability

The PowerBall lottery combines 31 states, the District of Columbia, and the Virgin Islands into a single lottery. To win the grand prize in this lottery, a player must select the five numbers from 1 to 59 and the one number from 1 to 39 that are chosen by the lottery commission. There are 195,249,054 different ways in which a player can make that selection. Thus the chances of winning the PowerBall lottery grand prize is 1 in 195,249,054. To put this in perspective, a person is about 400 times more likely to be struck by lightning than to win the PowerBall lottery grand prize.

When no person selects the winning lottery numbers, the grand prize increases. As of November 2010, the largest PowerBall jackpot was approximately $365 million, which was won on February 18, 2006.

This chapter focuses on the mathematics necessary to calculate the number of ways in which certain events, such as selecting a winning lottery number, can happen. This is part of the study of combinatorics. Once the number of ways in which an event can occur is calculated, it is possible to determine the probability of that event. Probability is another topic covered in this chapter.

ALEX GRIMM/Reuters/Corbis

section **12.1** The Counting Principle

Counting by Making a List

Combinatorics is the study of counting the different outcomes of some task. For example, if a coin is flipped, the side facing upward will be a head or a tail. The outcomes can be listed as {H, T}. There are two possible outcomes.

If a regular six-sided die is rolled, the possible outcomes are [dice], [dice], [dice], [dice], [dice], [dice]. The outcomes can also be listed as {1, 2, 3, 4, 5, 6}. There are six possible outcomes.

▼ example **1** Counting by Forming a List

List and then count the number of different outcomes that are possible when one letter from the word *Tennessee* is chosen.

Solution

The possible outcomes are {T, e, n, s}. There are four possible outcomes.

▼ check your progress **1** List and then count the number of different outcomes that are possible when one letter is chosen from the word *Mississippi*.

Solution *See page S44.* ◄

In combinatorics, an **experiment** is an activity with an observable outcome. The set of all possible outcomes of an experiment is called the **sample space** of the experiment. Flipping a coin, rolling a die, and choosing a letter from the word *Tennessee* are experiments. The sample spaces are {H, T}, {1, 2, 3, 4, 5, 6}, and {T, e, n, s}, respectively.

An **event** is one or more of the possible outcomes of an experiment. Flipping a coin and having a head show on the upward face, rolling a 5 when a die is tossed, and choosing a T from one of the letters in the word *Tennessee* are all examples of events. An event is a *subset* of the sample space.

▼ example **2** Listing the Elements of an Event

One number is chosen from the sample space

$$S = \{1, 2, 3, 4, 5, 6, 7, 8, 9, 10, 11, 12, 13, 14, 15, 16, 17, 18, 19, 20\}$$

List the elements in the following events.

a. The number is even.

b. The number is divisible by 5.

c. The number is a prime number.

Solution

a. {2, 4, 6, 8, 10, 12, 14, 16, 18, 20}

b. {5, 10, 15, 20}

c. {2, 3, 5, 7, 11, 13, 17, 19}

▼ check your progress 2   One digit is chosen from the digits 0 through 9. The sample space S is {0, 1, 2, 3, 4, 5, 6, 7, 8, 9}. List the elements in the following events.

a. The number is odd.

b. The number is divisible by 3.

c. The number is greater than 7.

Solution   *See page S44.*   ◄

Counting by Making a Table

Each of the experiments given above illustrates a *single-stage experiment*. A **single-stage experiment** is an experiment for which there is a single outcome. Some experiments have two, three, or more stages. Such experiments are called **multi-stage experiments**. To count the number of outcomes of such an experiment, a systematic procedure is helpful. Using a table to record results is one such procedure.

Consider the two-stage experiment of rolling two dice, one red and one green. How many different outcomes are possible? To determine the number of outcomes, make a table with the different outcomes of rolling the red die across the top and the different outcomes of rolling the green die down the side.

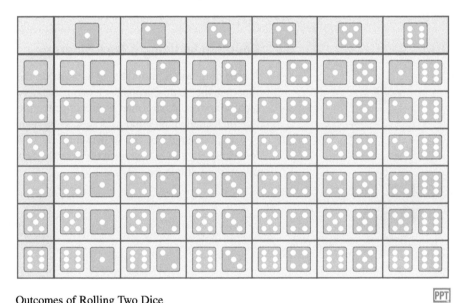

Outcomes of Rolling Two Dice

By counting the number of entries in the diagram above, we see that there are 36 different outcomes of the experiment of rolling two dice. The sample space is

{⚁⚁, ⚁⚁, ⚁⚁, ..., ⚁⚁, ⚁⚁}

From the table, several different events can be discussed.

- The sum of the pips (dots) on the upward faces is 7. There are six outcomes of this event. They are {⚁⚁, ⚁⚁, ⚁⚁, ⚁⚁, ⚁⚁, ⚁⚁}.

- The sum of the pips on the upward faces is 11. There are two outcomes of this event. They are {⚁⚁, ⚁⚁}.

- The number of pips on the upward faces are equal. There are six outcomes of this event. They are {⚁⚁, ⚁⚁, ⚁⚁, ⚁⚁, ⚁⚁, ⚁⚁}.

▼ example 3   Counting Using a Table

Two-digit numbers are formed from the digits 1, 3, and 8. Find the sample space and determine the number of elements in the sample space.

Solution
Use a table to list all the different two-digit numbers that can be formed by using the digits 1, 3, and 8.

	1	3	8
1	11	13	18
3	31	33	38
8	81	83	88

The sample space is {11, 13, 18, 31, 33, 38, 81, 83, 88}. There are nine two-digit numbers that can be formed from the digits 1, 3, and 8.

▼ check your progress 3   A die is tossed and then a coin is flipped. Find the sample space and determine the number of elements in the sample space.

Solution   *See page S44.*   ◄

Counting by Using a Tree Diagram

A **tree diagram** is another way to organize the outcomes of a multi-stage experiment. To illustrate the method, consider a computer store offering special prices on its most popular laptop models. A customer can choose from two sizes of RAM, three screen sizes, and two preloaded application packages. How many different laptops can customers choose?

We can organize the information by letting M_1 and M_2 represent the two sizes of RAM; S_1, S_2, and S_3 represent the three screen sizes; and A_1 and A_2 represent the two application packages (see Figure 12.1).

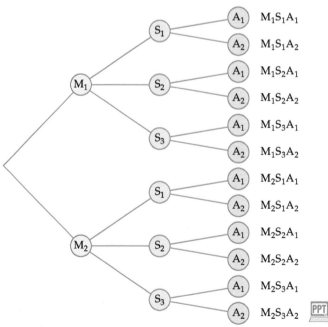

FIGURE 12.1

There are 12 possible laptops.

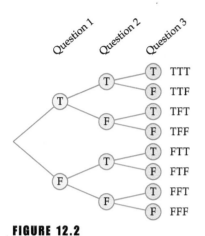

FIGURE 12.2

INSTRUCTOR NOTE

To convince students that explicitly listing all possible outcomes of an experiment can easily get out of hand, suggest that they draw a tree diagram for all 10 questions in Example 4. They should quickly see that it is not a reasonable request.

▼ **example** **4**   Counting Using a Tree Diagram

A true/false test consists of 10 questions. Draw a tree diagram to show the number of ways to answer the first three questions.

Solution
See the tree diagram in Figure 12.2. There are eight possible ways to answer the first three questions.

▼ **check your progress** **4**   Draw a tree diagram to determine the sample space for Example 3.

Solution   *See page S44.*   ◄

The Counting Principle

For each of the previous problems, the possible outcomes were listed and then counted to determine the number of different outcomes. However, it is not always possible or practical to list and count outcomes. For example, the number of different five-card poker hands that can be drawn from a standard deck of 52 playing cards is 2,598,960. Trying to create a list of these hands would be quite time consuming.

Consider again the problem of selecting a laptop. By using a tree diagram, we listed the 12 possible laptops. Another way to arrive at this result is to find the product of the numbers of choices available for RAM sizes, screen sizes, and application packages.

$$\begin{bmatrix} \text{number of} \\ \text{RAM sizes} \end{bmatrix} \times \begin{bmatrix} \text{number of} \\ \text{screen sizes} \end{bmatrix} \times \begin{bmatrix} \text{number of} \\ \text{application packages} \end{bmatrix} = \begin{bmatrix} \text{number of} \\ \text{laptops} \end{bmatrix}$$
$$2 \quad\times\quad 3 \quad\times\quad 2 \quad=\quad 12$$

For the example of tossing two dice, there were 36 possible outcomes. We can arrive at this result without listing the outcomes by finding the product of the number of possible outcomes of rolling the red die and the number of possible outcomes of rolling the green die.

$$\begin{bmatrix} \text{outcomes} \\ \text{of red die} \end{bmatrix} \times \begin{bmatrix} \text{outcomes} \\ \text{of green die} \end{bmatrix} = \begin{bmatrix} \text{number of} \\ \text{outcomes} \end{bmatrix}$$
$$6 \quad\times\quad 6 \quad=\quad 36$$

This method of determining the number of outcomes of a multi-stage experiment without listing them is called the **counting principle**.

▼ **Counting Principle**

Let E be a multi-stage experiment. If $n_1, n_2, n_3, ..., n_k$ are the number of possible outcomes of each of the k stages of E, then there are $n_1 \cdot n_2 \cdot n_3 \cdot \cdots \cdot n_k$ possible outcomes for E.

▼ **example** **5**   Counting by Using the Counting Principle

In horse racing, a *trifecta* consists of choosing the exact order of the first three horses across the finish line. If there are eight horses in a race, how many trifectas are possible, assuming there are no ties?

Solution

Any one of the eight horses can be first, so $n_1 = 8$. Because a horse cannot finish both first and second, there are seven horses that can finish second; thus $n_2 = 7$. Similarly, there are six horses that can finish third; $n_3 = 6$. By the counting principle, there are $8 \cdot 7 \cdot 6 = 336$ possible trifectas.

▼ **check your progress** **5** Nine runners are entered in a 100-meter dash for which a gold, silver, and bronze medal will be awarded for first, second, and third place finishes, respectively. In how many possible ways can the medals be awarded? (Assume that there are no ties.)

Solution *See page S44.*

MATHMATTERS   Coding Characters for Web Pages

POINT OF INTEREST

The ISO (International Organization for Standardization) develops standards used in science, engineering, and commerce. ISO Latin 1 is one such standard that web browsers such as Internet Explorer, Safari, Mozilla, Bing, Yahoo, Opera, and others use to display text in a browser. Because "International Organization for Standardization" would have different abbreviations in different languages ("IOS" in English; "OIN" in French, for Organisation internationale de normalisation), it was decided at the outset to use a word derived from the Greek *isos*, meaning "equal." (*Source*: www .iso.org)

So that web pages look the same on computers all over the world, certain standards must be adopted and adhered to by web page developers. ISO Latin 1 (also called ISO 8859-1) is a standard character set used by many web browsers. Each upper- and lower-case letter, numeral, punctuation mark, and special symbol (for instance, &, %, and @) has a representation as an eight-digit binary code. Here are some examples:

Partial ISO Latin 1 Character Set

Letter or symbol	Binary code	Letter or symbol	Binary code
A	0100 0001	7	0011 0111
a	0110 0001	&	0010 0110
B	0100 0010	=	0011 1101

Symbols used in foreign languages are also available in the ISO Latin 1 character set. For instance, to display the word *español* (Spanish for "Spanish"), it is necessary to code the letter ñ. This letter is given by 1111 0001.

Using the counting principle, we can count the number of possible characters that can be represented by this scheme. Each of the eight digits in the binary number can be a 0 or a 1. Thus there are two choices for the first digit, two choices for the second digit, and so on until the eighth digit, for which there are two choices. By the counting principle, there are $2 \cdot 2 \cdot 2 \cdot 2 \cdot 2 \cdot 2 \cdot 2 \cdot 2 = 2^8 = 256$ possible characters that can be represented using the ISO Latin 1 method.

You can see which character set your browser is using by selecting Preferences and looking at the font characteristics being used by the browser.

Counting With and Without Replacement

Consider an experiment in which balls colored red, blue, and green are placed in a box. A person reaches into the box and repeatedly pulls out a colored ball, keeping note of the color picked. The sequence of colors that can result depends on whether or not the balls are returned to the box after each pick. This is referred to as performing the experiment *with replacement* or *without replacement*.

Consider the following two situations.

1. How many four-digit numbers can be formed from the digits 1 through 9 if no digit can be repeated?

2. How many four-digit numbers can be formed from the digits 1 through 9 if a digit can be used repeatedly?

In the first case, there are nine choices for the first digit ($n_1 = 9$). Because a digit cannot be repeated, the first digit chosen cannot be used again. Thus there are only eight choices for the second digit ($n_2 = 8$). Because neither of the first two digits can be used as the third digit, there are only seven choices for the third digit ($n_3 = 7$). Similarly, there are six choices for the fourth digit ($n_4 = 6$). By the counting principle, there are $9 \cdot 8 \cdot 7 \cdot 6 = 3024$ four-digit numbers in which no digit is repeated.

In the second case, there are nine choices for the first digit ($n_1 = 9$). Because a digit can be used repeatedly, the first digit chosen can be used again. Thus there are nine choices for the second digit ($n_2 = 9$), and, similarly, there are nine choices for the third and fourth digits ($n_3 = 9, n_4 = 9$). By the counting principle, there are $9 \cdot 9 \cdot 9 \cdot 9 = 6561$ four-digit numbers when digits can be used repeatedly.

The set of four-digit numbers created without replacement includes numbers such as 3867, 7941, and 9128. For these numbers, no digit is repeated. However, numbers such as 6465, 9911, and 2222, each of which contains at least one repeated digit, can be created only with replacement.

question Does a multi-stage experiment performed with replacement generally have more or fewer possible outcomes than the same experiment performed without replacement?

▼ example 6 **Counting With and Without Replacement**

From the letters a, b, c, d, and e, how many four-letter groups can be formed if

a. a letter can be used more than once?

b. each letter can be used exactly once?

Solution

a. Because each letter can be repeated, there are $5 \cdot 5 \cdot 5 \cdot 5 = 625$ possible four-letter groups.

b. Because each letter can be used only once, there are $5 \cdot 4 \cdot 3 \cdot 2 = 120$ four-letter groups in which no letter is repeated.

▼ check your progress 6 In how many ways can three awards by given to five students if

a. each student may receive more than one award?

b. each student may receive no more than one award?

Solution *See page S44.*

answer More. Each stage of an experiment (after the first) performed without replacement will have fewer possible outcomes than the preceding stage. Performed with replacement, each stage of the experiment has the same number of outcomes.

EXCURSION

Decision Trees

Decision trees are tree diagrams that are used to solve problems that involve many choices. To illustrate, we will consider a particular puzzle. Suppose we are given eight coins, one of which is counterfeit and slightly heavier than the other seven. Using a balance scale, we must find the counterfeit coin.

INSTRUCTOR NOTE

There is a much more difficult version of this puzzle in which it is not known whether the counterfeit coin is heavier or lighter than the other coins. See http://mathforum.org/library/drmath/view/57947.html.

Designate the coins as c_1, c_2, c_3, c_4, c_5, c_6, c_7, and c_8. One way to determine the counterfeit coin is to weigh coins in pairs. This method is illustrated by the decision tree in Figure 12.3. In this case, it would take from one to four weighings to determine the counterfeit coin.

A second method is to divide the coins into two groups of four coins each, place each group on a pan of the balance scale, and take the coins from the side that goes down. Now divide these four coins into two groups of two coins each and weigh them on the balance scale. Again keep the coins from the heavier side. Weighing one of these coins against the other will reveal the counterfeit coin. The decision tree for this procedure is shown in Figure 12.4. Using this method, the counterfeit coin will be found in three weighings.

A third method is even more efficient. Divide the coins into three groups. Two of the groups contain three coins, and the third group contains two coins. Place each of the three-coin groups on the balance scale. If they balance, the counterfeit coin is in the third group. Placing a coin from the third group on each of the balance pans will determine the counterfeit coin. If the three coin groups do not balance, then take two of the coins from the pan that goes down and weigh them against each other. If these balance, the third coin is the counterfeit. If not, the counterfeit is the coin on the pan that goes down. This method requires only two weighings and is shown in the decision tree in Figure 12.5.

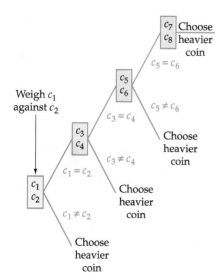

FIGURE 12.3

Using this method, it may take up to four weighings to determine which is the counterfeit coin.

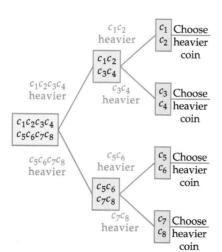

FIGURE 12.4

Using this method, it will always take three weighings to determine which is the counterfeit coin.

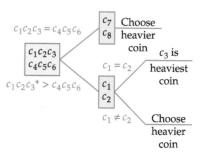

*We are assuming that $c_1 c_2 c_3$ is heavier than $c_4 c_5 c_6$. If $c_4 c_5 c_6$ is heavier, use those coins in the final weighing.

FIGURE 12.5

Using this method, it will take only two weighings to determine which is the counterfeit coin.

EXCURSION EXERCISES

For each of the following problems, draw a decision tree and determine the minimum number of weighings necessary to identify the counterfeit coin.

1. In a stack of 12 identical-looking coins, one is counterfeit and is lighter than the remaining 11 coins. Using a balance scale, identify the counterfeit coin in as few weighings as possible.

2. In a stack of 13 identical-looking coins, one is counterfeit and is heavier than the remaining 12 coins. Using a balance scale, identify the counterfeit coin in as few weighings as possible.

EXERCISE SET 12.1
(Suggested Assignment: The Enhanced WebAssign Exercises and Exercises 19 and 20)

■ In Exercises 1 to 10, list the elements of the sample space defined by each experiment.

1. Select an even single-digit whole number.

2. Select an odd single-digit whole number.

3. Select one day from the days of the week.

4. Select one month from the months of the year.

5. Toss a coin twice.

6. Toss a coin three times.

7. Roll a single die and then toss a coin.

8. Toss a coin and then choose a digit from the digits 1 through 4.

9. Choose a complete dinner from a dinner menu that allows a customer to choose from two salads, three entrees, and two desserts.

10. Choose a car during a new car promotion that allows a buyer to choose from three body styles, two radios, and two interior color schemes.

■ For Exercises 11 and 12, list the sample space of paths that start at A and pass through each vertex of the figure exactly once.

11. Square *ABCD*

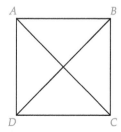

12. Pentagon *ABCDE*

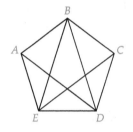

■ In Exercises 13 to 18, use the counting principle to determine the number of elements in the sample space.

13. Two digits are selected without replacement from the digits 1, 2, 3, and 4.

14. Two digits are selected with replacement from the digits 1, 2, 3, and 4.

15. The possible ways to complete a multiple-choice test consisting of 20 questions, with each question having four possible answers (a, b, c, or d)

16. The possible ways to complete a true–false examination consisting of 25 questions

17. The possible four-digit telephone number extensions that can be formed if 0, 8, and 9 are excluded as the first digit

18. The possible six-character passwords that can be formed using a letter from a to h and five numbers from 1 to 9. Assume a letter or number cannot be used more than once in any six-character password.

License Plates For Exercises 19 and 20, refer to the article below about North Carolina license plate numbers.

19. The article above describes the structure of a North Carolina license plate as "ABC-1234"—three letters, followed by four numbers. There are some restrictions on the letters and numbers:

- No letter can be I, O, Q, or U.
- The middle letter cannot be from the A through L sequence.
- The first number cannot be a 0.

What is the total number of license plate numbers possible given this structure?

20. About how many plate numbers were still available for issue when North Carolina switched back to blue lettering? Assume for the purposes of estimating that the switch took place at plate number ZNE-1000, the first plate in the ZNE series. *Note:* The plates of the ZNE series are issued in sequential order until all possible ZNE plates have been issued. The first ZNE plate is ZNE-1000, the second is ZNE-1001, and so on through ZNE-9999. The next plate issued is ZNF-1000, the beginning of the ZNF series. After reaching ZNZ-9999 (last of the ZNZ series), the next plate issued is ZPA-1000 (recall from Exercise 19 that the letter "O" is not used, nor are I, Q, or U).

■ In Exercises 21 to 26, use the following experiment. Two-digit numbers are formed, with replacement, from the digits 0–9.

21. How many two-digit numbers are possible?

22. How many two-digit even numbers are possible?

23. How many numbers are divisible by 5?

24. How many numbers are divisible by 3?

25. How many numbers are greater than 37?

26. How many numbers are less than 59?

■ In Exercises 27 to 30, use the following experiment. Four cards labeled A, B, C, and D are randomly placed in four boxes labeled A, B, C, and D. Each box receives exactly one card.

27. In how many ways can the cards be placed in the boxes?

28. Count the number of elements in the event that no box contains a card with the same letter as the box.

29. Count the number of elements in the event that *at least* one card is placed in the box with the corresponding letter.

30. If you add the answer for Exercise 28 and the answer for Exercise 29, is the sum the answer for Exercise 27? Why or why not?

■ **Lotteries** In Exercises 31 to 34, use the following experiment. A state lottery game consists of choosing one card from each of the four suits in a standard deck of playing cards. (There are 13 cards in each suit.)

31. Count the number of elements in the sample space.

32. Count the number of elements in the event that an ace, a king, a queen, and a jack are chosen.

33. Count the number of ways in which four aces can be chosen.

34. Count the number of ways in which four cards, each of a different face value, can be chosen.

35. Write a lesson that you could use to explain the meanings of the words *experiment, sample space,* and *event* as they apply to combinatorics.

36. Explain how a tree diagram is used to count the number of ways an experiment can be performed.

EXTENSIONS

Critical Thinking

37. Computer Programming   A main component of any computer programming language is its method of repeating a series of computations. Each programming language has its own syntax for performing those "loops." In one programming language, BASIC, the structure is similar to the display below.

```
FOR I = 1 TO 10        Start with I = 1.
    FOR J = 1 TO 15       Start with J = 1.
        SUM = I + J
    NEXT J          Increase J by 1 until J > 15.
NEXT I            Increase I by 1 until I > 10.
END
```

The program repeats each loop until the index variables, I and J in this case, exceed a certain value. After this program is executed, how many times will the instruction SUM = I + J have been executed? What is the final value of SUM?

38. Computer Programming   One way in which a software engineer can write a program to sort a list of numbers from smallest to largest is to use a *bubble sort*. The following code segment, written in BASIC, will perform a bubble sort of *n* numbers. We have left out the details of how the numbers are sorted in the list.

```
FOR J = 1 TO n − 1
    FOR K = J + 1 TO n
        <Sorting takes place here.>
    NEXT K
NEXT J
END
```

a. If the list contains 10 numbers ($n = 10$), how many times will the program loop before reaching the END statement?

b. Suppose there are 20 numbers in the list. How many times will the program loop before reaching the END statement?

Explorations

39. Use a tree diagram to display the relationships among the junior- and senior-level courses you must take for your major and their prerequisites.

40. Review the rules of the game checkers, and make a tree diagram that shows all of the first two moves that are possible by one player of a checker game. Assume that no moves are blocked by the opponent's checkers. (*Hint:* It may help to number the squares of the checkerboard.)

section **12.2** Permutations and Combinations

Factorial

Suppose four different colored squares are arranged in a row. One possibility is shown below.

How many different ways are there to order the colors? There are four choices for the first square, three choices for the second square, two choices for the third square, and one choice for the fourth square. By the counting principle, there are $4 \cdot 3 \cdot 2 \cdot 1 = 24$ different arrangements of the four squares. Note from this example that the number of arrangements equals the product of the natural numbers n through 1, where n is the number of objects. This product is called a *factorial*.

▼ *n* Factorial

n **factorial** is the product of the natural numbers n through 1 and is symbolized by *n!*.

$$n! = n \cdot (n - 1) \cdot (n - 2) \cdot \cdots \cdot 3 \cdot 2 \cdot 1$$

CALCULATOR NOTE

The factorial of a number becomes quite large for even relatively small numbers. For instance, 58! is the approximate number of atoms in the known universe. The number 70! is greater than 10 with 100 zeros after it. This number is larger than most scientific calculators can handle.

Here are some examples:

$5! = 5 \cdot 4 \cdot 3 \cdot 2 \cdot 1 = 120$

$8! = 8 \cdot 7 \cdot 6 \cdot 5 \cdot 4 \cdot 3 \cdot 2 \cdot 1 = 40{,}320$

$1! = 1$

On some occasions it will be necessary to use 0! (zero factorial). Because it is impossible to define zero factorial in terms of a product of natural numbers, a standard definition is used.

▼ Zero Factorial

$$0! = 1$$

A factorial can be written in terms of smaller factorials. This is useful when calculating large factorials. For example,

$10! = 10 \cdot 9!$

$10! = 10 \cdot 9 \cdot 8!$

$10! = 10 \cdot 9 \cdot 8 \cdot 7!$

▼ example **1** Simplify Factorials

Evaluate: **a.** $5! - 3!$ **b.** $\dfrac{9!}{6!}$

Solution

a. $5! - 3! = (5 \cdot 4 \cdot 3 \cdot 2 \cdot 1) - (3 \cdot 2 \cdot 1) = 120 - 6 = 114$

b. $\dfrac{9!}{6!} = \dfrac{9 \cdot 8 \cdot 7 \cdot \cancel{6!}}{\cancel{6!}} = 9 \cdot 8 \cdot 7 = 504$

▼ **check your progress 1** Evaluate: **a.** $7! + 4!$ **b.** $\dfrac{8!}{4!}$

Solution   *See page S44.*   ◄

Permutations

Determining the number of possible arrangements of distinct objects in a definite order, as we did with the squares earlier, is one application of the counting principle. Each arrangement of this type is called a *permutation*.

▼ **Permutation**

A **permutation** is an arrangement of objects in a definite order.

For example, abc and cba are two different permutations of the letters a, b, and c. As a second example, 122 and 212 are two different permutations of one 1 and two 2s.

The counting principle is used to count the number of different permutations of any set of objects. We will begin our discussion with *distinct* objects. For instance, the objects a, b, c, d are distinct, whereas the objects ♥ ♥ ♠ ♦ are not all distinct.

Suppose that there are two songs, *Two Step* and *Respect*, in a playlist, and you wish to play both on your music player. There are two ways to choose the first song; $n_1 = 2$. There is one way to choose the second song; $n_2 = 1$. By the counting principle, there are $2 \cdot 1 = 2! = 2$ permutations or orders in which you can play the two songs. Each permutation gives a different play list. These are

Permutation (playlist) 1	Permutation (playlist) 2
Two Step	*Respect*
Respect	*Two Step*

With three songs in a playlist, *Two Step, Respect,* and *Let It Be,* there are three choices for the first song, two choices for the second song, and one choice for the third song. By the counting principle, there are $3 \cdot 2 \cdot 1 = 3! = 6$ permutations in which you can play the three songs.

Permutation 1	Permutation 2	Permutation 3	Permutation 4	Permutation 5	Permutation 6
Two Step	*Two Step*	*Respect*	*Respect*	*Let It Be*	*Let It Be*
Respect	*Let It Be*	*Two Step*	*Let It Be*	*Two Step*	*Respect*
Let It Be	*Respect*	*Let It Be*	*Two Step*	*Respect*	*Two Step*

With four songs in a playlist, there are $4 \cdot 3 \cdot 2 \cdot 1 = 4! = 24$ orders in which you could play the songs. In general, if there are *n* songs in a playlist, then there are *n*! permutations, or orders, in which the songs could be played.

Suppose now that you have a playlist that consists of eight songs but you have time to listen to only three of the songs. You could choose any one of the eight songs to play first, then any one of the seven remaining songs to play second, and then any one of the remaining six songs to play third. By the counting principle, there are $8 \cdot 7 \cdot 6 = 336$ permutations in which the songs could be played.

The following formula can be used to determine the number of permutations of n distinct objects (songs in the examples above), of which k are selected.

▼ Permutation Formula for Distinct Objects

The number of permutations of n distinct objects selected k at a time is

$$P(n, k) = \frac{n!}{(n - k)!}$$

Applying this formula to the situation above in which there were eight songs ($n = 8$), of which there was time to play only three songs ($k = 3$), we have

$$P(8, 3) = \frac{8!}{(8 - 3)!} = \frac{8!}{5!} = \frac{8 \cdot 7 \cdot 6 \cdot 5!}{5!} = 8 \cdot 7 \cdot 6 = 336$$

There are 336 permutations of playing the songs. This is the same answer we obtained using the counting principle.

▼ example 2 Counting Permutations

A university tennis team consists of six players who are ranked from 1 through 6. If a tennis coach has 10 players from which to choose, how many different tennis teams can the coach select?

Solution

Because the players on the tennis team are ranked from 1 through 6, a team with player A in position 1 is different from a team with player A in position 2. Therefore, the number of different teams is the number of permutations of 10 players selected six at a time.

$$P(10, 6) = \frac{10!}{(10 - 6)!} = \frac{10!}{4!} = \frac{10 \cdot 9 \cdot 8 \cdot 7 \cdot 6 \cdot 5 \cdot 4!}{4!}$$
$$= 10 \cdot 9 \cdot 8 \cdot 7 \cdot 6 \cdot 5 = 151,200$$

There are 151,200 possible tennis teams.

▼ check your progress 2 A college golf team consists of five players who are ranked from 1 through 5. If a golf coach has eight players from which to choose, how many different golf teams can the coach select?

Solution *See page S45.* ◀

▼ example 3 Counting Permutations

In 2010, the Kentucky Derby had 20 horses entered in the race. How many different finishes of first, second, third, and fourth place were possible?

Solution

Because the order in which the horses finish the race is important, the number of possible finishes of first, second, third, and fourth place is $P(20, 4)$.

The Kentucky Derby

$$P(20, 4) = \frac{20!}{(20 - 4)!} = \frac{20!}{16!} = \frac{20 \cdot 19 \cdot 18 \cdot 17 \cdot 16!}{16!}$$

$$= 20 \cdot 19 \cdot 18 \cdot 17 = 116{,}280$$

There were 116,280 possible finishes of first, second, third, and fourth places.

▼ **check your progress** **3**   There were 43 cars entered in the 2010 Daytona 500 NASCAR race. How many different ways could first, second, and third place prizes have been awarded?

Solution   *See page S45.*   ◄

MATHMATTERS   How Many Shuffles?

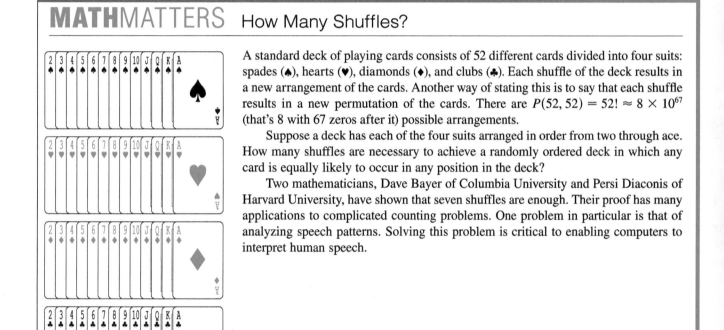

A standard deck of
playing cards

A standard deck of playing cards consists of 52 different cards divided into four suits: spades (♠), hearts (♥), diamonds (♦), and clubs (♣). Each shuffle of the deck results in a new arrangement of the cards. Another way of stating this is to say that each shuffle results in a new permutation of the cards. There are $P(52, 52) = 52! \approx 8 \times 10^{67}$ (that's 8 with 67 zeros after it) possible arrangements.

Suppose a deck has each of the four suits arranged in order from two through ace. How many shuffles are necessary to achieve a randomly ordered deck in which any card is equally likely to occur in any position in the deck?

Two mathematicians, Dave Bayer of Columbia University and Persi Diaconis of Harvard University, have shown that seven shuffles are enough. Their proof has many applications to complicated counting problems. One problem in particular is that of analyzing speech patterns. Solving this problem is critical to enabling computers to interpret human speech.

Applying Several Counting Techniques

The permutation formula is derived from the counting principle. This formula is just a convenient way to express the number of ways the items in an ordered list can be arranged. Some counting problems require using both the permutation formula and the counting principle.

▼ **example** **4**   Counting Using Several Methods

Five women and four men are to be seated in a row of nine chairs. How many different seating arrangements are possible if

a. there are no restrictions on the seating arrangements?

b. the women sit together and the men sit together?

Solution

Because seating arrangements have a definite order, they are permutations.

a. If there are no restrictions on the seating arrangements, then the number of seating arrangements is $P(9, 9)$.

$$P(9, 9) = \frac{9!}{(9 - 9)!} = \frac{9!}{0!} = 9! = 362,880$$

There are 362,880 seating arrangements.

b. This is a multi-stage experiment, so both the permutation formula and the counting principle will be used. There are 5! ways to arrange the women and 4! ways to arrange the men. We must also consider that either the women or the men could be seated at the beginning of the row. There are two ways to do this. By the counting principle, there are $2 \cdot 5! \cdot 4!$ ways to seat the women together and the men together.

$$2 \cdot 5! \cdot 4! = 5760$$

There are 5760 arrangements in which women sit together and men sit together.

▼ **check your progress** **4** There are seven tutors, three juniors and four seniors, who must be assigned to the 7 hours that a math center is open each day. If each tutor works 1 hour per day, how many different tutoring schedules are possible if

a. there are no restrictions?

b. the juniors tutor during the first 3 hours and the seniors tutor during the last 4 hours?

Solution *See page S45.* ◄

Permutations of Indistinguishable Objects

Up to this point we have been counting the number of permutations of *distinct* objects. We now look at the situation of arranging objects when some of them are identical. In the case of identical or indistinguishable objects, a modification of the permutation formula is necessary. The general idea for modifying the formula is to count the number of permutations as if all of the objects were distinct and then remove the permutations that are not different in appearance.

Consider the permutations of the letters *bbbcc*. We first assume the letters are all different—for example, $b_1b_2b_3c_1c_2$. Using the permutation formula, there are $5! = 120$ permutations. Now we need to remove repeated permutations. Consider for a moment the letter *b* in the permutation $bbbc_1c_2$. If the *b*'s are written as b_1, b_2, and b_3 (so that they are distinguishable), then

$$b_1b_2b_3c_1c_2 \quad b_1b_3b_2c_1c_2 \quad b_2b_1b_3c_1c_2 \quad b_2b_3b_1c_1c_2 \quad b_3b_2b_1c_1c_2 \quad b_3b_1b_2c_1c_2$$

are all distinct permutations that end with c_1c_2. There are six of these permutations. Note that $3! = 6$, where 3 is the number of *b*'s.

However, because the *b*'s are not distinct, each of these permutations of *b*'s should have been counted only once. Thus there are six times too many permutations of *b*'s for each arrangement of c_1 and c_2.

A similar argument applies to the *c*'s. If the *c*'s are identical, then c_1c_2bbb and c_2c_1bbb are the same permutation. For each arrangement of *b*'s, there are two arrangements of *c*'s that yield identical permutations. Note that there are two *c*'s and that $2! = 2$, the number of identical permutations. Thus there are two times too many permutations of *c*'s for each arrangement of *b*'s.

Combining the results above, the number of permutations of *bbbcc* is

$$\frac{5!}{3! \cdot 2!} = \frac{5 \cdot 4 \cdot 3 \cdot 2 \cdot 1}{(3 \cdot 2 \cdot 1) \cdot (2 \cdot 1)} = 10$$

There are 10 distinct permutations of *bbbcc*.

▼ **Permutations of Objects, Some of Which Are Identical**

The number of distinguishable permutations of *n* objects of *r* different types, where k_1 identical objects are of one type, k_2 of another, and so on, is given by

$$\frac{n!}{k_1! \cdot k_2! \cdot \cdots \cdot k_r!}$$

where $k_1 + k_2 + \cdots + k_r = n$.

▼ **example 5**   **Permutations of Identical Objects**

A password requires 7 characters. If a person who lives at 155 Nunn Road wants a password to be an arrangement of the characters 155NUNN, how many different passwords are possible?

Solution
We are looking for the number of permutations of the characters 155NUNN. With $n = 7$ (number of characters), $k_1 = 1$ (number of 1s), $k_2 = 2$ (number of 5s), $k_3 = 3$ (number of Ns), and $k_4 = 1$ (number of Us) we have

$$\frac{7!}{1! \cdot 2! \cdot 3! \cdot 1!} = \frac{7 \cdot 6 \cdot 5 \cdot 4 \cdot 3!}{2 \cdot 3!} = 420$$

There are 420 possible passwords.

▼ **check your progress 5**   Eight coins—3 pennies, 2 nickels, and 3 dimes—are placed in a single stack. How many different stacks are possible if

a. there are no restrictions on the placement of the coins?

b. the dimes must stay together?

Solution   *See page S45.*   ◀

INSTRUCTOR NOTE

Students often confuse permutations with combinations. Emphasize that if the particular order of objects makes a difference, then we are counting permutations. You can also point out that there are more permutations of objects than combinations.

Combinations

For some arrangements of objects, the order of the arrangement is important. These are permutations. If a telephone extension is 2537, then the digits must be dialed in exactly that order. On the other hand, if you were to receive a $1 bill, a $5 bill, and a $10 bill, you would have $16 regardless of the order in which you received the bills. A **combination** is a collection of objects for which the order is not important. The three-letter sequences acb and bca are *different* permutations but the *same* combination.

question   From a group of 45 applicants, five identical scholarships will be awarded. Is the number of ways in which the scholarships can be awarded determined by permutations or combinations?

answer   Combinations. The order in which the scholarship winners are chosen is not important.

The formula for finding the number of combinations is derived in much the same manner as the formula for finding the number of permutations of identical objects. Consider the problem of finding the number of combinations possible when choosing three letters from the letters a, b, c, d, and e, without replacement. For each choice of three letters, there are 3! permutations. For example, choosing the letters a, d, and e gives the following six permutations.

    ade   aed   dea   dae   ead   eda

Because there are six permutations and each permutation is the *same* combination, the number of permutations is six times the number of combinations. This is true each time three letters are selected. Therefore, to find the number of combinations of five objects chosen three at a time, divide the number of permutations by $3! = 6$. The number of combinations of five objects chosen three at a time is

$$\frac{P(5,3)}{3!} = \frac{5!}{3! \cdot (5-3)!} = \frac{5!}{3! \cdot 2!} = \frac{5 \cdot 4 \cdot 3!}{3! \cdot 2!} = \frac{5 \cdot 4}{2 \cdot 1} = 10$$

CALCULATOR NOTE

Some calculators can compute permutations and combinations directly. For instance, on a TI-83/84 calculator, enter 11 nCr 5 for $C(11, 5)$. The nCr operation is accessible in the probability menu after pressing the MATH key.

▼ **Combination Formula**

The number of combinations of n objects chosen k at a time is

$$C(n, k) = \frac{P(n, k)}{k!} = \frac{n!}{k! \cdot (n-k)!}$$

In Section 12.1, we stated that there were 2,598,960 possible 5-card poker hands. This number was calculated using the combination formula. Because the 5-card hand ace of hearts, king of diamonds, queen of clubs, jack of spades, 10 of hearts is exactly the same as the 5-card hand king of diamonds, jack of spades, queen of clubs, 10 of hearts, ace of hearts, the order of the cards is not important, and therefore the number of hands is a combination. The number of different 5-card poker hands is the combination of 52 cards chosen 5 at a time, which is given by $C(52, 5)$.

$$C(52, 5) = \frac{52!}{5! \cdot (52-5)!} = \frac{52!}{5! \cdot 47!} = \frac{52 \cdot 51 \cdot 50 \cdot 49 \cdot 48 \cdot 47!}{5! \cdot 47!}$$

$$= \frac{52 \cdot 51 \cdot 50 \cdot 49 \cdot 48}{5 \cdot 4 \cdot 3 \cdot 2 \cdot 1} = 2{,}598{,}960$$

▼ **example 6**   **Counting Using the Combination Formula**

An emergency room at a hospital has 11 nurses on staff. Each night a team of 5 nurses is on duty. In how many different ways can the team of 5 nurses be chosen?

Solution

This is a combination problem, because the order in which the nurses are chosen is not important. The 5 nurses N_1, N_2, N_3, N_4, N_5 are the same as the 5 nurses N_3, N_5, N_1, N_2, N_4.

$$C(11, 5) = \frac{11!}{5! \cdot (11-5)!} = \frac{11!}{5! \cdot 6!} = \frac{11 \cdot 10 \cdot 9 \cdot 8 \cdot 7 \cdot 6!}{5! \cdot 6!}$$

$$= \frac{11 \cdot 10 \cdot 9 \cdot 8 \cdot 7}{5 \cdot 4 \cdot 3 \cdot 2 \cdot 1} = 462$$

There are 462 possible teams of 5 nurses.

▼ **check your progress 6**   A restaurant employs 16 waiters and waitresses. In how many ways can a group of 9 waiters and waitresses be chosen for the lunch shift?

Solution   *See page S45.*   ◄

▼ **example 7**   Counting Using the Combination Formula and the Counting Principle

A committee of 5 is chosen from 5 mathematicians and 6 economists. How many different committees are possible if the committee must include 2 mathematicians and 3 economists?

Solution
Because a committee of professors A, B, C, D, and E is exactly the same as a committee of professors B, D, E, A, and C, choosing a committee is an example of choosing a combination. There are 5 mathematicians from whom 2 are chosen, which is equivalent to $C(5, 2)$ combinations. There are 6 economists from whom 3 are chosen, which is equivalent to $C(6, 3)$ combinations. Therefore, by the counting principle, there are $C(5, 2) \cdot C(6, 3)$ ways to choose 2 mathematicians and 3 economists.

$$C(5, 2) \cdot C(6, 3) = \frac{5!}{2! \cdot 3!} \cdot \frac{6!}{3! \cdot 3!} = 10 \cdot 20 = 200$$

There are 200 possible committees consisting of 2 mathematicians and 3 economists.

▼ **check your progress 7**   An IRS auditor randomly chooses 5 tax returns to audit from a stack of 10 tax returns, 4 of which are from corporations and 6 of which are from individuals. In how many different ways can the auditor choose the tax returns if the auditor wants to include 3 corporate and 2 individual returns?

Solution   *See page S45.*   ◄

MATHMATTERS   Buying Every Possible Lottery Ticket

A lottery prize in Pennsylvania reached $65 million. A resident of the state suggested that it might be worth buying a ticket for every possible combination of numbers. To win the $65 million, a person needed to correctly select 6 of 50 numbers. Each ticket costs $1. Because the order of the numbers drawn is not important, the number of different possible tickets is $C(50, 6) = 15,890,700$. Thus it would cost the resident $15,890,700 to purchase tickets for every possible combination of numbers.

It might seem that an approximately $16 million investment to win $65 million is reasonable. Unfortunately, when prize levels reach such lofty heights, many more people play the lottery. This increases the chances that more than one person will select the winning combination of numbers. In fact, there were eight people with the winning numbers and each received approximately $8 million. Now the $16 million investment does not look very appealing.

Some of the problems in this section require a basic knowledge of what comprises a standard deck of playing cards. In a standard deck, there are 4 suits: spades, hearts, diamonds, and clubs. Each suit has 13 cards: 2 through 10, jack, queen, king, and ace. See the Math Matters on page 747.

▼ **example 8** **Counting Problems with Cards**

From a standard deck of playing cards, 5 cards are chosen. How many 5-card combinations contain

a. 2 kings and three queens?

b. 5 hearts?

c. 5 cards of the same suit?

Solution

a. There are $C(4, 2)$ ways of choosing 2 kings from 4 kings and $C(4, 3)$ ways of choosing 3 queens from 4 queens. By the counting principle, there are $C(4, 2) \cdot C(4, 3)$ ways of choosing 2 kings and 3 queens.

$$C(4, 2) \cdot C(4, 3) = \frac{4!}{2! \cdot 2!} \cdot \frac{4!}{3! \cdot 1!} = 6 \cdot 4 = 24$$

There are 24 ways of choosing 2 kings and 3 queens.

b. There are $C(13, 5)$ ways of choosing 5 hearts from 13 hearts.

$$C(13, 5) = \frac{13!}{5! \cdot 8!} = 1287$$

There are 1287 ways to choose 5 hearts from 13 hearts.

c. From part b, there would also be 1287 ways of choosing 5 spades from 13 spades, 5 clubs from 13 clubs, or 5 diamonds from 13 diamonds. Because there are 4 suits from which to choose and $C(13, 5)$ ways of choosing 5 cards from a suit, by the counting principle there are $4 \cdot C(13, 5)$ ways to choose 5 cards of the same suit.

$$4 \cdot C(13, 5) = 4 \cdot 1287 = 5148$$

There are 5148 ways of choosing 5 cards of the same suit from a standard deck of playing cards.

Five cards of the same suit is called a *flush* in poker.

▼ **check your progress 8** From a standard deck of playing cards, 5 cards are chosen. How many 5-card combinations contain 4 cards of the same suit?

Solution *See page S45.*

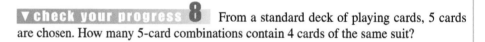

EXCURSION

Choosing Numbers in Keno

A popular gambling game called keno, first introduced in China over 2000 years ago, is played in many casinos. In keno, there are 80 balls numbered from 1 to 80. The casino randomly chooses 20 balls from the 80 balls. These are "lucky balls" because if a gambler chooses some of the numbers on these balls, there is a possibility of winning money. The amount that is won depends on the number of lucky numbers the gambler has selected. The number of ways in which a casino can choose 20 balls from 80 is

$$C(80, 20) = \frac{80!}{20! \cdot 60!} \approx 3,535,000,000,000,000,000$$

Photolibrary

A keno card is used to mark the numbers chosen.

Once the casino chooses the 20 lucky balls, the remaining 60 balls are unlucky for the gambler. A gambler who chooses 5 numbers will have from 0 to 5 lucky numbers.

Let's consider the case in which 2 of the 5 numbers chosen by the gambler are lucky numbers. Because 5 numbers were chosen, there must be 3 unlucky numbers among the 5 numbers. The number of ways of choosing 2 lucky numbers from 20 lucky numbers is $C(20, 2)$. The number of ways of choosing 3 unlucky numbers from 60 unlucky numbers is $C(60, 3)$. By the counting principle, there are $C(20, 2) \cdot C(60, 3) = 190 \cdot 34{,}220 = 6{,}501{,}800$ ways to choose 2 lucky and 3 unlucky numbers.

EXCURSION EXERCISES

For each of the following exercises, assume that a gambler playing keno has randomly chosen 4 numbers.

1. In how many ways can the gambler choose no lucky numbers?

2. In how many ways can the gambler choose exactly 1 lucky number?

3. In how many ways can the gambler choose exactly 2 lucky numbers?

4. In how many ways can the gambler choose exactly 3 lucky numbers?

5. In how many ways can the gambler choose 4 lucky numbers?

EXERCISE SET 12.2
(Suggested Assignment: The Enhanced WebAssign Exercises and Exercise 51)

■ In Exercises 1 to 30, evaluate each expression.

1. $8!$

2. $5!$

3. $9! - 5!$

4. $(9 - 5)!$

5. $(8 - 3)!$

6. $8! - 3!$

7. $P(8, 5)$

8. $P(7, 2)$

9. $P(9, 7)$

10. $P(10, 5)$

11. $P(8, 0)$

12. $P(7, 0)$

13. $P(8, 8)$

14. $P(10, 10)$

15. $P(8, 2) \cdot P(5, 3)$

16. $\dfrac{P(10, 4)}{P(8, 4)}$

17. $\dfrac{P(6, 0)}{P(6, 6)}$

18. $\dfrac{P(6, 3) \cdot P(5, 2)}{P(4, 3)}$

19. $C(9, 2)$

20. $C(8, 6)$

21. $C(12, 0)$

22. $C(11, 11)$

23. $C(6, 2) \cdot C(7, 3)$

24. $C(8, 5) \cdot C(9, 4)$

25. $\dfrac{C(10, 4) \cdot C(5, 2)}{C(15, 6)}$

26. $\dfrac{C(4, 3) \cdot C(5, 2)}{C(9, 5)}$

27. $\dfrac{C(9, 7) \cdot C(5, 3)}{C(14, 10)}$

28. $3! \cdot C(8, 5)$

29. $4! \cdot C(10, 3)$

30. $5! \cdot C(18, 0)$

■ In Exercises 31 to 34, how many combinations are possible? Assume that the items are distinct.

31. 7 items chosen 5 at a time

32. 8 items chosen 3 at a time

33. 12 items chosen 7 at a time

34. 11 items chosen 11 at a time

35. Is it possible to calculate $C(7, 9)$? Think of your answer in terms of 7 items chosen 9 at a time.

36. Is it possible to calculate $C(n, k)$ where $k > n$? See Exercise 35 for some help.

37. Music Downloads A student downloaded 5 music files to a portable MP3 player. In how many different orders can the songs be played?

38. Elections The board of directors of a corporation must select a president, a secretary, and a treasurer. In how many possible ways can this be accomplished if there are 20 members on the board of directors?

39. Elections A committee of 16 students must select a president, a vice-president, a secretary, and a treasurer. In how many possible ways can this be accomplished?

40. The Olympics A gold, a silver, and a bronze medal are awarded in an Olympic event. In how many possible ways can the medals be awarded for a 200-meter sprint in which there are 9 runners?

41. Music Festival   Six country music bands and 3 rock bands are signed up to perform at an all-day festival. How many different orders can the bands play in if

   a. there are no restrictions on the order?

   b. all the bands of each type must perform in a row?

42. Radio Show   Five rock songs and 6 rap songs are on a disc jockey's playlist for a radio show. How many different orders can the songs be played in if

   a. there are no restrictions on the order?

   b. all the rap songs are played consecutively?

43. Passwords   A password requires 8 characters. If a soccer player whose jersey number is 77 wants his password to be an arrangement of the numbers and letters in SOCCER77, how many passwords are possible?

44. Gardening   A gardener is planting a row of tulip bulbs. She has a mix of bulbs for 10 yellow tulips and 6 red tulips. How many color arrangements are possible for the row of tulips?

45. Firefighters   At a certain fire station, one team of firefighters consists of 8 firefighters. If there are 24 firefighters qualified for a team, how many different teams are possible?

46. Platoons   A typical platoon consists of 20 soldiers. If there are 30 soldiers available to create a platoon, how many different platoons are possible?

47. Exam Questions   A professor gives his students 7 essay questions to prepare for an exam. Only 3 of the questions will actually appear on the exam. How many different exams are possible?

48. Test Banks   A math quiz is generated by randomly choosing 5 questions from a test bank consisting of 50 questions. How many different quizzes are possible?

49. Committee Selection   A committee of 6 people is chosen from 8 women and 8 men. How many different committees are possible that consist of 3 women and 3 men?

50. Quality Control   In a shipment of 20 smart phones, 2 are defective. How many ways can a quality control inspector randomly test 5 smart phones, of which 2 are defective?

51. Football   In the National Football Conference (NFC), the NFC East division has 4 teams. During the regular season, each NFC East team plays each of the other NFC East teams twice. How many games are played between teams of the NFC East during a regular season?

52. Basketball   In the Eastern Conference of the National Basketball Association (NBA), the Atlantic Division has 7 teams and the Central Division has 8 teams. During the regular season, each NBA team plays each of the other teams in its division 4 times. How many games are played between teams of the Atlantic Division during a regular season? How many games are played between teams of the Central Division during a regular season?

53. Geometry   A hexagon is a 6-sided plane figure. A diagonal is a line segment connecting any 2 nonadjacent vertices. How many diagonals are possible?

54. Geometry   Seven distinct points are drawn on a circle. How many different triangles can be drawn in which each vertex of the triangle is at 1 of the 7 points?

55. Softball   Eighteen people decide to play softball. In how many ways can the 18 people be divided into 2 teams of 9 people?

56. Bowling   Fifteen people decide to form a bowling league. In how many ways can the 15 people be divided into 3 teams of 5 people each?

57. Signal Flags   The Coast Guard uses signal flags as a method of communicating between ships. If 4 different flags are available and the order in which the flags are raised is important, how many different signals are possible? Assume that 4 flags are raised.

58. Pizza Toppings   A restaurant offers a special pizza with any 5 toppings. If the restaurant has 12 toppings from which to choose, how many different special pizzas are possible?

59. Letter Arrangements   How many different letter arrangements are possible using all the letters of the word *committee*?

60. Color Arrangements   A set of 12 plates came with 4 red, 4 green, and 4 yellow plates, but 2 red plates have been broken. The remaining plates are stacked on a shelf. How many different color arrangements are possible for the stack of plates?

61. Coin Tosses   Ten identical coins are tossed. How many possible arrangements of the coins include 5 heads and 5 tails?

62. Coin Tosses   Twelve identical coins are tossed. How many possible arrangements of the coins include 8 heads and 4 tails?

63. Concerts Three groups will perform at a choral concert. One group will sing 3 pieces, one group will sing 4 pieces, and one group will sing 2 pieces. In how many possible orders can the 9 pieces be performed, assuming each group performs all its songs consecutively?

64. Morse Code In 1835, Samuel Morse, a professor of art at New York University, devised a code that could be transmitted over a wire by using an electric current. This was the invention of the telegraph. The code used is called Morse code. It consists of a dot or a dash or a combination of up to 5 dots and/or dashes. For instance, the letter c is represented by — • — •, and • — represents the letter a. The numeral 0 is represented by 5 dashes, — — — — —. How many different symbols can be represented using Morse code?

■ Exercises 65 to 70 refer to a standard deck of playing cards. Assume that 5 cards are randomly chosen from the deck.

65. How many hands contain 4 aces?

66. How many hands contain 2 aces and 2 kings?

67. How many hands contain exactly 3 jacks?

68. How many hands contain exactly 3 jacks and 2 queens?

69. How many hands contain exactly two 7s?

70. How many hands contain exactly two 7s and two 8s?

71. Write a few sentences that explain the difference between a permutation and a combination of distinct objects. Give examples of each.

72. Francis Bacon, a contemporary of William Shakespeare, invented a cipher (a secret code) based on permutations of the letters a and b. Write an essay on Bacon's method and the intended use of his scheme.

EXTENSIONS

Critical Thinking

73. If the expression $(x + y)^5$ is expanded, the result is initially 32 terms. The terms consist of every possible product of the form $a_1a_2a_3a_4a_5$, where each a_i is either x or y. After combining like terms, the expansion simplifies to 6 terms with variable parts x^5, x^4y, x^3y^2, x^2y^3, xy^4, and y^5. Use the formula for the arrangements of nondistinct objects to find the coefficient of each term and then write the expansion of $(x + y)^5$.

74. If the expression $(x + y)^6$ is expanded, the result is initially 64 terms. The terms consist of every possible product of the form $a_1a_2a_3a_4a_5a_6$, where each a_i is either x or y. After combining like terms, the expansion simplifies to 7 terms with variable parts x^6, x^5y, x^4y^2, x^3y^3, x^2y^4, xy^5, and y^6. Use the formula for the arrangements of nondistinct objects to find the coefficient of each term and then write the expansion of $(x + y)^6$.

Explorations

75. Hamiltonian Circuits In Chapter 5, we examined the problem of determining certain paths through a network (or graph). One particular path, called a *Hamiltonian circuit,* visited each vertex (dot) of a graph exactly once and returned to the starting vertex. A Hamiltonian circuit is shown as the green path in the following network. The number associated with each line (edge) of the graph indicates the distance between two vertices. (The indicated distance does not match the physical distance in the drawing.) In most applications, the object is to find the *shortest* path. The green path shown here is the shortest path that visits each vertex exactly once and returns to the starting vertex.

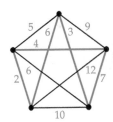

One method of searching for the shortest Hamiltonian circuit is by trial and error. For a network with a small number of vertices, this procedure will work fine. As the number of vertices increases, the likelihood of finding the shortest path using trial-and-error solution becomes extremely remote.

The counting principle can be used to find the number of possible paths through a network in which every pair of vertices possible is connected by a line. For the complete network with 5 vertices shown in the figure on the next page, beginning at the home vertex labeled H, there are 4 choices for the next vertex to visit, 3 choices for the next, 2 choices for the next, and finally 1 choice that returns to H. By the counting principle, there are $4 \cdot 3 \cdot 2 \cdot 1 = 24$ possible circuits. Traveling circuit *HABCDH* is the same as traveling the circuit in the reverse order, *HDCBAH*. Thus there are $\frac{24}{2} = 12$ possible circuits. In general, there are $\frac{(n-1)!}{2}$ possible Hamiltonian circuits through a network of n vertices, where $n \geq 3$.

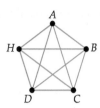

a. A network has 8 vertices, and each vertex is connected to the other vertices. How many Hamiltonian circuits are possible?

b. For the network below, find the number of possible Hamiltonian circuits.

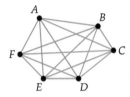

c. Suppose a network has 20 vertices (with every pair of vertices connected by a line segment) and a computer is available that can analyze 1 million paths per second. How many years would it take this computer to find all the possible Hamiltonian circuits of this network? (Assume that a year is 365 days.)

d. Suppose a network has 40 vertices (with every pair of vertices connected by a line segment) and a computer is available that can analyze 1 trillion paths per second. How many years would it take this computer to find all the possible Hamiltonian circuits? Assume a year is 365 days.

76. **Lotteries** The PowerBall multistate lottery is played by choosing 5 numbers between 1 and 59 (regular numbers) and 1 number between 1 and 39 (bonus number). A player wins a prize under any of the following conditions:

i) Only the bonus number is chosen correctly.

ii) The bonus number and 1 regular number are chosen correctly.

iii) The bonus number and 2 regular numbers are chosen correctly.

iv) Three regular numbers are chosen correctly.

v) The bonus number and 3 regular numbers are chosen correctly.

vi) Four regular numbers are chosen correctly.

vii) The bonus number and 4 regular numbers are chosen correctly.

viii) Five regular numbers are chosen correctly.

ix) The bonus number and 5 regular numbers are chosen correctly. This combination results in winning the jackpot.

In how many different ways can a person who plays this game win a prize?

---

s e c t i o n 12.3 Probability and Odds

Introduction to Probability

In California, the likelihood of selecting the winning lottery numbers in the Super Lotto Plus game is approximately 1 chance in 41,000,000. In contrast, the likelihood of being struck by lightning is about 1 chance in 500,000. Comparing the likelihood of winning the Super Lotto Plus to the likelihood of being struck by lightning, you are about 80 times more likely to be struck by lightning than to pick the winning California lottery numbers.

The likelihood of the occurrence of a particular event is described by a number between 0 and 1. (You can think of this as a percentage between 0% and 100%, inclusive.) This number is called the **probability** of the event. An event that is not very likely has a probability close to 0; an event that is very likely has a probability close to 1 (100%). For instance, the probability of being struck by lightning is close to 0. However, if you randomly choose a basketball player from the National Basketball Association, it is very likely that the player is over 6 feet tall, so the probability is close to 1.

Because any event has from a 0% to 100% chance of occurring, probabilities are always between 0 and 1, inclusive. If an event *must* occur, its probability is 1. If an event *cannot* occur, its probability is 0.

Probabilities can be calculated by considering the outcome of experiments. Here are some examples of experiments.

- Flip a coin and observe the outcome as a head or a tail.
- Select a company and observe its annual profit.
- Record the time a person spends at the checkout line in a supermarket.

The sample space of an experiment is the set of all possible outcomes of the experiment. For example, consider tossing a coin three times and observing the outcome as a head or a tail. Using H for head and T for tail, the sample space is

$$S = \{HHH, HHT, HTH, HTT, THH, THT, TTH, TTT\}$$

Note that the sample space consists of *every* possible outcome of tossing three coins.

▼ example 1   Find a Sample Space

A single die is rolled once. What is the sample space for this experiment?

Solution
The sample space is the set of possible outcomes of the experiment.

$$S = \{\boxdot, \boxdot, \boxdot, \boxdot, \boxdot, \boxdot\}$$

▼ check your progress 1   A coin is tossed twice. What is the sample space for this experiment?

Solution   *See page S45.*   ◀

Formally, an event is a subset of a sample space. Using the sample space of Example 1, here are some possible events:

- There are an even number of pips (dots) facing up. The event is

$$E_1 = \{\boxdot, \boxdot, \boxdot\}.$$

- The number of pips facing up is greater than 4. The event is $E_2 = \{\boxdot, \boxdot\}$.
- The number of pips facing up is less than 20. The event is

$$E_3 = \{\boxdot, \boxdot, \boxdot, \boxdot, \boxdot, \boxdot\}.$$

Because the number of pips facing up is always less than 20, this event will always occur. The event and the sample space are the same.

- The number of pips facing up is greater than 15. The event is $E_4 = \varnothing$, the empty set. This is an impossible event; the number of pips facing up cannot be greater than 15.

Outcomes of some experiments are **equally likely**, which means that the chance of any one outcome is just as likely as the chance of another. For instance, if 4 balls of the same size but different colors—red, blue, green, and white—are placed in a box and a blindfolded person chooses 1 ball, the chance of choosing a green ball is the same as the chance of choosing any other color ball.

In the case of equally likely outcomes, the probability of an event is based on the number of elements in the event and the number of elements in the sample space. We will use $n(E)$ to denote the number of elements in the event E and $n(S)$ to denote the number of elements in the sample space S.

▼ **Probability of an Event**

For an experiment with sample space S of *equally likely outcomes,* the probability $P(E)$ of an event E is given by

$$P(E) = \frac{n(E)}{n(S)} = \frac{\text{number of elements in } E}{\text{total number of elements in sample space } S}$$

Because each outcome of rolling a fair die is equally likely, the probability of the events E_1 through E_4 described on page 757 can be determined from the formula for the probability of an event.

$$P(E_1) = \frac{3}{6} \quad \longleftarrow \text{Number of elements in } E_1$$
$$\phantom{P(E_1) = \frac{3}{6}} \quad \longleftarrow \text{Number of elements in the sample space}$$
$$= \frac{1}{2}$$

The probability of rolling an even number of pips on a single roll of one die is $\frac{1}{2}$ (or 50%).

$$P(E_2) = \frac{2}{6} \quad \longleftarrow \text{Number of elements in } E_2$$
$$\phantom{P(E_2) = \frac{2}{6}} \quad \longleftarrow \text{Number of elements in the sample space}$$
$$= \frac{1}{3}$$

The probability of rolling a number greater than 4 on a single roll of one die is $\frac{1}{3}$.

$$P(E_3) = \frac{6}{6} \quad \longleftarrow \text{Number of elements in } E_3$$
$$\phantom{P(E_3) = \frac{6}{6}} \quad \longleftarrow \text{Number of elements in the sample space}$$
$$= 1$$

The probability of rolling a number less than 20 on a single roll of one die is 1 (or 100%). Recall that the probability of any event that is certain to occur is 1.

$$P(E_4) = \frac{0}{6} \quad \longleftarrow \text{Number of elements in } E_4$$
$$\phantom{P(E_4) = \frac{0}{6}} \quad \longleftarrow \text{Number of elements in the sample space}$$
$$= 0$$

The probability of rolling a number greater than 15 on a single roll of one die is 0. (It is not possible to roll any number greater than 6.)

▼ example **2**   Probability of Equally Likely Outcomes

A fair coin—one for which it is equally likely that heads or tails will result from a single toss—is tossed 3 times. What is the probability that 2 heads and 1 tail are tossed?

Solution
Determine the number of elements in the sample space. The sample space must include every possible toss of a head or a tail (in order) in 3 tosses of the coin.

$$S = \{\text{HHH, HHT, HTH, HTT, THH, THT, TTH, TTT}\}$$

In Example 2, we calculated the probability as $\frac{3}{8}$. However, we could have expressed the probability as a decimal, 0.375, or as a percent, 37.5%. A probability can always be expressed as a fraction, a decimal, or a percent.

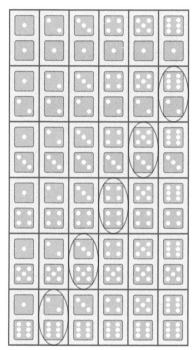

FIGURE 12.6 Outcomes of the roll of 2 dice

Point up	15
Side	85
Total	100

The elements in the event are $E = \{HHT, HTH, THH\}$.

$$P(E) = \frac{n(E)}{n(S)} = \frac{3}{8}$$

The probability is $\frac{3}{8}$.

▼ check your progress 2 If a fair die is rolled once, what is the probability that an odd number will show on the upward face?

Solution   *See page S45.* ◄

question Is it possible that the probability of some event could be 1.23?

▼ example 3 **Calculating Probabilities with Dice**

Two fair dice are tossed, one after the other. What is the probability that the sum of the pips on the upward faces of the 2 dice equals 8?

Solution
The dice must be considered as distinct, so there are 36 possible outcomes. ([ ][ ] and [ ][ ] are considered different outcomes.) Therefore, $n(S) = 36$. The sample space is shown at the left. (See also Section 12.1.) Let E represent the event that the sum of the pips on the upward faces is 8. These outcomes are circled in Figure 12.6. By counting the number of circled pairs, $n(E) = 5$.

$$P(E) = \frac{n(E)}{n(S)} = \frac{5}{36}$$

The probability that the sum of the pips is 8 is $\frac{5}{36}$.

▼ check your progress 3 Two fair dice are tossed. What is the probability that the sum of the pips on the upward faces of the 2 dice equals 7?

Solution   *See page S45.* ◄

Empirical Probability

Probabilities such as those calculated in the preceding examples are sometimes referred to as **theoretical probabilities**. In Example 2, we assume that, in theory, we have a perfectly balanced coin, and we calculate the probability based on the fact that each outcome is equally likely. Similarly, we assume the dice in Example 3 are equally likely to land with any of the 6 faces upward.

When a probability is based on data gathered from an experiment, it is called an **experimental probability** or an **empirical probability**. For instance, if we tossed a thumbtack 100 times and recorded the number of times it landed "point up," the results might be as shown in the table at the left. From this experiment, the empirical probability of "point up" is

$$P(\text{point up}) = \frac{15}{100} = 0.15$$

answer No. All probabilities must be between 0 and 1, inclusive.

▼ Empirical Probability of an Event

If an experiment is performed repeatedly and the occurrence of the event E is observed, the probability $P(E)$ of the event E is given by

$$P(E) = \frac{\text{number of times event } E \text{ occurred}}{\text{number of times the experiment was performed}}$$

▼ **example 4**   Calculate an Empirical Probability

A survey of the registrar of voters office in a certain city showed the following information on the ages and party affiliations of registered voters. If one voter is chosen from this survey, what is the probability that the voter is a Republican?

Age	Republican	Democrat	Independent	Other	Total
18–28	205	432	98	112	847
29–38	311	301	109	83	804
39–49	250	251	150	122	773
≥50	272	283	142	107	804
Total	1038	1267	499	424	3228

Solution

Let R be the event that a Republican is selected. Then

$$P(R) = \frac{1038}{3228} \quad \longleftarrow \text{ Number of Republicans in the survey}$$
$$\qquad\qquad \longleftarrow \text{ Total number of people surveyed}$$
$$\approx 0.32$$

The probability that the selected person is a Republican is approximately 0.32.

▼ **check your progress 4**   Using the data from Example 4, what is the probability that a randomly selected person is between the ages of 39 and 49?

Solution   *See page S45.*

HISTORICAL NOTE

Gregor Mendel
(mĕn′dl) (1822–1884) was an Augustinian monk and teacher. His study of the transmission of traits from one generation to another was actually started to confirm the prevailing theory of the day that environment influenced the traits of a plant. However, his research seemed to suggest that plants have certain "determiners" that dictate the characteristics of the next generation. Thus began the study of heredity. It took over 30 years for Mendel's conclusions to be accepted in the scientific community.

Application to Genetics

Completed in April 2003, the Human Genome Project was a 13-year-long project designed to completely map the genetic make-up of *Homo sapiens*. Researchers hope to use this information to treat and prevent certain hereditary diseases.

The concept behind this project began with Gregor Mendel and his work on flower color and how it was transmitted from generation to generation. From his studies, Mendel concluded that flower color seems to be predictable in future generations by making certain assumptions about a plant's color "determiner." He concluded that red was a *dominant* determiner of color and that white was a *recessive* determiner. Today, geneticists talk about the *gene* for flower color and the *allele* of the gene. A gene consists of two dominant alleles (two red), a dominant and a recessive allele (red and white), or two recessive alleles (two white). Because red is the dominant allele, a flower will be white only if no dominant allele is present.

Later work by Reginald Punnett (1875–1967) showed how to determine the probability of certain flower colors by using a **Punnett square**. Using a capital letter for a dominant allele (R for red) and the corresponding lower-case letter for recessive allele (r for white), Punnett arranged the alleles of the parents in a square. A parent could be RR, Rr, or rr.

Suppose that the genotype (genetic composition) of one parent is rr and the genotype of the other is Rr. The first parent is represented in the left column of the square, and the second parent is represented in the top row. The genotypes of the offspring are shown in the body of the table and are the result of combining one allele from each parent.

Parents	R	r
r	Rr	rr
r	Rr	rr

Because each of the genotypes of the offspring are equally likely, the probability that a flower is red is $\frac{1}{2}$ (two Rr genotypes of the four possible genotypes), and the probability that a flower is white is $\frac{1}{2}$ (two rr genotypes of the four possible genotypes).

▼ example **5**   Probability Using a Punnett Square

A child will have cystic fibrosis if the child inherits the recessive gene from both parents. Using F for the normal allele and f for the mutant recessive allele, suppose a parent who is Ff (said to be a *carrier*) and a parent who is FF (does not have the mutant allele) decide to have a child.

a. What is the probability that the child will have cystic fibrosis? (To have the disease, the child must be ff.)

b. What is the probability that the child will be a carrier?

Solution
Make a Punnett square.

Parents	F	F
F	FF	FF
f	Ff	Ff

a. To have the disease, the child must be ff. From the table, there is no combination of the alleles that will produce ff. Therefore, the child cannot have the disease, and the probability is 0.

b. To be a carrier, exactly one allele must be f. From the table, there are two cases out of four of a genotype with one f. The probability that the child will be a carrier is $\frac{2}{4} = \frac{1}{2}$.

▼ check your progress **5**   For a certain type of hamster, the color cinnamon, C, is dominant and the color white, c, is recessive. If both parents are Cc, what is the probability that an offspring will be white?

Solution   *See page S46.*   ◀

TAKE NOTE

The probability that a parent will pass a certain genetic characteristic on to a child is one-half. However, the probability that a parent has a certain genetic characteristic is not one-half. For instance, the probability of passing on the allele for cystic fibrosis by a parent who has the mutant allele is 0.5. However, the probability that a person randomly selected from the population has the mutant allele is less than 0.025. We will discuss this further in Section 12.5.

Peyton Manning

© PCN Photography/Alamy

INSTRUCTOR NOTE

Make sure that students understand that a fraction such as $\frac{1}{4}$ when used to express odds is being used in a very different way from when $\frac{1}{4}$ is used to express a probability.

Calculating Odds

Based on his pass completion average over the years 1998 to 2010, when quarterback Peyton Manning throws the football, the *odds in favor* of the pass being completed are 13 to 7. These odds indicate that out of 20 (i.e., 13 + 7) passes thrown by Manning, you can expect 13 passes to be completed and 7 passes to be incomplete.

When the racehorse Ice Box raced in the Belmont Stakes, the last stage of the Triple Crown, the *odds against* Ice Box winning the race were 3 to 1. In this case, bettors estimated that in 4 (i.e., 3 + 1) races, Ice Box would lose 3 races and win 1 race.

A **favorable outcome** of an experiment is one that satisfies some event. For instance, in the case of a Peyton Manning pass, you can assume that out of 20 passes, there will be 13 favorable outcomes (13 completed passes). The opposite event, an incomplete pass, is an **unfavorable outcome**. Odds are frequently expressed in terms of favorable and unfavorable outcomes.

▼ Odds of an Event

Let E be an event in a sample space of equally likely outcomes. Then

$$\textbf{Odds in favor of } E = \frac{\text{number of favorable outcomes}}{\text{number of unfavorable outcomes}}$$

$$\textbf{Odds against } E = \frac{\text{number of unfavorable outcomes}}{\text{number of favorable outcomes}}$$

When the odds of an event are written in fractional form, the fraction bar is read as the word *to*. Thus odds of $\frac{3}{2}$ are read as "3 to 2." We can also write odds of $\frac{3}{2}$ as 3:2. This form is also read as "3 to 2."

▼ example 6 **Calculate Odds**

If a pair of fair dice is rolled once, what are the odds in favor of rolling a sum of 7?

Solution

Let E be the event of rolling a sum of 7. From Figure 12.6 on page 759, the 6 favorable outcomes are $E = \{$ ⚀⚅, ⚁⚄, ⚂⚃, ⚃⚂, ⚄⚁, ⚅⚀ $\}$. The unfavorable outcomes are the remaining 30 possibilities. (Because there are 36 possible outcomes when tossing 2 dice and 6 of them are favorable, then $36 - 6 = 30$ are unfavorable.)

$$\text{Odds in favor of } E = \frac{\text{number of favorable outcomes}}{\text{number of unfavorable outcomes}} = \frac{6}{30} = \frac{1}{5}$$

The odds in favor of rolling a sum of 7 are 1 to 5.

▼ check your progress 6 If 3 red, 4 white, and 5 blue balls are placed in a box and 1 ball is randomly selected from the box, what are the odds against the ball being blue?

Solution *See page S46.*

◄

Odds express the likelihood of an event and are therefore related to probability. When the odds of an event are known, the probability of the event can be determined. Conversely, when the probability of an event is known, the odds of the event can be determined.

▼ **The Relationship between Odds and Probability**

1. Suppose E is an event in a sample space and that the *odds in favor* of E are $\frac{a}{b}$. Then $P(E) = \frac{a}{a+b}$.

2. Suppose E is an event in a sample space. Then the *odds in favor* of E are $\frac{P(E)}{1 - P(E)}$.

▼ **example 7** **Determine Probability from Odds**

In 2010, the racehorse Drosselmeyer won the Belmont Stakes, beating the favorite Ice Box. The odds against Drosselmeyer winning the race were 12 to 1. What was the probability of Drosselmeyer winning the race?

Solution

Because the odds *against* Drosselmeyer winning the race were 12 to 1, the odds *in favor* of Drosselmeyer winning the race were 1 to 12, or, as a fraction, $\frac{1}{12}$. Now use the formula for calculating the probability of an event when the odds in favor are known.

$$P(E) = \frac{a}{a+b}$$

$$P(E) = \frac{1}{1+12} = \frac{1}{13} \qquad \bullet\, a = 1, b = 12$$

The probability of Drosselmeyer winning the race was $\frac{1}{13}$ or about 7.7%.

▼ **check your progress 7** A report issued by the Southern California Earthquake Data Center estimates that the probability of an earthquake of magnitude 6.7 or greater within 30 years in the San Francisco Bay Area is about 60%. What are the odds in favor of that type of earthquake occurring in that region in the next 30 years?

Solution *See page S46.*

MATHMATTERS   The Birth of Probability Theory

Pierre Fermat and Blaise Pascal are generally considered to be the founders of the study of probability. For many historians, the starting point of the formal discussion of probability is contained in a letter from Pascal to Fermat written in July 1654.

The Chevalier de Mere said to me that he found a falsehood in the theory of numbers for the following reason. If one wants to throw a six with a single die, there is an advantage in undertaking to do it in four throws, as the odds are 671 to 625. If one throws two sixes with a pair of dice, there is a disadvantage in having only 24 throws. However, 24 to 36 (which is the number of cases for two dice) is 4 to 6 (which is the number of cases on one die). This is the great "scandal" which makes him proclaim loftily that the theorems are not constant and Arithmetic is self-contradictory.

Basically (although it certainly isn't obvious from this letter), the Chevalier was claiming that there is better than a 50% chance of tossing a 6 in 4 rolls of a single die. By his reasoning, he concluded that there should be a better than 50% chance of rolling a pair of 6s in 24 tosses of 2 dice. However, the Chevalier tested his theory and found that 25 tosses were needed. From his tests, he concluded a "falsehood in the theory of numbers." In this letter, Pascal was mocking the inability of the Chevalier to correctly determine the probabilities involved.

EXCURSION

Photolibrary

The Value of Pi by Simulation

A **simulation** is an activity designed to approximate a given situation. For instance, pilots fly a simulator to practice maneuvers that would be dangerous to try in an airplane. Chemists use computer programs to simulate chemical processes. This Excursion uses a simulation to approximate the value of π.

This Excursion will work better if four or five people work together. Get 25 toothpicks, and then tape several sheets of blank paper together to form a large rectangular shape. Use a large ruler to draw parallel lines on the paper such that the distance between the lines is the length of one toothpick. Then drop all of the toothpicks from approximately knee height onto the paper. (Make sure that none of the toothpicks lands off the paper.) Record the number of toothpicks that cross any of the parallel lines. Repeat this process 10 times for a total of 250 toothpicks dropped.[1]

EXCURSION EXERCISES

1. Using the recorded data, calculate the empirical probability that a dropped toothpick will cross a line.

2. The theoretical probability that a toothpick will cross a line is $\frac{2}{\pi}$. Use a calculator to find the value of $\frac{2}{\pi}$, rounded to four decimal places.

3. The experimental value you calculated in Exercise 1 should be close to the theoretical value you calculated in Exercise 2 (at least it should be if you dropped a toothpick around 1000 times). Calculate the percent error between the experimental value and the theoretical value. (The percent error can be determined by finding the difference between the two values and dividing it by the theoretical value.)

4. Explain how performing this experiment thousands of times could give you an approximate value for π.

5. Using your data, what approximate value for π do you calculate?

6. This experiment is based on a famous 18th-century problem called the *Buffon needle problem*. Look up "Buffon needle problem" in the library or on the Internet and write a short essay about this problem and how it applies to this Excursion.

EXERCISE SET **12.3** (Suggested Assignment: The Enhanced WebAssign Exercises and Exercises 20, 41, 43, 45, 47, and 49)

■ In Exercises 1 to 6, list the elements of the sample space for each experiment.

1. A coin is flipped 3 times.

2. An even number between 1 and 11 is selected at random.

3. One day in the first 2 weeks of November is selected.

4. A current U.S. coin is selected from a piggybank.

5. A state is selected from the states in the U.S. whose name begins with the letter A.

6. A month is selected from the months that have exactly 30 days.

---

[1] See our website at www.cengage.com/math/aufmann for a computer program that can be used to simulate the dropping of the toothpicks. This program will enable you to perform the experiment many thousands of times.

■ In Exercises 7 to 15, assume that it is equally likely for a child to be born a boy or a girl and that the Lin family is planning on having 3 children.

7. List the elements of the sample space for the genders of the 3 children.

8. List the elements of the event that the Lins have 2 boys and 1 girl.

9. List the elements of the event that the Lins have at least 2 girls.

10. List the elements of the event that the Lins have no girls.

11. List the elements of the event that the Lins have at least 1 girl.

12. Compute the probability that the Lins will have 2 boys and 1 girl.

13. Compute the probability that the Lins will have at least 2 girls.

14. Compute the probability that the Lins will have no girls.

15. Compute the probability that the Lins will have at least 1 girl.

■ In Exercises 16 to 19, a coin is tossed 4 times. Assuming that the coin is equally likely to land on heads or tails, compute the probability of each event occurring.

16. 2 heads and 2 tails

17. 1 head and 3 tails

18. All 4 coin tosses are identical

19. At least 2 tails

20. **Asteroids**   Read the following article about asteroid 1999 RQ36. The two values given for the probability that the asteroid will collide with Earth are not exactly equal. State whether the decimal value given for the probability is greater than or less than "one chance in a thousand."

IN THE NEWS

Collision Course?

An asteroid known to the scientists who study it as 1999 RQ36 could collide with Earth sometime before the year 2200. Scientists have recently revised their estimates of how likely impact is, putting the overall probability at 0.00092, or about one chance in a thousand.

SOURCE: http://www.csmonitor.com

■ In Exercises 21 to 24, a dodecahedral die (one with 12 sides numbered from 1 to 12) is tossed once. Find each of the following probabilities.

21. The number on the upward face is 12.

22. The number on the upward face is not 10.

23. The number on the upward face is divisible by 4.

24. The number on the upward face is less than 5 or greater than 9.

■ In Exercises 25 to 34, 2 regular 6-sided dice are tossed. Compute the probability that the sum of the pips on the upward faces of the 2 dice is each of the following. (See Figure 12.6 on page 759 for the sample space of this experiment.)

25. 6   26. 11

27. 2   28. 12

29. 1   30. 14

31. At least 10   32. At most 5

33. An even number   34. An odd number

35. If 2 dice are rolled, compute the probability of rolling doubles (both dice show the same number of pips).

36. If 2 dice are rolled, compute the probability of *not* rolling doubles.

■ In Exercises 37 to 40, a card is selected at random from a standard deck of playing cards.

37. Compute the probability that the card is a 9.

38. Compute the probability that the card is a face card (jack, queen, or king).

39. Compute the probability that the card is between 5 and 9, inclusive.

40. Compute the probability that the card is between 3 and 6, inclusive.

■ **Voter Characteristics**   In Exercises 41 to 46, use the data given in Example 4, page 760, to compute the probability that a randomly chosen voter from the survey will satisfy the following.

41. The voter is a Democrat.

42. The voter is not a Republican.

43. The voter is 50 years old or older.

44. The voter is under 39 years old.

45. The voter is between 39 and 49 and is registered as an Independent.

46. The voter is under 29 and is registered as a Democrat.

■ **Education Levels**   In Exercises 47 to 50, a survey asked 850 respondents about their highest levels of completed education. The results are given in the following table.

Education completed	Number of respondents
No high school diploma	52
High school diploma	234
Associate's degree or 2 years of college	274
Bachelor's degree	187
Master's degree	67
Ph.D. or professional degree	36

If a respondent from the survey is selected at random, compute the probability of each of the following.

47. The respondent did not complete high school.

48. The respondent has an associate's degree or 2 years of college (but not more).

49. The respondent has a Ph.D. or professional degree.

50. The respondent has a degree beyond a bachelor's degree.

■ **Annual Salaries**   A random survey asked respondents about their current annual salaries. The results are given in the following table. Use the table for Exercises 51 to 54.

Salary range	Number of respondents
Below $18,000	24
$18,000–$27,999	41
$28,000–$35,999	52
$36,000–$45,999	58
$46,000–$59,999	43
$60,000–$79,999	39
$80,000–$99,999	22
$100,000 or more	14

If a respondent from the survey is selected at random, compute the probability of the following.

51. The respondent earns from $36,000 to $45,999 annually.

52. The respondent earns from $60,000 to $79,999 per year.

53. The respondent earns at least $80,000 per year.

54. The respondent earns less than $36,000 annually.

55. Genotypes   The following Punnett square for flower color shows two parents of genotype Rr, where R corresponds to the dominant red flower allele and r represents the recessive white flower allele. (See Example 5.)

Parents	R	r
R	RR	Rr
r	Rr	rr

What is the probability that the offspring of these parents will have white flowers?

56. Genotypes   One parent plant with red flowers has genotype RR and the other with white flowers has genotype rr, where R is the dominant allele for a red flower and r is the recessive allele for a white flower. Compute the probability of one of the offspring having white flowers. *Hint:* Draw a Punnett square.

57. Genotypes   The eye color of mice is determined by a dominant allele E, corresponding to black eyes, and a recessive allele e, corresponding to red eyes. If two mice, one of genotype EE and the other of genotype ee, have offspring, compute the probability of one of the offspring having red eyes. *Hint:* Draw a Punnett square.

58. Genotypes   The height of a certain plant is determined by a dominant allele T corresponding to tall plants, and a recessive allele t corresponding to short (or dwarf) plants. If both parent plants have genotype Tt, compute the probability that the offspring plants will be tall. *Hint:* Draw a Punnett square.

59. Explain the difference between the probability of an event and the odds of the same event.

60. Give an example of an event that has probability 0 and one that has probability 1.

■ In Exercises 61 to 66, the odds in favor of an event occurring are given. Compute the probability of the event occurring.

61. 1 to 2   **62.** 1 to 4

63. 3 to 7   **64.** 3 to 5

65. 8 to 5   **66.** 11 to 9

■ In Exercises 67 to 72, the probability of an event occurring is given. Find the odds in favor of the event occurring.

67. 0.2   **68.** 0.6

69. 0.375   **70.** 0.28

71. 0.55   **72.** 0.81

73. Game Shows   The game board for the television show "Jeopardy" is divided into 6 categories with each category containing 5 answers. In the Double Jeopardy round, there are 2 hidden Daily Double squares. What are the odds in favor of choosing a Daily Double square on the first turn?

74. Game Shows   The spinner for the televsion game show "Wheel of Fortune" is shown below. On a single spin of the wheel, what are the odds against stopping on $300?

"Wheel of Fortune" courtesy Califon Productions, Inc.

75. If a single fair die is rolled, what are the odds in favor of rolling an even number?

76. If a card is randomly pulled from a standard deck of playing cards, what are the odds in favor of pulling a heart?

77. A coin is tossed 4 times. What are the odds against the coin showing heads all 4 times?

Earthquakes   For Exercises 78 to 80, use the table, which is based on information in an article discussing the probabilities of earthquakes occurring in California.

Earthquake of magnitude . . .	Will occur in California within . . .	Probability or odds
7.5 or greater	next 30 years	Probability: 46%
7.5 or greater	any one year	Odds against: 48 to 49
6.7 or greater	next 30 years	Odds in favor: 332 to 1

Source: http://www.digitjournal.com

78. What are the odds in favor of an earthquake of magnitude 7.5 or greater occurring in California within the next 30 years?

79. What is the probability of an earthquake of magnitude 7.5 or greater occurring in California within any 1 year?

80. What is the probability of an earthquake of magnitude 6.7 or greater occurring in California within the next 30 years?

81. Football   A bookmaker has placed 8 to 3 odds *against* a particular football team winning its next game. What is the probability, in the bookmaker's view, of the team winning?

82. Candy Colors   A snack-size bag of M&Ms candies contains 12 red candies, 12 blue, 7 green, 13 brown, 3 orange, and 10 yellow. If a candy is randomly picked from the bag, compute

Rachel Epstein/The Image Works, Inc.

a. the odds of getting a green M&M.

b. the probability of getting a green M&M.

EXTENSIONS

Critical Thinking

83. If 4 cards labeled A, B, C, and D are randomly placed in 4 boxes also labeled A, B, C, and D, 1 to each box, find the probability that no card will be in a box with the same letter.

84. Determine the probability that if 10 coins are tossed, 5 heads and 5 tails will result.

85. In a family of 3 children, all of whom are girls, a family member new to probability reasons that the probability that each child would be a girl is 0.5. Therefore, the probability that the family would have 3 girls is $0.5 + 0.5 + 0.5 = 1.5$. Explain why this reasoning is not valid.

■ In Exercises 86 and 87, a hand of 5 cards is dealt from a standard deck of playing cards. You may want to review the material on combinations before doing these exercises.

86. Find the probability that the hand will contain all 4 aces.

87. Find the probability that the hand will contain 3 jacks and 2 queens.

Explorations

Roulette Exercises 88 to 93 use the casino game roulette. Roulette is played by spinning a wheel with 38 numbered slots. The numbers 1 through 36 appear on the wheel, half of them colored black and half colored red. Two slots, numbered 0 and 00, are colored green. A ball is placed on the spinning wheel and allowed to come to rest in one of the slots. Bets are placed on where the ball will land.

Vuk Nenezic/Shutterstock.com

88. You can place a bet that the ball will stop in a black slot. If you win, the casino will pay you $1 for each dollar you bet. What is the probability of winning this bet?

89. You can bet that the ball will land on an odd number. If you win, the casino will pay you $1 for each dollar you bet. What is the probability of winning this bet?

90. You can bet that the ball will land on any number from 1 to 12. If you win, the casino will pay you $2 for each dollar you bet. What is the probability of winning this bet?

91. You can bet that the ball will land on any particular number. If you win, the casino will pay you $35 for each dollar you bet. What is the probability of winning this bet?

92. You can bet that the ball will land on one of 0 or 00. If you win, the casino will pay you $17 for each dollar you bet. What is the probability of winning this bet?

93. You can bet that the ball will land on certain groups of 6 numbers (such as 1 to 6). If you win, the casino will pay you $5 for each dollar you bet. What is the probability of winning this bet?

section 12.4 Addition and Complement Rules

The Addition Rule for Probabilities

Suppose you draw a single card from a standard deck of playing cards. The sample space S consists of the 52 cards of the deck. Therefore, $n(S) = 52$. Now consider the events

E_1 = a 4 is drawn = {♠4, ♥4, ♦4, ♣4}

E_2 = a spade is drawn

= {♠A, ♠2, ♠3, ♠4, ♠5, ♠6, ♠7, ♠8, ♠9, ♠10, ♠J, ♠Q, ♠K}

It is possible, on one draw, to satisfy the conditions of both events: the ♠4 could be drawn. This card is an element of both E_1 and E_2.

Now consider the events

E_3 = a 5 is drawn = {♠5, ♥5, ♦5, ♣5}

E_4 = a king is drawn = {♠K, ♥K, ♦K, ♣K}

In this case, it is not possible to draw one card that satisfies the conditions of both events. There are no elements common to both sets. Two events that cannot both occur at the same time are called **mutually exclusive events**. The events E_3 and E_4 are mutually exclusive events, whereas E_1 and E_2 are not.

▼ **Mutually Exclusive Events**

Two events A and B are mutually exclusive if they cannot occur at the same time. That is, A and B are mutually exclusive when $A \cap B = \varnothing$.

---

question A die is rolled once. Let E be the event that an even number is rolled and let O be the event that an odd number is rolled. Are the events E and O mutually exclusive?

The probability of either of two mutually exclusive events occurring can be determined by adding the probabilities of the individual events.

▼ **Probability of Mutually Exclusive Events**

If A and B are two mutually exclusive events, then the probability of A or B occurring is

$$P(A \text{ or } B) = P(A) + P(B)$$

▼ **example 1**   Probability of Mutually Exclusive Events

Suppose a single card is drawn from a standard deck of playing cards. Find the probability of drawing a 5 or a king.

Solution
Let $A = \{\spadesuit 5, \heartsuit 5, \diamondsuit 5, \clubsuit 5\}$ and $B = \{\spadesuit K, \heartsuit K, \diamondsuit K, \clubsuit K\}$. There are 52 cards in a standard deck of playing cards; thus $n(S) = 52$. Because the events are mutually exclusive, we can use the formula for the probability of mutually exclusive events.

$P(A \text{ or } B) = P(A) + P(B)$ • Formula for the probability of mutually
   exclusive events

$= \dfrac{1}{13} + \dfrac{1}{13} = \dfrac{2}{13}$ • $P(A) = \dfrac{4}{52} = \dfrac{1}{13}, P(B) = \dfrac{4}{52} = \dfrac{1}{13}$

The probability of drawing a 5 or a king is $\dfrac{2}{13}$.

▼ **check your progress 1**   Two fair dice are tossed once. What is the probability of rolling a 7 or an 11? For the sample space for this experiment, see page 735.

Solution *See page S46.* ◄

Consider the experiment of rolling two dice. Let A be the event of rolling a sum of 8 and let B be the event of rolling a double (the same number on both dice).

$A = \{ \square\square, \square\square, \square\square, \square\square, \square\square \}$

$B = \{ \square\square, \square\square, \square\square, \square\square, \square\square, \square\square \}$

These events are *not* mutually exclusive because it is possible to satisfy the conditions of each event on one toss of the dice—a $\square$ $\square$ could be rolled. Therefore, $P(A \text{ or } B)$, the probability of a sum of 8 or a double, cannot be calculated using the formula for the probability of mutually exclusive events. However, a modification of that formula can be used.

---

answer Yes. It is not possible to roll an even number and an odd number on a single roll of the die.

The $P(A \text{ and } B)$ term in the Addition Rule for Probabilities is subtracted to compensate for the overcounting of the first two terms of the formula. If two events are mutually exclusive, then $A \cap B = \emptyset$. Therefore, $n(A \cap B) = 0$ and $P(A \text{ and } B) = \frac{n(A \cap B)}{n(S)} = 0$. For mutually exclusive events, the Addition Rule for Probabilities is the same as the formula for the probability of mutually exclusive events.

Recall that the probability of an event A is $P(A) = \frac{n(A)}{n(S)}$. Therefore, $P(A \text{ and } B) = \frac{n(A \cap B)}{n(S)}$.

▼ Addition Rule for Probabilities

If A and B are two events in a sample space S, then

$$P(A \text{ or } B) = P(A) + P(B) - P(A \text{ and } B)$$

Using this formula with

$A = \{$ ⚄⚀, ⚃⚁, ⚂⚂, ⚁⚃, ⚀⚄ $\}$

$B = \{$ ⚀⚀, ⚁⚁, ⚂⚂, ⚃⚃, ⚄⚄, ⚅⚅ $\}$

$A \cap B = \{$ ⚂⚂ $\}$

the probability of A or B can be calculated.

$$P(A \text{ or } B) = P(A) + P(B) - P(A \text{ and } B)$$

$$= \frac{5}{36} + \frac{6}{36} - \frac{1}{36} \qquad • P(A) = \frac{5}{36}, P(B) = \frac{6}{36}, P(A \cap B) = \frac{1}{36}$$

$$= \frac{10}{36} = \frac{5}{18}$$

On a single roll of two dice, the probability of rolling a sum of 8 or a double is $\frac{5}{18}$.

▼ example 2   Use the Addition Rule for Probabilities

The table at the left shows data from an experiment conducted to test the effectiveness of a flu vaccine. If one person is selected from this population, what is the probability that the person was vaccinated or contracted the flu?

Solution

Let $V = \{$people who were vaccinated$\}$ and $F = \{$people who contracted the flu$\}$. These events are not mutually exclusive because there are 21 people who were vaccinated and who contracted the flu. The sample space S consists of the 490 people who participated in the experiment. From the table, $n(V) = 219$, $n(F) = 97$, and $n(V \text{ and } F) = 21$.

$$P(V \text{ or } F) = P(V) + P(F) - P(V \text{ and } F)$$

$$= \frac{219}{490} + \frac{97}{490} - \frac{21}{490}$$

$$= \frac{295}{490} \approx 0.602$$

The probability of selecting a person who was vaccinated or who contracted the flu is approximately 60.2%.

	F	**No F**	**Total**
V	21	198	219
No V	76	195	271
Total	97	393	490

V: Vaccinated
F: Contracted the flu

▼ check your progress 2   The data in the table on the next page show the starting salaries of college graduates with selected degrees. If one person is chosen from this population, what is the probability that the person has a degree in business or has a starting salary between $20,000 and $24,999?

Salary (in $)	Degree			
	Engineering	**Business**	**Chemistry**	**Psychology**
Less than 20,000	0	4	1	12
20,000–24,999	4	16	3	16
25,000–29,999	7	21	5	15
30,000–34,999	12	35	5	7
35,000 or more	12	22	4	5

Solution   *See page S46.*

The Complement of an Event

Consider the experiment of tossing a single die once. The sample space is

$$S = \{\ \boxed{\cdot}\ ,\ \boxed{\cdot\cdot}\ ,\ \boxed{\cdot\cdot\cdot}\ ,\ \boxed{::}\ ,\ \boxed{:\cdot:}\ ,\ \boxed{:::}\ \}$$

Now consider the event $E = \{\ \boxed{\cdot}\ \}$, that is, the event of tossing a $\boxed{\cdot}$. The probability of E is

$$P(E) = \frac{1}{6}\quad \begin{array}{l}\longleftarrow \text{Number of elements in } E \\ \longleftarrow \text{Number of elements in the sample space}\end{array}$$

The **complement** of an event E, symbolized by E^c, is the "opposite" event of E. The complement includes all those outcomes of S that are not in E and excludes the outcomes in E. For the event E above, E^c is the event of not tossing a $\boxed{\cdot}$. Thus

$$E^c = \{\ \boxed{\cdot\cdot}\ ,\ \boxed{\cdot\cdot\cdot}\ ,\ \boxed{::}\ ,\ \boxed{:\cdot:}\ ,\ \boxed{:::}\ \}$$

Note that because E and E^c are opposite events, they are mutually exclusive, and their union is the entire sample space S. Thus $P(E) + P(E^c) = P(S)$. But $P(S) = 1$, so $P(E^c) = 1 - P(E)$.

▼ Probability of the Complement of an Event

If E is an event and E^c is the complement of the event, then

$$P(E^c) = 1 - P(E)$$

Continuing our example, the probability of not tossing a $\boxed{\cdot}$ is given by

$$P\left(\text{not a } \boxed{\cdot}\right) = 1 - P\left(\boxed{\cdot}\right)$$

$$= 1 - \frac{1}{6} = \frac{5}{6}$$

You can also verify the probability of E^c directly:

$$P\left(\text{not a } \boxed{\cdot}\right) = \frac{5}{6}\quad \begin{array}{l}\longleftarrow \text{Number of elements in } E^c \\ \longleftarrow \text{Number of elements in the sample space}\end{array}$$

INSTRUCTOR NOTE

It may be helpful to have students practice formulating a probability in terms of the complement of an event. Ask the students how they might use the complement to determine the probability of the following situations.

- Not rolling doubles in the game of Monopoly
- A randomly selected student on campus is enrolled in at least 2 classes
- A randomly selected student in the class has received at least 1 traffic violation

▼ **example 3** Find a Probability by the Complement Rule

The probability of tossing a sum of 11 on the toss of 2 dice is $\frac{1}{18}$. What is the probability of not tossing a sum of 11 on the toss of 2 dice?

Solution
Use the formula for the probability of the complement of an event. Let $E = \{$toss a sum of 11$\}$. Then $E^c = \{$toss a sum that is not 11$\}$.

$$P(E^c) = 1 - P(E)$$

$$= 1 - \frac{1}{18} = \frac{17}{18} \qquad \bullet P(E) = \frac{1}{18}$$

The probability of not tossing a sum of 11 is $\frac{17}{18}$.

▼ **check your progress 3** The probability that a person has type A blood is 34%. What is the probability that a person does not have type A blood?

Solution *See page S46.* ◄

TAKE NOTE ✓

The phrase "at least one" means 1 or more. Tossing a coin 3 times and asking the probability of getting at least 1 head means to calculate the probability of getting 1, 2, or 3 heads.

Suppose we toss a coin 3 times and want to calculate the probability of having heads occur *at least once*. We could list all the possibilities of tossing 3 coins, as shown below, and then find the ones that contain at least 1 head.

$$\{\underbrace{\text{HHH, HHT, HTH, HTT, THH, THT, TTH,}}_{\text{at least one head}} \text{TTT}\}$$

The probability of at least 1 head is $\frac{7}{8}$.

Another way to calculate this result is to use the formula for the probability of the complement of an event. Let $E = \{$at least 1 head$\}$. From the list above, note that E contains every outcome except TTT (no heads). Thus $E^c = \{$TTT$\}$ and we have

$$P(E) = 1 - P(E^c)$$

$$= 1 - \frac{1}{8} = \frac{7}{8} \qquad \bullet P(E^c) = \frac{n(E^c)}{n(S)} = \frac{1}{8}$$

This is the same result that we calculated above. As we will see, sometimes working with a complement is much less work than proceeding directly.

Combinatoric Formulas and Probability

In many cases, the principles of counting that were discussed in Sections 12.1 and 12.2 are part of the process of calculating a probability.

▼ **example 4** Find a Probability Using the Complement Rule

A die is tossed 4 times. What is the probability that a will show on the upward face at least once?

Solution
Let $E = \{$at least one 6$\}$. Then $E^c = \{$no 6s$\}$. To calculate the number of elements in the sample space (all possible outcomes of tossing a die 4 times) and the number of items in E^c, we will use the counting principle.

Because on each toss of the die there are 6 possible outcomes,

$$n(S) = 6 \cdot 6 \cdot 6 \cdot 6 = 1296$$

On each toss of the die there are 5 numbers that are not 6s. Therefore,

$$n(E^c) = 5 \cdot 5 \cdot 5 \cdot 5 = 625$$
$$P(E) = 1 - P(E^c)$$
$$= 1 - \frac{625}{1296} = \frac{671}{1296}$$
$$\approx 0.518$$

When a die is tossed 4 times, the probability that a ⚅ will show on the upward face at least once is approximately 0.518.

▼**check your progress 4** A pair of dice is rolled 3 times. What is the probability that a sum of 7 will occur at least once?

Solution   *See page S46.*   ◄

MATHMATTERS   The Monty Hall Problem

A famous probability puzzle began with the game show *Let's Make a Deal,* of which Monty Hall was the host, and goes something like the following. (See also the Chapter 1 opener on page 1.) Suppose you appear on the show and are shown three closed doors. Behind one of the doors is the grand prize; behind the other two doors are less desirable prizes (like a goat!). If you select the door hiding the grand prize, you win that prize. The probability of randomly choosing the grand prize, of course, is 1/3. After you choose a door, the show's host (who knows where the grand prize is) does not immediately open it to show you what you have won. Instead, he opens one of the other two doors and reveals a goat. Obviously you are relieved that you did not choose that particular door, but he then asks if you would like to switch your choice to the third door. Should you stay with your original choice, or switch? Most people would say at first that it makes no difference. However, computer simulations that play the game over and over have shown that you *should* switch. In fact, you double your chances of winning the grand prize if you give up your first choice! See Exercise 54 on page 777 for a mathematical investigation, and then try a simulation of your own with Exercises 56 and 57.

We end this section with another example of using counting principles in the calculation of a probability. In this case, we will use the combination formula $C(n, r) = \frac{n!}{r!(n-r)!}$, which gives the number of ways r objects can be chosen from n objects.

▼**example 5**   Find a Probability Using the Combination Formula

Every manufacturing process has the potential to produce a defective article. Suppose a manufacturing process for tableware produces 40 dinner plates, of which 3 are defective. (For instance, there is a paint flaw.) If 5 plates are randomly selected from the 40, what is the probability that at least 1 is defective?

Solution

Let $E = \{$at least one plate is defective$\}$. It is easier to work with the complement event, $E^c = \{$no plates are defective$\}$. To calculate the number of elements in the sample space (all possible outcomes of choosing 5 plates from 40), use the combination formula with $n = 40$ (the number of plates) and $r = 5$ (the number of plates that are chosen). Then

$$n(S) = C(40, 5) = \frac{40!}{5!\,(40 - 5)!} \qquad \bullet\, n = 40, r = 5$$

$$= \frac{40!}{5!\,35!} = 658,008$$

To find the number of outcomes that contain no defective plates, all of the plates chosen must come from the 37 nondefective plates. Therefore, we must calculate the number of ways 5 objects can be chosen from 37. Thus $n = 37$ (the number of nondefective plates) and $r = 5$ (the number of plates chosen).

$$n(E^c) = C(37, 5) = \frac{37!}{5!\,(37 - 5)!} \qquad \bullet\, n = 37, r = 5$$

$$= \frac{37!}{5!\,32!} = 435,897$$

$$P(E) = 1 - P(E^c)$$

$$= 1 - \frac{435,897}{658,008} = \frac{222,111}{658,008}$$

$$\approx 0.338$$

The probability is approximately 0.338, or 33.8%.

▼ **check your progress** **5**   The winner of a contest will be blindfolded and then allowed to reach into a hat containing 31 \$1 bills and 4 \$100 bills. The winner can remove 4 bills from the hat and keep the money. Find the probability that the winner will pull out at least 1 \$100 bill.

Solution   *See page S46.* ◀

EXCURSION

Photolibrary

Keno Revisited

In Section 12.2 (see pages 752–753), we looked at the popular casino game keno, in which a player chooses numbers from 1 to 80 and hopes that the casino will draw balls with the same numbers.

A player can choose only 1 number or as many as 20. The casino will then pick 20 numbered balls from the 80 possible; if enough of the player's numbers match the lucky numbers the casino chooses, the player wins money. The amount won varies according to how many numbers were chosen and how many match.

EXCURSION EXERCISES

1. A gambler playing keno has randomly chosen 5 numbers. What is the probability that the gambler will match at least 1 lucky number?

2. If 5 numbers are chosen, compute the probability of matching fewer than 5 lucky numbers.

3. If the keno player chooses 15 numbers, bets $1, and matches 13 of the lucky numbers, the gambler will be paid $12,000. What is the probability of this occurring?

4. If the keno player chooses 15 numbers and matches 5 or 6 of the lucky numbers, the gambler gets the bet back but is not paid any extra. What is the probability of this occurring?

5. Some casinos will let you choose up to 20 numbers. In this case, if you don't match any of the lucky numbers, the casino pays you! Although this may seem unusual, it is actually more difficult not to match any of the lucky numbers than it is to match a few of them. Compute the probability of not matching any of the lucky numbers at all, and compare it to the probability of matching 5 lucky numbers.

6. If 20 numbers are chosen, find the probability of matching *at least 1* lucky number.

EXERCISE SET 12.4

(Suggested Assignment: The Enhanced WebAssign Exercises and Exercises 33 and 34)

1. What are mutually exclusive events? How do you calculate the probability of mutually exclusive events?

2. Give an example of two mutually exclusive events and an example of two events that are not mutually exclusive.

■ In Exercises 3 to 6, first verify that the compound event consists of two mutually exclusive events, and then compute the probability of the compound event occurring.

3. A single card is drawn from a standard deck of playing cards. Find the probability of drawing a 4 or an ace.

4. A single card is drawn from a standard deck of playing cards. Find the probability of drawing a heart or a club.

5. Two dice are rolled. Find the probability of rolling a 2 or a 10.

6. Two dice are rolled. Find the probability of rolling a 7 or an 8.

7. If $P(A) = 0.2$, $P(B) = 0.5$, and $P(A \text{ and } B) = 0.1$, find $P(A \text{ or } B)$.

8. If $P(A) = 0.6$, $P(B) = 0.4$, and $P(A \text{ and } B) = 0.2$, find $P(A \text{ or } B)$.

9. If $P(A) = 0.3$, $P(B) = 0.8$, and $P(A \text{ or } B) = 0.9$, find $P(A \text{ and } B)$.

10. If $P(A) = 0.7$, $P(A \text{ and } B) = 0.4$, and $P(A \text{ or } B) = 0.8$, find $P(B)$.

■ In Exercises 11 to 14, suppose you ask a friend to randomly choose an integer between 1 and 10, inclusive.

11. What is the probability that the number will be more than 6 or odd?

12. What is the probability that the number will be less than 5 or even?

13. What is the probability that the number will be even or prime?

14. What is the probability that the number will be prime or greater than 7?

■ In Exercises 15 to 20, two dice are rolled. Determine the probability of each of the following. ("Doubles" means that both dice show the same number.)

15. Rolling a 6 or doubles

16. Rolling a 7 or doubles

17. Rolling an even number or doubles

18. Rolling a number greater than 7 or doubles

19. Rolling an odd number or a number less than 6

20. Rolling an even number or a number greater than 9

■ In Exercises 21 to 26, a single card is drawn from a standard deck. Find the probability of each of the following events.

21. Drawing an 8 or a spade

22. Drawing an ace or a red card

23. Drawing a jack or a face card

24. Drawing a red card or a face card

25. Drawing a diamond or a black card

26. Drawing a spade or a red card

■ **Employment** In Exercises 27 to 30, use the data in the table below, which shows the employment status of individuals in a particular town by age group.

Age	Full-time	Part-time	Unemployed
0–17	24	164	371
18–25	185	203	148
26–34	348	67	27
35–49	581	179	104
50+	443	162	173

27. If a person is randomly chosen from the town's population, what is the probability that the person is aged 26 to 34 or is employed part-time?

28. If a person is randomly chosen from the town's population, what is the probability that the person is at least 50 years old or unemployed?

29. If a person is randomly chosen from the town's population, what is the probability that the person is under 18 or employed part-time?

30. If a person is randomly chosen from the town's population, what is the probability that the person is 18 or older or employed full-time?

31. Contests If the probability of winning a particular contest is 0.04, what is the probability of not winning the contest?

32. Weather Suppose the probability that it will rain tomorrow is 0.38. What is the probability that it will not rain tomorrow?

Professional Sports The National Collegiate Athletic Association (NCAA) keeps statistics on the number of seniors playing on men's NCAA teams who are drafted to play on professional teams. Exercises 33 and 34 use some of these statistics.

33. There is a 1 in 75 chance that a senior on an NCAA basketball team will be drafted by a National Basketball Association (NBA). What is the probability that a senior NCAA basketball player will not be drafted to play on an NBA team?

34. The odds in favor of a senior on an NCAA ice hockey team being drafted by a National Hockey League (NHL) are 1 to 25. What is the probability that a senior NCAA ice hockey player will not be drafted to play on an NHL team?

■ In Exercises 35 to 46, use the formula for the probability of the complement of an event.

35. Two dice are tossed. What is the probability of not tossing a 7?

36. Two dice are tossed. What is the probability of not getting doubles?

37. Two dice are tossed. What is the probability of getting a sum of at least 4?

38. Two dice are tossed. What is the probability of getting a sum of at most 11?

39. A single card is drawn from a deck. What is the probability of not drawing an ace?

40. A single card is drawn from a deck. What is the probability of not drawing a face card?

41. A coin is flipped 4 times. What is the probability of getting at least 1 tail?

42. A coin is flipped 4 times. What is the probability of getting at least 2 heads?

43. A single die is rolled 3 times. What is the probability that a 1 will show on the upward face at least once?

44. A single die is rolled 4 times. Find the probability that a 5 will be rolled at least once.

45. A pair of dice is rolled 3 times. What is the probability that a sum of 8 on the 2 dice will occur at least once?

46. A pair of dice is rolled 4 times. Compute the probability that a sum of 11 on the 2 dice will occur at least once.

47. A magician shuffles a standard deck of playing cards and allows an audience member to pull out a card, look at it, and replace it in the deck. Two additional people do the same. Find the probability that of the 3 cards drawn, at least 1 is a face card.

48. If a person draws 3 cards from a standard deck (without replacing them), what is the probability that at least 1 of the cards is a face card?

49. E-Readers An electronics store receives a shipment of 30 e-readers. Unbeknownst to the store, 4 of the e-readers are defective. If the store sells 12 of these e-readers the first day, what is the probability that at least 1 of the 12 buyers will get a defective e-reader?

AP Photo/Michael Probst

50. Blu-ray Players An electronics store currently has 28 new Blu-ray players in stock, of which 5 are defective. If customers buy 3 Blu-ray players, what is the probability that at least 1 of them will be defective?

51. Prize Drawing Three employees of a restaurant each contributed 1 business card for a random drawing to win a prize. Forty-two business cards were received in all, and 3 cards will be drawn for prizes. Determine the probability that at least 1 of the restaurant employees will win a prize.

52. Coins A bag contains 44 U.S. quarters and 6 Canadian quarters. (The coins are identical in size.) If 5 quarters are randomly picked from the bag, what is the probability of getting at least 1 Canadian quarter?

EXTENSIONS

Critical Thinking

53. Blackjack In the game blackjack, a player is dealt 2 cards from a standard deck of playing cards. The player has a blackjack if one card is an ace and the other card is a 10, a jack, a queen, or a king. In some casinos, blackjack is played with more than 1 standard deck of playing cards. Does using more than 1 deck of cards change the probability of getting a blackjack?

54. Monty Hall Problem The *Monty Hall problem* is described in the Math Matters on page 773. The question arises, "If the contestant changes his or her original choice of door, what is the probability of choosing the door hiding the grand prize?" To answer this question, complete the following.

    a. What is the probability that the contestant will choose the grand prize on the first try?

    b. What is the probability that the contestant will not choose the grand prize on the first try?

    c. What is the probability that the person will choose the grand prize by switching?

55. Door Codes A planned community has 300 homes, each with an automatic garage door opener operated by a code of 8 numbers. The homeowner sets each of the 8 numbers to be 0 or 1. For example, a door opener code might be 01101001. Assuming all the homes in the community are sold, what is the probability that at least 2 homeowners will set their door openers to use the same code and will therefore be able to open each other's garage doors?

Cooperative Learning

56. Monty Hall Problem When someone is first presented with the Monty Hall problem in Exercise 54, there is a tendency for that person to think that switching his or her choice of door does not make any difference. You can actually simulate the Monty Hall game using playing cards and show that one is twice as likely to win by switching from the first choice. To do this you will need at least 2 people and 3 cards, say the ace of spades to represent the grand prize, and the 2 of hearts and the 2 of diamonds, which represent the other 2 prizes. Shuffle the cards and place them face down on the table. Choose 1 of the cards. The other player picks up the remaining 2 cards, removes 1 of the cards that is not the ace (it is possible that both cards will not be the ace), and places the other card face down on the table. Now you may either stay with your original selection or change to the remaining card. Perform this experiment 30 times staying with your original selection and 30 times where you change to the other card. Keep a record of how many times staying with your original selection resulted in selecting the ace and how many times switching cards resulted in choosing the ace. On the basis of your results, is staying or switching the better strategy?

57. Monty Hall Problem The benefit of switching doors in the Monty Hall problem is even more dramatic if there are more hidden prizes from which to choose. Instead of using 3 cards as in Exercise 56, use 5 cards, the four 2s and the ace of spades. Shuffle them well, place them face down on a table, and choose 1 card. Another person then picks up the remaining 4 cards, removes three 2s, and places the remaining card face down on the table. Now you may either stay with your original selection or switch to the remaining card. Perform this experiment 30 times staying with your original selection and 30 where you change to the other card. Keep a record of how many times staying with your original selection resulted in selecting the ace and how many times changing resulted in choosing the ace. On the basis of your results, is staying or switching the better strategy? About how many more times did switching result in choosing the ace than did staying with your original choice?

Explorations

58. Poker In 5-card stud poker, a hand containing 5 cards of the same suit from a standard deck is called a flush. In this Exploration, you will compute the probability of getting a flush.

    a. How many 5-card poker hands are possible?

    b. How many 5-card poker hands are possible that contain all spades?

    c. What is the probability of getting a 5-card poker hand containing all spades?

    d. What is the probability of getting a 5-card poker hand containing all hearts? All diamonds? All clubs?

    e. Are the events of getting 5 spades, 5 hearts, 5 diamonds, or 5 clubs mutually exclusive?

    f. What is the probability of getting a flush in 5-card stud poker?

Conditional Probability

	F	No *F*	Total
V	21	198	219
No *V*	76	195	271
Total	97	393	490

V: Vaccinated
F: Contracted the flu

	F	No *F*	Total
V	21	198	219

In the preceding section, we discussed the effectiveness of a flu vaccination in preventing the onset of the flu. The table from that discussion is shown again at the left. From the table, we can calculate the probability that one person randomly selected from this population will have the flu.

$$P(F) = \frac{n(F)}{n(S)} = \frac{97}{490} \approx 0.198$$ • $n(F) = 97, n(S) = 490$
(*S* denotes the sample space)

Now consider a slightly different situation. We could ask, "What is the probability that a person randomly chosen from this population will contract the flu *given* that the person received the flu vaccination?"

In this case, we know that the person received a flu vaccination, and we want to determine the probability that the person will contract the flu. Therefore, the only part of the table that is of concern to us is the top row. In this case, we have

$$P(F \text{ given } V) = \frac{21}{219} \approx 0.096$$

Thus the probability that an individual will contract the flu given that the individual has been vaccinated is about 0.096.

The probability of an event *B* occurring given that we know some other event *A* has already occurred is called a **conditional probability** and is denoted $P(B|A)$.

▼ **Conditional Probability Formula**

If *A* and *B* are two events in a sample space *S*, then the conditional probability of *B* given that *A* has occurred is

$$P(B|A) = \frac{P(A \text{ and } B)}{P(A)}$$

The symbol $P(B|A)$ is read "the probability of *B* given *A*."

To see how this formula applies to the flu data above, let

  S = {all people participating in the test}

  F = {people who contracted the flu}

  V = {people who were vaccinated}

Then $F \cap V$ = {people who contracted the flu *and* were vaccinated}.

$$P(F|V) = \frac{P(F \text{ and } V)}{P(V)} = \frac{\frac{21}{490}}{\frac{219}{490}}$$ • $P(F \text{ and } V) = \frac{n(F \cap V)}{n(S)} = \frac{21}{490}$

  • $P(V) = \frac{n(V)}{n(S)} = \frac{219}{490}$

$$= \frac{21}{219} \approx 0.096$$

The probability that one person selected from this population contracted the flu given that the person received the vaccination, $P(F|V)$, is about 0.096. Our answer agrees with the calculation we performed directly from the table, but the Conditional Probability Formula enables us to find conditional probabilities even when we cannot compute them directly.

question In the preceding example, what is the interpretation of $P(V\,|\,F)$?

▼ example 1 **Determine a Conditional Probability**

The data in the table below show the results of a survey used to determine the number of adults who received financial help from their parents for certain purchases.

Age	College tuition	Buy a car	Buy a house	Total
18–28	405	253	261	919
29–39	389	219	392	1000
40–49	291	146	245	682
50–59	150	71	112	333
≥ 60	62	15	98	175
Total	1297	704	1108	3109

If one person is selected from this survey, what is the probability that the person received financial help to purchase a home, given that the person is between the ages of 29 and 39?

Solution
Let $B = \{$adults who received financial help for a home purchase$\}$ and
$A = \{$adults between 29 and 39$\}$. From the table, $n(A \cap B) = 392$, $n(A) = 1000$, and
$n(S) = 3109$. Using the Conditional Probability Formula, we have

$$P(B\,|\,A) = \frac{P(A \text{ and } B)}{P(A)} = \frac{\dfrac{392}{3109}}{\dfrac{1000}{3109}}$$

- $P(A \text{ and } B) = \dfrac{n(A \cap B)}{n(S)} = \dfrac{392}{3109}$

- $P(A) = \dfrac{n(A)}{n(S)} = \dfrac{1000}{3109}$

$$= \frac{392}{1000} = 0.392$$

The probability that a person received financial help to purchase a home, given that the person is between the ages of 29 and 39, is 0.392.

▼ check your progress 1 Two dice are tossed, one after the other. What is the probability that the result is a sum of 6, given that the first die is not a 3?

Solution *See page S46.* ◀

MATHMATTERS Bayes Theorem and Bertrand's Box Paradox

Bayes Theorem, named after Thomas Bayes (1702–1761), gives a method of computing conditional probabilities by knowing the reverse conditional probabilities. Consider the following problem, known as Bertrand's Box Paradox. Three boxes each contain 2 coins. One box has 2 gold coins, another has 2 silver coins, and the third box has 1 gold coin and 1 silver coin. A box is chosen at random, and a coin is pulled out (without looking at the other coin). If the coin that is taken out is gold, what is the probability that the other coin in the same box is also gold? Many people initially guess that the probability is $\frac{1}{2}$. To check, let B_1 be the event that the first box (with 2 gold coins) is chosen, B_2 the event that the second box (with 2 silver coins) is chosen, and B_3 the event that the third box (with 1 silver and 1 gold) is chosen. In addition, let G represent the event that a gold coin is pulled from a box, and S the event that a silver coin is chosen. The conditional probability of pulling a gold coin from a box, given that we know which box was chosen, is simple to compute. What we want to find, however, is the reverse: the conditional probability that we have chosen the first box, given that a gold coin was chosen, $P(B_1 | G)$. Bayes Theorem gives us a way to find this probability. In this case, the theorem states

$$P(B_1 | G) = \frac{P(G | B_1)P(B_1)}{P(G | B_1)P(B_1) + P(G | B_2)P(B_2) + P(G | B_3)P(B_3)}$$

$$= \frac{1 \cdot \frac{1}{3}}{1 \cdot \frac{1}{3} + 0 \cdot \frac{1}{3} + \frac{1}{2} \cdot \frac{1}{3}} = \frac{2}{3}$$

Thus, there is actually a $\frac{2}{3}$ chance that if a gold coin is pulled from a box, the other coin is also gold.

Product Rule for Probabilities

Suppose that 2 cards are drawn, without replacement, from a standard deck of playing cards. Let A be the event that an ace is drawn on the first draw and B the event that an ace is drawn on the second draw. Then the probability that an ace is drawn on the first *and* second draw is $P(A \text{ and } B)$. To find this probability, we can solve the Conditional Probability Formula for $P(A \text{ and } B)$.

$$\frac{P(A \text{ and } B)}{P(A)} = P(B | A)$$

$$P(A) \cdot \frac{P(A \text{ and } B)}{P(A)} = P(A) \cdot P(B | A) \qquad \bullet \text{ Multiply each side by } P(A).$$

$$P(A \text{ and } B) = P(A) \cdot P(B | A)$$

This is called the Product Rule for Probabilities.

▼ Product Rule for Probabilities

If A and B are two events in a sample space S, then

$$P(A \text{ and } B) = P(A) \cdot P(B | A)$$

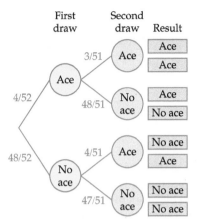

First draw | Second draw | Result

For the problem of drawing an ace from a standard deck of playing cards on the first and second draws, $P(A \text{ and } B)$ is the product of $P(A)$, the probability that the first drawn card is an ace, and $P(B \mid A)$, the probability of an ace on the second draw *given* that the first card drawn was an ace.

The tree diagram at the left shows the possible outcomes of drawing 2 cards from a deck without replacement. On the first draw, there are 4 aces in the deck of 52 cards. Therefore, $P(A) = \frac{4}{52} = \frac{1}{13}$. On the second draw, there are only 51 cards remaining and only 3 aces (an ace was drawn on the first draw). Therefore, $P(B \mid A) = \frac{3}{51} = \frac{1}{17}$. Putting these calculations together, we have

$$P(A \cap B) = P(A) \cdot P(B \mid A)$$
$$= \frac{1}{13} \cdot \frac{1}{17} = \frac{1}{221}$$

The probability of drawing an ace on the first and second draws is $\frac{1}{221}$.

The Product Rule for Probabilities can be extended to more than two events. The probability that a certain sequence of events will occur in succession is the product of the probabilities of each of the events *given* that the preceding events have occurred.

▼ Probability of Successive Events

The probability of two or more events occurring in succession is the product of the conditional probabilities of each of the events.

▼ **example 2**   **Find the Probability of Successive Events**

A box contains 4 red, 3 white, and 5 green balls. Suppose that 3 balls are randomly selected from the box in succession, without replacement.

a. What is the probability that first a red, then a white, and then a green ball are selected?

b. What is the probability that 2 white balls followed by 1 green ball are selected?

Solution

a. Let $A = \{$a red ball is selected first$\}$, $B = \{$a white ball is selected second$\}$, and $C = \{$a green ball is selected third$\}$. Then

$$P(A \text{ followed by } B \text{ followed by } C) = P(A) \cdot P(B \mid A) \cdot P(C \mid A \text{ and } B)$$
$$= \frac{4}{12} \cdot \frac{3}{11} \cdot \frac{5}{10}$$
$$= \frac{1}{22}$$

The probability of choosing a red, then a white, then a green ball is $\frac{1}{22}$.

b. Let $A = \{$a white ball is selected first$\}$, $B = \{$a white ball is selected second$\}$, and $C = \{$a green ball is selected third$\}$. Then

$$P(A \text{ followed by } B \text{ followed by } C) = P(A) \cdot P(B \mid A) \cdot P(C \mid A \text{ and } B)$$
$$= \frac{3}{12} \cdot \frac{2}{11} \cdot \frac{5}{10}$$
$$= \frac{1}{44}$$

The probability of choosing 2 white balls followed by 1 green ball is $\frac{1}{44}$.

TAKE NOTE

In part a, there are originally 12 balls in the box. After a red ball is selected, there are only 11 balls remaining, of which 3 are white. Thus $P(B \mid A) = \frac{3}{11}$. After a red ball and a white ball are selected, there are 10 balls left, of which 5 are green. Thus $P(C \mid A \text{ and } B) = \frac{5}{10}$.

In part b, we have a similar situation. However, after a white ball is selected, there are 11 balls remaining, of which only 2 are white. Therefore, $P(B \mid A) = \frac{2}{11}$.

▼ **check your progress** **2** A standard deck of playing cards is shuffled and 3 cards are dealt. Find the probability that the cards dealt are a spade followed by a heart followed by another spade.

Solution *See page S46.*

Independent Events

Earlier in this section we considered the probability of drawing 2 aces in a row from a standard deck of playing cards. Because the cards were drawn without replacement, the probability of an ace on the second draw *depended* on the result of the first draw.

Now consider the case of tossing a coin twice. The outcome of the first coin toss has no effect on the outcome of the second toss. So the probability of the coin flipping to a head or a tail on the second toss is not affected by the result of the first toss. When the outcome of a first event does not affect the outcome of a second event, the events are called *independent*.

> ▼ **Independent Events**
>
> If A and B are two events in a sample space and $P(B|A) = P(B)$, then A and B are called **independent events**.

For a mathematical verification, consider tossing a coin twice. We can compute the probability that the second toss comes up heads, given that the first coin toss came up heads. If A is the event of a head on the first toss, then $A = \{\text{HH, HT}\}$. Let B be the event of a head on the second toss. Then $B = \{\text{HH, TH}\}$. The sample space is $S = \{\text{HH, HT, TH, TT}\}$. The conditional probability $P(B|A)$ (the probability of a head on the second toss given a head on the first toss) is

$$P(B|A) = \frac{P(A \text{ and } B)}{P(A)} = \frac{\dfrac{1}{4}}{\dfrac{1}{2}} = \frac{1}{2}$$

- $P(A \text{ and } B) = \dfrac{n(A \cap B)}{n(S)} = \dfrac{1}{4}$
- $P(A) = \dfrac{n(A)}{n(S)} = \dfrac{2}{4} = \dfrac{1}{2}$

Thus $P(B|A) = \frac{1}{2}$. Note, however, that $P(B) = \frac{n(B)}{n(S)} = \frac{2}{4} = \frac{1}{2}$. Therefore, in the case of tossing a coin twice, the probability of the second event does not depend on the outcome of the first event, and we have $P(B|A) = P(B)$.

In general, this result enables us to simplify the product rule when two events are independent; the probability of two independent events occurring in succession is simply the product of the probabilities of each of the individual events.

TAKE NOTE

The Product Rule for Independent Events can be extended to more than 2 events. If E_1, E_2, E_3, and E_4 are independent events, then the probability that all 4 events will occur is $P(E_1) \cdot P(E_2) \cdot P(E_3) \cdot P(E_4)$.

> ▼ **Product Rule for Independent Events**
>
> If A and B are two independent events from the sample space S, then $P(A \text{ and } B) = P(A) \cdot P(B)$.

▼ **example** **3** Find the Probability of Independent Events

A pair of dice is tossed twice. What is the probability that the first roll is a sum of 7 and the second roll is a sum of 11?

Solution

The rolls of a pair of dice are independent; the probability of a sum of 11 on the second roll does not depend on the outcome of the first roll. Let $A = \{$sum of 7 on the first roll$\}$ and $B = \{$sum of 11 on the second roll$\}$. Then

$$P(A \text{ and } B) = P(A) \cdot P(B) = \frac{6}{36} \cdot \frac{2}{36} = \frac{1}{108}$$

▼ check your progress 3   A coin is tossed 3 times. What is the probability that heads appears on all 3 tosses?

Solution   *See page S47.*   ◄

Applications of Conditional Probability

Conditional probability is used in many real-world situations, such as to determine the efficacy of a drug test, to verify the accuracy of genetic testing, and to analyze evidence in legal proceedings.

▼ example 4   **Drug Testing and Conditional Probability**

Suppose that a company claims it has a test that is 95% effective in determining whether an athlete is using a steroid. That is, if an athlete is using a steroid, the test will be positive 95% of the time. In the case of a negative result, the company says its test is 97% accurate. That is, even if an athlete is not using steroids, it is possible that the test will be positive in 3% of the cases. Such an occurrence is called a **false positive**. Suppose this test is given to a group of athletes in which 10% of the athletes are using steroids. What is the probability that a randomly chosen athlete actually uses steroids, given that the athlete's test is positive?

Solution

Let S be the event that an athlete uses steroids and let T be the event that the test is positive. Then the probability we wish to determine is $P(S \mid T)$. Using the Conditional Probability Formula, we have

$$P(S \mid T) = \frac{P(S \text{ and } T)}{P(T)}$$

A tree diagram, shown at the left, can be used to calculate this probability. A positive test result can occur in two ways: either an athlete using steroids correctly tests positive, or an athlete not using steroids incorrectly tests positive. The probability of a positive test result, $P(T)$, corresponds to an athlete following path ST or path $S'T$ in the tree diagram. (S' symbolizes no steroid use and T' symbolizes a negative result.) $P(S \text{ and } T)$, the probability of using steroids and getting a positive test result, is path ST. Thus,

$$P(S \mid T) = \frac{P(S \text{ and } T)}{P(T)}$$

$$= \frac{(0.1)(0.95)}{(0.1)(0.95) + (0.9)(0.03)} \approx 0.779$$

Given that an athlete tests positive, the probability that the athlete actually uses steroids is approximately 77.9%.

This may be lower than you would expect, especially considering that the manufacturer claims that its test is 95% accurate. In fact, the prevalence of false-positive test results can be much more dramatic in cases of conditions that are present in a small percentage of the population. (See Exercise 61, page 787.)

▼ check your progress 4 A pharmaceutical company has a test that is 95% effective in determining whether a person has a certain genetic defect. It is possible, however, that the test may give a false positive result in 4% of cases. Suppose this particular genetic defect occurs in 2% of the population. Given that a person tests positive, what is the probability that the person actually has the defect?

Solution *See page S47.*

EXCURSION

Elena Elisseeva/Shutterstock.com

Photolibrary

Sharing Birthdays

Have you ever been introduced to someone at a party or other social gathering and discovered that you share the same birthday? It seems like an amazing coincidence when it occurs. In fact, how rare is this?

As an example, suppose 4 people have gathered for a dinner party. We can determine the probability that at least 2 of the guests have the same birthday. (For simplicity, we will ignore the February 29th birthday from leap years.) Let E be the event that at least 2 people share a birthday. It is easier to look at the complement E^C, the event that no one shares the same birthday.

If we start with 1 of the guests, then the second guest cannot share the same birthday, so that person has 364 possible dates for his or her birthday from a total of 365. Thus the conditional probability that the second guest has a different birthday, given that we know the first person's birthday, is $\frac{364}{365}$. Similarly, the third person has 363 possible birthday dates that do not coincide with those of the first two guests. So the conditional probability that the third person does not share a birthday with either of the first 2 guests, given that we know the birthdays of the first 2 people, is $\frac{363}{365}$. The probability of the fourth guest having a distinct birthday is, similarly, $\frac{362}{365}$. We can use the Product Rule for Probabilities to find the probability that all of these conditions are met; that is, none of the 4 guests share a birthday.

$$P(E^c) = \frac{364}{365} \cdot \frac{363}{365} \cdot \frac{362}{365} \approx 0.984$$

Then $P(E) = 1 - P(E^c) \approx 0.016$, so there is about a 1.6% chance that in a group of 4 people, 2 or more will have the same birthday.

It would require 366 people gathered together to *guarantee* that 2 people in the group will have the same birthday. But how many people would be required to guarantee that the chance that at least 2 of them share the same birthday is at least 50/50? Make a guess before you proceed through the exercises. The results may surprise you!

INSTRUCTOR NOTE

As suggested, you can introduce this Excursion by first asking the class how likely they think it is that 2 of the students in the class would have the same birthday. Then ask each student when his or her birthday is and determine if any of the students share a birthday. (If you have a large number of students, you can proceed month by month.) If 2 or more students do have a common birthday, it can serve as evidence that perhaps it doesn't require as many people present as one might think to have a good chance of a shared birthday.

EXCURSION EXERCISES

1. If 8 people are present at a meeting, find the probability that at least 2 share a common birthday.

2. Compute the probability that at least 2 people among a group of 15 have the same birthday.

3. If 23 people are in attendance at a party, what is the probability that at least 2 share a birthday?

4. In a group of 40 people, what would you estimate to be the probability that at least 2 people share a birthday? If you have the patience, compute the probability to check your guess.

EXERCISE SET 12.5 (Suggested Assignment: The Enhanced WebAssign Exercises and Exercise 63)

1. What is a conditional probability?

2. Explain the difference between independent events and dependent events.

■ In Exercises 3 to 6, compute the conditional probabilities $P(A|B)$ and $P(B|A)$.

3. $P(A) = 0.7$, $P(B) = 0.4$, $P(A \text{ and } B) = 0.25$

4. $P(A) = 0.45$, $P(B) = 0.8$, $P(A \text{ and } B) = 0.3$

5. $P(A) = 0.61$, $P(B) = 0.18$, $P(A \text{ and } B) = 0.07$

6. $P(A) = 0.2$, $P(B) = 0.5$, $P(A \text{ and } B) = 0.2$

■ **Employment** In Exercises 7 to 10, use the data in the table below, which shows the employment status of individuals in a particular town by age group.

	Full-time	Part-time	Unemployed
0–17	24	164	371
18–25	185	203	148
26–34	348	67	27
35–49	581	179	104
≥ 50	443	162	173

7. If a person in this town is selected at random, find the probability that the individual is employed part-time, given that he or she is between the ages of 35 and 49.

8. If a person in the town is randomly selected, what is the probability that the individual is unemployed, given that he or she is 50 years old or older?

9. A person from the town is randomly selected; what is the probability that the individual is employed full-time, given that he or she is between 18 and 49 years of age?

10. A person from the town is randomly selected; what is the probability that the individual is employed part-time, given that he or she is at least 35 years old?

■ **Video Games** In Exercises 11 to 14, use the data in the following table, which shows the results of a survey of 2000 gamers concerning their favorite home video game systems, organized by age group. If a survey participant is selected at random, determine the probability of each of the following. Round to the nearest hundredth.

	Nintendo Wii	Microsoft Xbox 360	Sega Dreamcast	Sony Playstation 3
0–12	63	84	55	51
13–18	105	139	92	113
19–24	248	217	83	169
≥ 25	191	166	88	136

11. The participant prefers the Nintendo Wii system.

12. The participant prefers the Xbox 360, given that the person is between the ages of 13 and 18.

13. The participant prefers Sega Dreamcast, given that the person is between the ages of 13 and 24.

14. The participant is under 12 years of age, given that the person prefers the Playstation 3 system.

15. A pair of dice is tossed. Find the probability that the sum on the 2 dice is 8, given that the sum is even.

16. A pair of dice is tossed. Find the probability that the sum on the 2 dice is 12, given that doubles are rolled.

17. A pair of dice is tossed. What is the probability that doubles are rolled, given that the sum on the 2 dice is less than 7?

18. A pair of dice is tossed. What is the probability that the sum on the 2 dice is 8, given that the sum is more than 6?

19. What is the probability of drawing 2 cards in succession (without replacement) from a standard deck and having them both be face cards? Round to the nearest thousandth.

20. Two cards are drawn from a standard deck without replacement. Find the probability that both cards are hearts. Round to the nearest thousandth.

21. Two cards are drawn from a standard deck without replacement. What is the probability that the first card is a spade and the second card is red? Round to the nearest thousandth.

22. Two cards are drawn from a standard deck without replacement. What is the probability that the first card is a king and the second card is not? Round to the nearest thousandth.

■ **Candy Colors** In Exercises 23 to 26, a snack-size bag of M&Ms candies is opened. Inside, there are 12 red candies, 12 blue, 7 green, 13 brown, 3 orange, and 10 yellow. Three candies are pulled from the bag in succession, without replacement.

23. Determine the probability that the first candy drawn is blue, the second is red, and the third is green.

24. Determine the probability that the first candy drawn is brown, the second is orange, and the third is yellow.

25. What is the probability that the first two candies drawn are green and the third is red?

26. What is the probability that the first candy drawn is orange, the second is blue, and the third is orange?

■ In Exercises 27 to 30, 3 cards are dealt from a shuffled standard deck of playing cards.

27. Find the probability that the first card dealt is red, the second is black, and the third is red.

28. Find the probability that the first 2 cards dealt are clubs and the third is a spade.

29. What is the probability that the 3 cards dealt are, in order, an ace, a face card, and an 8? (A face card is a jack, queen, or king.)

30. What is the probability that the 3 cards dealt are, in order, a red card, a club, and another red card?

■ **Student Attendance** In Exercises 31 to 34, the probability that a student enrolled at a local high school will be absent on a particular day is 0.04, assuming that the student was in attendance the previous school day. However, if a student is absent, the probability that he or she will be absent again the following day is 0.11. For each exercise, assume that the student was in attendance the previous day.

31. What is the probability that a student will be absent 3 days in a row?

32. What is the probability that a student will be absent 2 days in a row but then show up on the third day?

33. Find the probability that a student will be absent, attend the next day, but then be absent again the third day.

34. Find the probability that a student will be absent 4 days in a row.

■ In Exercises 35 to 38, determine whether the events are independent.

35. A single die is rolled and then rolled a second time.

36. Numbered balls are pulled from a bin one by one to determine the winning lottery numbers.

37. Numbers are written on slips of paper in a hat; 1 person pulls out a slip of paper without replacing it, then a second person pulls out a slip of paper.

38. In order to determine who goes first in a game, 1 person picks a number between 1 and 10, then a second person picks a number from the remaining 9 numbers.

■ In Exercises 39 to 44, a pair of dice is tossed twice.

39. Find the probability that both rolls give a sum of 8.

40. Find the probability that the first roll is a sum of 6 and the second roll is a sum of 12.

41. Find the probability that the first roll is a total of at least 10 and the second roll is a total of at least 11.

42. Find the probability that both rolls result in doubles.

43. Find the probability that both rolls give even sums.

44. Find the probability that both rolls give at most a sum of 4.

45. A fair coin is tossed 4 times in succession. Find the probability of getting 2 heads followed by 2 tails.

46. **Monopoly** In the game of Monopoly, a player is sent to jail if he or she rolls doubles with a pair of dice 3 times in a row. What is the probability of rolling doubles 3 times in succession?

Michael Newman/PhotoEdit, Inc.

47. Find the probability of tossing a pair of dice 3 times in succession and getting a sum of at least 10 on all 3 tosses.

48. Find the probability of tossing a pair of dice 3 times in succession and getting a sum of at most 3 on all 3 tosses.

■ In Exercises 49 to 54, a card is drawn from a standard deck and replaced. After the deck is shuffled, another card is pulled.

49. What is the probability that both cards pulled are aces?

50. What is the probability that both cards pulled are face cards?

51. What is the probability that the first card drawn is a spade and the second card is a diamond?

52. What is the probability that the first card drawn is an ace and the second card is not an ace?

53. Find the probability that the first card drawn is a heart and the second card is a spade.

54. Find the probability that the first card drawn is a face card and the second card is black.

55. A standard deck of playing cards is shuffled, and 3 people each choose a card. Find the probability that the first 2 cards chosen are diamonds and the third card is black if

a. the cards are chosen *with* replacement.

b. the cards are chosen *without* replacement.

56. A standard deck of playing cards is shuffled and 3 people each choose a card. Find the probability that all 3 cards are face cards if

a. the cards are chosen *with* replacement.

b. the cards are chosen *without* replacement.

57. A bag contains 5 red marbles, 4 green marbles, and 8 blue marbles. Find the probability of pulling 2 red marbles followed by a green marble if the marbles are pulled from the bag

a. with replacement.

b. without replacement.

58. A box contains 3 medium t-shirts, 5 large t-shirts, and 4 extra-large t-shirts. If someone randomly chooses 3 t-shirts from the box, find the probability that the first t-shirt is large, the second is medium, and the third is large if the shirts are chosen

a. with replacement.

b. without replacement.

59. Drug Testing A company that performs drug testing guarantees that its test determines a positive result with 97% accuracy. However, the test also gives 6% false positives. If 5% of those being tested actually have the drug present in their bloodstream, find the probability that a person testing positive has actually been using drugs. Round to the nearest thousandth.

60. Genetic Testing A test for a genetic disorder can detect the disorder with 94% accuracy. However, the test will incorrectly report positive results for 3% of those without the disorder. If 12% of the population has the disorder, find the probability that a person testing positive actually has the genetic disorder. Round to the nearest thousandth.

61. Disease Testing A pharmaceutical company has developed a test for a rare disease that is present in 0.5% of the population. The test is 98% accurate in determining a positive result, and the chance of a false positive is 4%. What is the probability that someone who tests positive actually has the disease? Round to the nearest thousandth.

62. HIV Testing When used together, the ELISA and Western Blot tests for HIV are more than 99% accurate in determining a positive result. The chance of a false positive is between 1 and 5 for every 100,000 tests. If we assume a 99% accuracy rate for correctly identifying positive results and 5/100,000 false positives, find the probability that someone who tests positive has HIV. [The Centers for Disease Control and Prevention (CDC) estimate that about 0.3% of U.S. residents are infected with HIV.] Round to the nearest thousandth.

EXTENSIONS

Critical Thinking

63. Suppose you are standing at a street corner and flip a coin to decide whether you will go north or south from your current position. When you reach the next intersection, you repeat the procedure. (This problem is a simplified version of what is called a *random walk* problem. Problems of this type are important in economics, physics, chemistry, biology, and other disciplines.)

a. After performing this experiment 3 times, what is the probability that you will be 3 blocks north of your original position?

b. After performing this experiment 4 times, what is the probability that you will be 2 blocks north of your original position?

c. After performing this experiment 4 times, what is the probability that you will be back at your original position?

Cooperative Learning

64. From a standard deck of playing cards, choose 4 red cards and 4 black cards. Deal the 8 cards, face up, in 2 rows of 4 cards each. A good event is that each column contains a red and a black card. (The column can be red/black or black/red.) A bad event is any other situation. Although this problem is stated in terms of cards, it has a very practical application. If several proteins in a cell break (say, from radiation therapy) and then reattach, it is possible that the new protein is harmful to the cell.

a. Perform this experiment 50 times and keep a record of the number of good events and bad events.

b. Use your data to predict the probability of a good event.

c. Repeat parts a and b again.

d. Use your data from the complete 100 trials to predict the probability of a good event.

e. Calculate the theoretical probability of a good event.

Explorations

65. Monty Hall Problem This is another explanation, using conditional probability, of the Monty Hall problem discussed in Exercise 54 of the last section. Let the door chosen by the contestant be labeled 1 and the other two doors be labeled 2 and 3. In the following exercises, we will use A to represent the event that the grand prize is behind door 1, B to represent the prize is behind door 2,

and C to represent the prize is behind door 3. We will use $\overline{A}$ to represent that Monty Hall opens door 1, $\overline{B}$ to represent he opens door 2, and $\overline{C}$ that he opens door 3.

a. What is the probability that Monty Hall opens door 2 given that the grand prize is behind door 1? This is $P(\overline{B}|A)$.

b. What is the probability that Monty Hall opens door 2 given that the grand prize is behind door 2? This is $P(\overline{B}|B)$.

c. What is the probability that Monty Hall opens door 2 given that the grand prize is behind door 3? This is $P(\overline{B}|C)$.

d. The probability that the grand prize is behind door 1 given that door 2 is opened (you have not switched choices) is given by Bayes Theorem in the following form.

$$P(A|\overline{B}) = \frac{P(\overline{B}|A)P(A)}{P(\overline{B}|A)P(A) + P(\overline{B}|B)P(B) + P(\overline{B}|C)P(C)}$$

What is the probability?

e. Find the probability of choosing the grand prize if you switch doors. That is, find

$$P(C|\overline{B}) = \frac{P(\overline{B}|C)P(C)}{P(\overline{B}|A)P(A) + P(\overline{B}|B)P(B) + P(\overline{B}|C)P(C)}$$

f. Is switching the better strategy?

---

section **12.6** **Expectation**

Expectation

Suppose a barrel contains a large number of balls, half of which have the number 1000 painted on them and the other half of which have the number 500 painted on them. As the grand prize winner of a contest, you get to reach into the barrel (blindfolded, of course) and select 10 balls. Your prize is the sum of the numbers on the balls in cash.

If you are very lucky, all of the balls will have 1000 painted on them, and you will win $10,000. If you are very unlucky, all of the balls will have 500 painted on them, and you will win $5000. Most likely, however, approximately one-half of the balls will have 1000 painted on them and one-half will have 500 painted on them. The amount of your winnings in this case will be 5(1000) + 5(500), or $7500. Because 10 balls were drawn, your amount of winnings per ball is $\frac{\$7500}{10} = \$750$.

The number $750 is called the *expected value* or the *expectation* of the game. You cannot win $750 on one draw, but if given the opportunity to draw many times, you will win, on average, $750 per ball.

---

question Can an expectation be negative?

---

answer Yes. For instance, if the expected value of a gambling game is negative, it simply means that, in the long run, a person will lose that amount of money, on average, on each play.

Another way we can calculate expectation is to use probabilities. For the game above, one-half of the balls have the number 1000 painted on them and one-half have the number 500 painted on them. Therefore, $P(1000) = \frac{1}{2}$ and $P(500) = \frac{1}{2}$. The expectation is calculated as follows.

Expectation

$$= \text{(probability of winning \$1000)} \cdot \$1000 + \text{(probability of winning \$500)} \cdot \$500$$

$$= P(1000) \cdot \$1000 + P(500) \cdot \$500$$

$$= \frac{1}{2} \cdot \$1000 + \frac{1}{2} \cdot \$500 = \$500 + \$250 = \$750$$

The general result for experiments with numerical outcomes follows.

INSTRUCTOR NOTE

You can point out to students that expectation is like a weighted average. We are finding the average value of the outcomes, but each outcome is weighted by the probability of it occurring.

▼ Expectation

Let $S_1, S_2, S_3, \ldots, S_n$ be the possible numerical outcomes of an experiment, and let $P(S_1), P(S_2), P(S_3), \ldots, P(S_n)$ be the probabilities of those outcomes. Then the **expectation** of the experiment is

$$P(S_1) \cdot S_1 + P(S_2) \cdot S_2 + P(S_3) \cdot S_3 + \cdots + P(S_n) \cdot S_n$$

That is, to find the expectation of an experiment, multiply the probability of each outcome of the experiment by the outcome and then add the results.

PhotoDisc Green/Getty Images

▼ example **Expectation in Gambling**

One of the wagers in roulette is to place a bet on 1 of the numbers from 0 to 36 or on 00. If that number comes up, the player wins 35 times the amount bet (and keeps the original bet). Suppose a player bets \$1 on a number. What is the player's expectation?

Solution

Let S_1 be the event that the player's number comes up and the player wins \$35. Because there are 38 numbers from which to choose, $P(S_1) = \frac{1}{38}$. Let S_2 be the event that the player's number does not come up and the player therefore loses \$1. Then $P(S_2) = 1 - \frac{1}{38} = \frac{37}{38}$.

$$\text{Expectation} = P(S_1) \cdot S_1 + P(S_2) \cdot S_2$$

$$= \frac{1}{38}(35) + \frac{37}{38}(-1) = -\frac{1}{19}$$

$$\approx -0.053$$

• The amount the player can win is entered as a positive number. The amount that can be lost is entered as a negative number.

The player's expectation is approximately −\$0.053. This means that, on average, the player will lose about \$0.05 every time this bet is made.

TAKE NOTE ✓

In Example 1, suppose the player bets \$5 instead of \$1. The payoff is then \$175 (35 · 5) if the player wins and −\$5 if the player loses. The expectation is $\frac{1}{38}(175) + \frac{37}{38}(-5) = -\frac{5}{19}$. Note that the bet is 5 times greater and the expectation, $-\frac{5}{19}$, is 5 times the expectation when \$1 is bet. Thus a player who makes \$5 bets can expect to lose 5 times as much money as a player who makes \$1 bets.

▼ check your progress **1** In roulette it is possible to place a wager that 1 of the numbers between 1 and 12 (inclusive) will come up. If it does, the player wins twice the amount bet. Suppose a player bets \$5 that a number between 1 and 12 will come up. What is the player's expectation?

Solution *See page S47.* ◀

In Example 1, the fact that the player is losing approximately 5 cents on each dollar bet means that the casino's expectation is positive 5 cents; it is earning (on average) 5 cents for every dollar spent making that particular bet at the roulette wheel. An individual player may get lucky, but over time the casino can plan on a predictable profit.

MATHMATTERS A Bargain at Any Price?

HISTORICAL NOTE

Daniel Bernoulli (bər-nōo′lē) (1700–1782) was the son of Jean Bernoulli I and the nephew of Jacques Bernoulli. For a time, he was a professor in St. Petersburg, Russia, where he collaborated with Leonhard Euler. There he wrote a paper on probability and expectation in which he discussed the game described at the right, now known as the St. Petersburg Paradox.

The Swiss mathematician Daniel Bernoulli discussed the following game in a paper he published in 1738. A fair coin is repeatedly tossed until the coin comes up tails. Let n = the number of total coin flips. When the coin comes up tails, you are paid 2^n dollars. Thus, if the first flip is tails, you are paid \$2. If the coin comes up heads 5 times in a row and the sixth flip comes up tails, you are paid $2^6 = \$64$. How much would you pay to play such a game? \$5? \$20? The game becomes interesting when we compute the expected value:

Expectation
$$= P(\text{T}) \cdot 2^1 + P(\text{HT}) \cdot 2^2 + P(\text{HHT}) \cdot 2^3 + P(\text{HHHT}) \cdot 2^4 + \cdots$$

$$= \frac{1}{2} \cdot 2 + \frac{1}{4} \cdot 4 + \frac{1}{8} \cdot 8 + \frac{1}{16} \cdot 16 + \cdots$$

$$= 1 + 1 + 1 + 1 + \cdots$$

There is no maximum number of times the coin can be flipped, and the expectation is infinite! Theoretically, it would be worthwhile to play no matter how high the fee is.

Business Applications of Expectation

When an insurance company sells a life insurance policy, the premium (the cost to purchase the policy) is based on many factors, but one of the most important is the probability that the insured person will outlive the term of the policy. Such probabilities are found in *mortality tables,* which give the probability that a person of a certain age will live 1 more year. For the life insurance company, it is very much like gambling. The company wants to know its expectation on a policy—that is, how much it will pay out, on average, for each policy it writes. Here is an example.

HISTORICAL NOTE

Edmond Halley (hăl′ē) (1656–1742), of Halley's comet fame, also created, for the city of Breslau, Germany, one of the first mortality tables, which he published in 1693. This was one of the first attempts to relate mortality and age in a population.

▼ example 2 **Expectation in Insurance**

According to mortality tables published in the National Vital Statistics Report, the probability that a 21-year-old will die within 1 year is approximately 0.000962. Suppose that the premium for a 1-year, \$25,000 life insurance policy for a 21-year-old is \$32. What is the insurance company's expectation for this policy?

Solution

Let S_1 be the event that the person dies within 1 year. Then $P(S_1) = 0.000962$, and the company must pay out \$25,000. Because the company charged \$32 for the policy, the company's actual loss is \$24,968. Let S_2 be the event that the policy holder does not die during the year of the policy. Then $P(S_2) = 0.999038$, and the company keeps the premium of $32. The expectation is

$$\begin{aligned}\text{Expectation} &= P(S_1) \cdot S_1 + P(S_2) \cdot S_2 \\ &= 0.000962(-24,968) + 0.999038(32) \\ &= 7.95\end{aligned}$$

• The amount the company pays out is entered as a negative number. The amount the company receives is entered as a positive number.

The company's expectation is $7.95, so the company earns, on average, $7.95 for each policy sold.

▼ check your progress 2   The probability that an 18-year-old will die within 1 year is approximately 0.000753. Suppose that the premium for a 1-year, $10,000 life insurance policy for an 18-year-old is $45. What is the insurance company's expectation for this policy?

Solution   *See page S47.*

Expectation is also used when a company bids on a project. The company must try to predict the costs and amount of work involved to give a bid that allows it to make a profit from completing the project. At the same time, if the bid is too high, the client may reject the offer. Because it is impossible to predict in advance the exact requirements of the job, probabilities can be used to analyze the likelihood of making a profit, as the next example demonstrates.

▼ example 3   Expected Company Profits

Suppose a software company bids on a project to update the database program for an accounting firm. The software company assesses its potential profit as shown in the following table.

Profit/Loss	Probability
$75,000	0.10
$50,000	0.25
$20,000	0.50
−$10,000	0.10
−$25,000	0.05

What is the profit expectation for the company?

Solution
The company's expected profit is

Expectation
$= 0.10(75,000) + 0.25(50,000) + 0.50(20,000) + 0.10(-10,000) + 0.05(-25,000)$
$= 7500 + 12,500 + 10,000 - 1000 - 1250 = 27,750$

The company's expected profit is $27,750.

▼ check your progress 3   A road construction company bids on a project to build a new freeway. The company estimates its potential profit as shown in the table at the left. What is the profit expectation for the company?

Solution   *See page S47.*

Profit/Loss	Probability
$500,000	0.05
$250,000	0.30
$150,000	0.35
−$100,000	0.20
−$350,000	0.10

EXCURSION

Yale Joel/Time & Life Pictures/Getty Images

Photolibrary

Chuck-a-luck

Chuck-a-luck is a game of chance in which 3 dice in a cage are tumbled. Bets can be placed on the values that the 3 dice will show. The table below shows the different bets and the amount won if $1 is wagered.

Numbers bet	Bet on a number from 1 through 6	
	if 1 die matches	Pays $1
	if 2 dice match	Pays $2
	if all 3 dice match	Pays $10
Field bet	Bet that the sum of the numbers showing on all 3 dice will be 5, 6, 7, 8, 13, 14, 15, or 16	Pays $1
Over 10	Bet that the sum of the numbers showing on all 3 dice will be more than 10	Pays $1
Under 11	Bet that the sum of the numbers showing on all 3 dice will be less than 11	Pays $1

HISTORICAL NOTE

Chuck-a-luck, also known (sometimes with slight variations) as "Bird Cage," "Sic Bo," or "Sweat," is one of the oldest dice games. It was probably the most popular dice game played by the soldiers of the Civil War. The term *tinhorn gambler* is derived from the practice of using a metal chute in place of a cage to tumble the dice.

EXCURSION EXERCISES

1. Compute the probability that if a bet is placed on the number 5, all 3 dice will show a 5.
2. What is the probability that *exactly* 1 die will show a 5?
3. What is the probability that 2 (but not 3) dice will show a 5?
4. Find the probability that none of the dice will show a 5.
5. What is the expectation for a $1 bet placed on the number 5?
6. Determine the expectation for wagering $1 on the Over 10 bet.
7. Determine the expectation for wagering $1 on the Under 11 bet.
8. Which bet is most favorable for the player? Which is most favorable for the casino? (The expectation for a $1 Field bet is −4 cents.)

EXERCISE SET 12.6

(Suggested Assignment: The Enhanced WebAssign Exercises and Exercise 26)

1. The outcomes of an experiment and the probability of each outcome are given in the table below. Compute the expectation for this experiment.

Outcome	Probability
30	0.15
40	0.2
50	0.4
60	0.05
70	0.2

2. The outcomes of an experiment and the probability of each outcome are given in the table below. Compute the expectation for this experiment.

Outcome	Probability
5	0.4
6	0.3
7	0.1
8	0.08
9	0.07
10	0.05

3. Roulette One of the wagers in the game of roulette is to place a bet that the ball will land on a black number. (Eighteen of the numbers are black, 18 are red, and 2 are green.) If the ball lands on a black number, the player wins the amount bet. If a player bets $1, find the player's expectation.

4. Roulette One of the wagers in roulette is to bet that the ball will stop on a number that is a multiple of 3. (Both 0 and 00 are not included.) If the ball stops on such a number, the player wins double the amount bet. If a player bets $1, compute the player's expectation.

■ **Casino Games** Many casinos have a game called the Big Six Money Wheel that has 54 slots in which are displayed a Joker, the casino logo, and various dollar amounts, as shown in the table below. Players may bet on the Joker, the casino logo, or one or more dollar denominations. The wheel is spun and if the wheel stops on the same place as the player's bet, the player wins that amount for each dollar bet. Exercises 5 to 8 use this game.

Denomination	Number of slots
$40 (Joker)	1
$40 (Casino logo)	1
$20	2
$10	4
$5	7
$2	15
$1	24

Barbara Alper/Stock Boston, LLC

5. If a player bets $1 on the Joker denomination, find the player's expectation.

6. If a player bets $1 on the $20 denomination, find the player's expectation.

7. If a player bets $1 on the $5 denomination, find the player's expectation.

8. If a player bets $1 on the $2 denomination, find the player's expectation.

Life Insurance Exercises 9 to 14 use data taken from mortality tables published in the National Vital Statistics Report.

9. The probability that a 22-year-old female in the United States will die within 1 year is approximately 0.000487.

If an insurance company sells a 1-year, $25,000 life insurance policy to such a person for $75, what is the company's expectation?

10. The probability that a 28-year-old male in the United States will die within 1 year is approximately 0.001362. If an insurance company sells a 1-year, $20,000 life insurance policy to such a person for $155, what is the company's expectation?

11. The probability that an 80-year-old male in the United States will die within 1 year is approximately 0.070471. If an insurance company sells a 1-year, $10,000 life insurance policy to such a person for $495, what is the company's expectation?

12. The probability that an 80-year-old female in the United States will die within 1 year is approximately 0.050409. If an insurance company sells a 1-year, $15,000 life insurance policy to such a person for $860, what is the company's expectation?

13. The probability that a 30-year-old male in the United States will die within 1 year is about 0.001406. An insurance company is preparing to sell a 30-year-old male a 1-year, $30,000 life insurance policy. How much should it charge for its premium in order to have a positive expectation for the policy?

14. The probability that a 25-year-old female in the United States will die within 1 year is about 0.000509. An insurance company is preparing to sell a 25-year-old female a 1-year, $75,000 life insurance policy. How much should it charge for its premium in order to have a positive expectation for the policy?

15. Construction A construction company has been hired to build a custom home. The builder estimates the probabilities of potential profit (or loss) as shown in the table below. What is the profit expectation for the company?

Profit/Loss	Probability
$100,000	0.10
$60,000	0.40
$30,000	0.25
$0	0.15
−$20,000	0.08
−$40,000	0.02

Marquis/Shutterstock.com

16. Painting   A professional painter has been hired to paint a commercial building for $18,000. From this fee, the painter must buy supplies and pay employees. The painter estimates the potential profit as shown in the table below. What is the profit expectation for the painter?

Profit/Loss	Probability
$10,000	0.15
$8,000	0.35
$5,000	0.2
$3,000	0.2
$1,000	0.1

17. Design Consultant   A consultant has been hired to redesign a company's production facility. The consultant estimates the probabilities of her potential profit as shown in the table below. What is her profit expectation?

Profit/Loss	Probability
$40,000	0.05
$30,000	0.2
$20,000	0.5
$10,000	0.2
$5,000	0.05

18. Office Rentals   A real estate company has purchased an office building with the intention of renting office space to small businesses. The company estimates the probabilities of potential profit (or loss) as shown in the table below. What is the profit expectation for the company?

Profit/Loss	Probability
$700,000	0.15
$400,000	0.25
$200,000	0.25
$50,000	0.20
−$100,000	0.10
−$250,000	0.05

19. Lotteries   The Florida lottery game Mega Money is played by choosing 4 numbers from 1 to 44 and 1 number from 1 to 22. The table below shows the probability of winning one day's $2 million jackpot along with the amounts and probabilities of other prizes on that day.

Match 4+Mega	$2,000,000	$\dfrac{1}{2,986,522}$
Match 4	$2164	$\dfrac{3}{426,646}$
Match 3+Mega	$440	$\dfrac{80}{1,493,261}$
Match 3	$70	$\dfrac{240}{213,323}$
Match 2+Mega	$27.50	$\dfrac{2340}{1,493,261}$
Match 2	$2.50	$\dfrac{7020}{213,323}$
Match 1+Mega	$3	$\dfrac{19,760}{1,493,261}$
Mega	$1	$\dfrac{45,695}{1,493,261}$

Assuming that the jackpot was not split among multiple winners, find the expectation for buying a $1 lottery ticket. Round to the nearest tenth of a cent.

20. Lotteries   The Florida lottery game Lotto is played by choosing 6 numbers from 1 to 53. If these numbers match the ones drawn by the lottery commission, you win the jackpot. The table below shows the probability of winning one day's $26 million jackpot, along with the amounts and probabilities of other prizes on that day.

Match 6	$26,000,000	$\dfrac{1}{22,957,480}$
Match 5	$4273	$\dfrac{141}{11,478,740}$
Match 4	$64	$\dfrac{3243}{4,591,496}$
Match 3	$5	$\dfrac{16,215}{1,147,874}$

Assuming that the jackpot was not split among multiple winners, find the expectation for buying a $1 lottery ticket. Round to the nearest tenth of a cent.

EXTENSIONS

Critical Thinking

21. If a pair of regular dice is tossed once, use the expectation formula to determine the expected sum of the numbers on the upward faces of the 2 dice.

22. Consider rolling a pair of unusual dice, for which the faces have the number of pips indicated.

Die 1: {1, 2, 3, 4, 5, 6}

Die 2: {0, 0, 0, 6, 6, 6}

   a. List the sample space for the experiment.

   b. Compute the probability of each possible sum of the upward faces on the dice.

   c. What is the expected value of the sum of the numbers on the upward faces of the 2 dice?

23. Two dice, one labeled 1, 2, 2, 3, 3, 4 and the other labeled 1, 3, 4, 5, 6, 8, are rolled once. Use the formula for expectation to determine the expected sum of the numbers on the upward faces of the 2 dice. Dice such as these are called *Sicherman dice.*

24. Suppose you purchase a ticket for a prize and your expectation is −$1. What is the meaning of this expectation?

Cooperative learning

25. Efron's dice Suppose you are offered 1 of 2 pairs of dice, a red pair or a green pair, that are labeled as follows.

Red die 1: 0, 0, 4, 4, 4, 4

Red die 2: 2, 3, 3, 9, 10, 11

Green die 1: 3, 3, 3, 3, 3, 3

Green die 2: 0, 1, 7, 8, 8, 8

After you choose, your friend will receive the other pair. Which pair should you choose if you are going to play a game in which each of you rolls your dice and the player with the higher sum wins? Dice such as these are part of a set of 4 pairs of dice called *Efron's dice.* Which pair should you choose? Explain why you chose the dice you did.

26. **Lotteries** The PowerBall lottery commission chooses 5 white balls from a drum containing 59 balls marked with the numbers 1 through 59, and 1 red ball from a separate drum containing 39 balls. The following table shows the approximate probability of winning a certain prize if the numbers you choose match those chosen by the lottery commission.

Match	Prize	Approximate probability
○○○○○ + ●	Jackpot	1:195,249,054
○○○○○	$200,000	1:5,138,133
○○○○ + ●	$10,000	1:723,145
○○○○	$100	1:19,030
○○○ + ●	$100	1:13,644
○○○	$7	1:359
○○ + ●	$7	1:787
○ + ●	$4	1:123
○	$3	1:62

Probabilities based on $1 play.

SOURCE: http://www.powerball.com/pb_prizes.asp

   a. Assuming that the jackpot for a certain drawing is $36 million, what is your expectation if you purchase 1 ticket for $1? Round to the nearest cent. Assume that the jackpot is not split among multiple winners.

   b. What is unusual about the 2 prizes for $7?

CHAPTER 12 SUMMARY

The following table summarizes essential concepts in this chapter. The references given in the right-hand column list Examples and Exercises that can be used to test your understanding of a concept.

12.1 The Counting Principle

Sample Spaces An experiment is an activity with an observable outcome. The sample space of an experiment is the set of all possible outcomes. A table or a tree diagram can be used to list all the outcomes in the sample space of a multi-stage experiment.	See **Examples 3 and 4** on pages 736 and 737, and then try Exercises 3 and 4 on page 798.

The Counting Principle Let E be a multi-stage experiment. If $n_1, n_2, n_3, \ldots, n_k$ are the number of possible outcomes of each of the k stages of E, then there are $n_1, n_2, n_3, \ldots, n_k$ possible outcomes for E.	See **Example 5** on page 737, and then try Exercises 5 and 7 on page 798.
Counting With and Without Replacement Some multi-stage experiments involve repeatedly choosing an element from a given set. If an element is returned to the set and can be chosen again, the experiment is performed *with replacement*. If an element is not returned to the set and cannot be chosen again, the experiment is performed *without replacement*.	See **Example 6** on page 739, and then try Exercises 1 and 2 on page 798.

12.2 Permutations and Combinations

n factorial n factorial is the product of the natural numbers n through 1. $n! = (n - 1) \cdot (n - 2) \cdots \cdots 3 \cdot 2 \cdot 1$ $0! = 1$	See **Example 1** on page 744, and then try Exercises 9 to 11 on page 798.
Permutation Formula for Distinct Objects A permutation is an arrangement of objects in a definite order. The number of permutations of n distinct objects selected k at a time is $P(n, k) = \dfrac{n!}{(n - k)!}$	See **Examples 2 and 3** on page 746, and then try Exercises 15 and 20 on pages 798 and 799.
Permutations of Objects, Some of Which Are Identical The number of permutations of n objects of r different types, where k_1 identical objects are of one type, k_2 of another, and so on, is $\dfrac{n!}{k_1! \cdot k_2! \cdots \cdots k_r!}$, where $k_1 + k_2 + \cdots + k_r = n$.	See **Example 5** on page 749, and then try Exercise 18 on page 798.
Combination Formula A combination is a collection of objects for which the order is not important. The number of combinations of n objects chosen k at a time is $C(n, k) = \dfrac{P(n, k)}{k!} = \dfrac{n!}{k! \cdot (n - k)!}$	See **Examples 2 and 3** on page 746, and then try Exercises 21 and 22 on page 799.
Applying Several Counting Techniques Some counting problems require using the counting principle along with a permutation or combination formula.	See **Examples 4, 7, and 8** on pages 747, 751, and 752, and then try Exercises 23 to 25 on page 799.

12.3 Probability and Odds

Theoretical Probability of an Event For an experiment with sample space S of equally likely outcomes, the theoretical probability $P(E)$ of an event E is given by $P(E) = \dfrac{n(E)}{n(S)} = \dfrac{\text{number of elements in } E}{\text{number of elements in } S}$	See **Examples 2 and 3** on pages 758 and 759, and then try Exercises 27 and 31 on page 799.
Empirical Probability of an Event If an experiment is performed repeatedly and the occurrence of the event E is observed, the empirical probability $P(E)$ of the event E is given by $P(E) = \dfrac{\text{number of times event } E \text{ occurred}}{\text{number of times the experiment was performed}}$	See **Example 4** on page 760, and then try Exercise 29 on page 799.

Punnet Square Using a capital letter to represent a dominant allele and a lower-case letter to represent a recessive allele, a Punnet square shows all possible genotypes of the offspring of two parents with given genotypes.	See **Example 5** on page 761, and then try Exercise 44 on page 799.
Odds of an Event Let E be an event in a sample space of equally likely outcomes. Then $$\text{Odds in favor of } E = \frac{\text{number of favorable outcomes}}{\text{number of unfavorable outcomes}}$$ $$\text{Odds against } E = \frac{\text{number of unfavorable outcomes}}{\text{number of favorable outcomes}}$$	See **Example 6** on page 762, and then try Exercises 41 and 42 on page 799.
Relationship between Odds and Probability If the odds in favor of event E are $\frac{a}{b}$, then $P(E) = \frac{a}{a+b}$. This relationship is also expressed by the equation $$\text{Odds in favor of } E = \frac{P(E)}{1 - P(E)}$$	See **Example 7** on page 763, and then try Exercise 43 on page 799.

12.4 Addition and Complement Rules

Probability of Mutually Exclusive Events Two events A and B are mutually exclusive if they cannot occur at the same time. In this case, the probability of A or B occurring is $$P(A \text{ or } B) = P(A) + P(B)$$	See **Example 1** on page 769, and then try Exercises 27 and 31 on page 799.
Addition Rule for Probabilities If A and B are two events in a sample space S, then $$P(A \text{ or } B) = P(A) + P(B) - P(A \text{ and } B)$$	See **Example 2** on page 770, and then try Exercise 38 on page 799.
Probability of the Complement of an Event The complement of an event E in a sample space S includes all those outcomes of S that are not in E and excludes the outcomes in E. The symbol for the complement of E is E^c. To find the probability of E^c, use the relationship $$P(E^c) = 1 - P(E)$$	See **Example 3** on page 772, and then try Exercise 39 on page 799.
Using Combinatorics Formulas to Find Probabilities In many cases, the process of finding a probability involves using the counting principle and/or a permutation or combination formula.	See **Examples 4 and 5** on pages 772 and 773, and then try Exercises 60 and 63 on page 800.

12.5 Conditional Probability

Conditional Probability Formula If A and B are two events in a sample space S, then the conditional probability of B given that A has occurred is $$P(B \mid A) = \frac{P(A \text{ and } B)}{P(A)}$$	See **Example 1** on page 779, and then try Exercises 35 and 36 on page 799.
Product Rule for Probabilities If A and B are two events in a sample space S, then $$P(A \text{ and } B) = P(A) \cdot P(B \mid A)$$ **Probability of Successive Events** The probability of two or more events occurring in succession is the product of the conditional probabilities of each of the events.	See **Example 2** on page 781, and then try Exercises 51 and 57 on page 800.

Product Rule for Independent Events If A and B are two events in a sample space S and $P(B\mid A) = P(B)$, then A and B are independent events. In this case $$P(A \text{ and } B) = P(A) \cdot P(B)$$	See **Example 3** on page 782, and then try Exercise 39 on page 799.
Using Several Formulas to Find Probabilities The process of finding a conditional probability may involve using other probability formulas.	See **Example 4** on page 783, and then try Exercise 61 on page 800.

12.6 Expectation

Expectation If $S_1, S_2, S_3, \ldots, S_n$ are the possible numerical outcomes of an experiment and $P(S_1), P(S_2), P(S_3), \ldots, P(S_n)$ are the probabilities of those outcomes, then the expectation of the experiment is $$P(S_1) \cdot S_1 + P(S_2) \cdot S_2 + P(S_3) \cdot S_3 + \cdots + P(S_n) \cdot S_n$$	See **Example 1** on page 789, and then try Exercises 64 and 65 on page 800.
Business Applications of Expectation Insurance companies can use expectation to help determine life insurance premiums. Other businesses can use expectation to determine potential profits.	See **Examples 2 and 3** on pages 790 and 791, and then try Exercises 68 and 70 on page 800.

CHAPTER 12 REVIEW EXERCISES

■ In Exercises 1 and 2, list the elements of the sample space for the given experiment.

1. Two-digit numbers are formed, with replacement, from the digits 1, 2, and 3.

2. Two-digit numbers are formed, without replacement, from the digits 2, 6, and 8.

3. Use a tree diagram to list all possible outcomes that result from tossing 4 coins.

4. Use a table to list all possible 2-character codes that can be formed from one of the digits 7, 8, or 9 followed by one of the letters A or B.

5. An athletic shoe store sells jogging shoes in 3 styles that come in 4 colors. Each color comes in 6 sizes. How many distinct shoes are available?

6. The combination for a lock to a bicycle chain contains 4 numbers chosen from the numbers 0 through 9. How many different lock combinations are possible? Assume that a number can be used more than once.

7. **Serial Numbers** In the 1970s and 1980s, the Conn music company assigned the first 4 characters in serial numbers of instruments in the following way. The first character is one of the letters G or H and indicates the decade in which the instrument was made: "G" for 1970s and "H" for 1980s. The second character indicates the month of the year: "A" for January, "B" for February, and so on. The third character is a number from 0 to 9 indicating the year in the decade a instrument is made, and the fourth character is a number from 0 to 9 indicating the type of instrument: 1 = cornet, 2 = trumpet, 3 = alto horn, 4 = French horn, 5 = mellophone, 6 = valve trombone, 7 = slide trombone, 8 = euphonium, 9 = tuba, 0 = sousaphone. How many 4-character serial numbers are possible?

8. **Codes** A *biquinary code* is a code that consists of 2 different binary digits (a binary digit is a 0 or 1) followed by 5 binary digits for which there are no restrictions. How many biquinary codes are possible?

■ In Exercises 9 to 14, evaluate each expression.

9. $7!$

10. $8! - 4!$

11. $\dfrac{9!}{2!\,3!\,4!}$

12. $P(10, 6)$

13. $P(8, 3)$

14. $\dfrac{C(6, 2) \cdot C(8, 3)}{C(14, 5)}$

15. In how many different ways can 7 people arrange themselves in a line to receive service from a bank teller?

16. A matching test has 7 definitions that are to be paired with 7 words. Assuming each word corresponds to exactly 1 definition, how many different matches are possible by random matching?

17. A matching test has 7 definitions to be matched with 5 words. Assuming each word corresponds to exactly 1 definition, how many different matches are possible by random matching?

18. How many distinct arrangements are possible using the letters of the word *letter*?

19. Twelve identical coins are tossed. How many distinct arrangements are possible consisting of 4 heads and 8 tails?

20. Work Shifts Three positions are open at a manufacturing plant: the day shift, the swing shift, and the night shift. In how many different ways can 5 people be assigned to the 3 shifts?

21. A professor assigns 25 homework problems, of which 10 will be graded. How many different sets of 10 problems can the professor choose to grade?

22. Stock Portfolios A stockbroker recommends 11 stocks to a client. If the client will invest in 3 of the stocks, how many different 3-stock portfolios can be selected?

23. Quality Control A quality control inspector receives a shipment of 15 computer monitors, of which 3 are defective. If the inspector randomly chooses 5 monitors, how many different sets can be formed that consist of 3 nondefective monitors and 2 defective monitors?

24. How many ways can 9 people be seated in 9 chairs if 2 of the people refuse to sit next to each other?

25. How many 5-card poker hands consist of 4 of a kind (4 aces, 4 kings, 4 queens, and so on)?

26. If it is equally likely that a child will be born a boy or a girl, compute the probability that a family of 4 children will have 1 boy and 3 girls.

27. If a coin is tossed 3 times, what is the probability of getting 1 head and 2 tails?

28. A large company currently employs 5739 men and 7290 women. If an employee is selected at random, what is the probability that the employee is a woman?

■ **Enrollment** In Exercises 29 and 30, use the table below, which shows the number of students at a university who are currently in each class level.

Class level	Number of students
First year	642
Sophomore	549
Junior	483
Senior	445
Graduate student	376

29. If a student is selected at random, what is the probability that the student is an upper-division undergraduate student (junior or senior)?

30. If a student is selected at random, what is the probability that the student is not a graduate student?

■ In Exercises 31 to 36, a pair of dice is tossed.

31. Find the probability that the sum of the pips on the 2 upward faces is 9.

32. Find the probability that the sum on the 2 dice is not 11.

33. Find the probability that the sum on the 2 dice is at least 10.

34. Find the probability that the sum on the 2 dice is an even number or a number less than 5.

35. What is the probability that the sum on the 2 dice is 9, given that the sum is odd?

36. What is the probability that the sum on the 2 dice is 8, given that doubles were rolled?

■ In Exercises 37 to 40, a single card is selected from a standard deck of playing cards.

37. What is the probability that the card is a heart or a black card?

38. What is the probability that the card is a heart or a jack?

39. What is the probability that the card is not a 3?

40. What is the probability that the card is red, given that it is not a club?

41. If a pair of dice is rolled, what are the odds in favor of getting a sum of 6?

42. If 1 card is drawn from a standard deck of playing cards, what are the odds that the card is a heart?

43. If the odds against an event occurring are 4 to 5, compute the probability of the event occurring.

44. Genotypes The hair length of a particular rodent is determined by a dominant allele H, corresponding to long hair, and a recessive allele h, corresponding to short hair. Draw a Punnett square for parents of genotypes Hh and hh, and compute the probability that the offspring of the parents will have short hair.

45. Two cards are drawn, without replacement, from a standard deck of playing cards. The probability that exactly 1 card is an ace is 0.145. The probability that exactly 1 card is a face card (jack, queen, or king) is 0.362, and the probability that a selection of 2 cards will contain an ace or a face card is 0.471. Find the probability that the 2 cards are an ace and a face card.

46. Surveys A recent survey asked 1000 people whether they liked cheese-flavored corn chips (642 people), jalapeño-flavored chips (487 people), or both (302 people). If 1 person is chosen from this survey, what is the probability that the person does not like either of the 2 flavors?

■ In Exercises 47 to 51, a box contains 24 different colored chips that are identical in size. Five are black, 4 are red, 8 are white, and 7 are yellow.

47. If a chip is selected at random, what is the probability that the chip will be yellow or white?

48. If a chip is selected at random, what are the odds in favor of getting a red chip?

49. If a chip is selected at random, find the probability that the chip is yellow given that it is not white.

50. If 5 chips are randomly chosen, without replacement, what is the probability that none of them is red?

51. If 3 chips are chosen without replacement, find the probability that the first one is yellow, the second is white, and the third is yellow.

■ **Voting**   In Exercises 52 to 56, use the table below, which shows the number of voters in a city who voted for or against a proposition (or abstained from voting) according to their party affiliations. Round answers to the nearest hundredth.

	For	**Against**	**Abstained**
Democrat	8452	2527	894
Republican	2593	5370	1041
Independent	1225	712	686

52. If a voter is chosen at random, compute the probability that the person voted against the proposition.

53. If a voter is chosen at random, compute the probability that the person is a Democrat or an Independent.

54. If a voter is randomly chosen, what is the probability that the person abstained from voting on the proposition and is not a Republican?

55. A voter is randomly selected. What is the probability that the individual voted for the proposition, given that the voter is a registered Independent?

56. A voter is randomly selected. What is the probability that the individual is registered as a Democrat, given that the person voted against the proposition?

57. A single die is rolled 3 times in succession. What is the probability that each roll gives a 6?

58. A single die is rolled 5 times in succession. Find the probability that a 6 will be rolled at least once.

59. A single die is rolled 5 times in succession. Find the probability that exactly 2 of the rolls give a 6.

60. A person draws a card from a standard deck and replaces it; she then does this 3 more times. What is the probability that she drew a spade at least once?

61. Disease Testing   A veterinarian uses a test to determine whether a dog has a disease that affects 7% of the dog population. The test correctly gives a positive result for 98% of dogs that have the disease, but gives false positives for 4% of dogs that do not have the disease. If a dog tests positive, what is the probability that the dog has the disease?

62. Weather   Suppose that in your area in the wintertime, if it rains one day, there is a 65% chance that it will rain the next day. If it does not rain on a given day, there is only a 15% chance that it will rain the following day. What is the probability that, if it didn't rain today, it will rain the next 2 days but not the following 2?

63. Batteries   About 1.2% of AA batteries produced by a particular manufacturer are defective. If a consumer buys a box of 12 of these batteries, what is the probability that at least 1 battery is defective?

64. Suppose it costs $4 to play a game in which a single die is rolled and you win the amount of dollars that the die shows. What is the expectation for the game?

65. I will flip 2 coins. If both coins come up tails, I will pay you $5. If one shows heads and one shows tails, you will pay me $2. If both coins come up heads, we will call it a draw. What is your expectation for this game?

66. Raffle Tickets   For a fundraiser, an elementary school is selling 800 raffle tickets for $1 each. From these, 5 tickets will be drawn. One of the winners gets $200 and the others each get $75. If you buy 1 raffle ticket, what is your expectation?

67. If a pair of dice is rolled 65 times, on how many rolls can we expect to get a total of 4?

68. **Life Insurance**   The probability that a 29-year-old female in the United States will die within 1 year is approximately 0.000595. If an insurance company sells a 1-year, $40,000 life insurance policy to a 29-year-old for $320, what is the company's expectation?

69. **Life Insurance**   The probability that a 19-year-old male in the United States will die within 1 year is approximately 0.001188. If an insurance company sells a 1-year, $25,000 life insurance policy to a 19-year-old for $795, what is the company's expectation?

70. Construction   A construction company has bid on a building renovation project. The company estimates the probabilities of potential profit (or loss) as shown in the table below. What is the profit expectation for the company?

Profit/loss	**Probability**
$25,000	0.20
$15,000	0.25
$10,000	0.20
$5,000	0.15
$0	0.10
−$5,000	0.10

CHAPTER 12 TEST

1. **Driver's License Numbers** A certain driver's license number begins with one of the letters A, D, G, or K and is followed by one of the digits 2, 3, or 4. List the elements in the sample space of the first two digits of this driver's license number.

2. **Computer Systems** A computer system can be configured using 1 of 3 processors of different speeds, 1 of 4 disk drives of different sizes, 1 of 3 monitors, and 1 of 2 graphics cards. How many different computer systems are possible?

3. **Transistors** In a very simple computer chip, 10 wires go from one transistor to a second transistor. For the computer to function, instructions must be sent between these 2 transistors. In how many ways can 4 different instructions be sent between the 2 transistors if no 2 instructions can be sent along the same wire?

4. A matching test asks students to match 10 words with 15 definitions. Assuming each word corresponds to exactly 1 definition, how many different pairs are possible?

5. **Softball** How many games are necessary in a softball league consisting of 8 teams if each team must play each of the other teams once?

6. A coin and a regular 6-sided die are tossed together once. What is the probability that the coin shows a head or the die has a 5 on the upward face?

7. A person draws 4 cards in succession from a standard deck of playing cards, replacing each card before drawing the next. What is the probability that the person draws at least 1 ace?

8. What is the probability of drawing 2 cards in succession (without replacement) from a standard deck of playing cards and having them both be hearts?

9. Four cards are drawn from a standard deck of playing cards without replacement. What is the probability that none of them are 9s?

10. Three coins are tossed once. What are the odds in favor of the coins showing all heads?

11. **Disease Testing** A new medical test can determine whether a human has a disease that affects 5% of the population. The test correctly gives a positive result for 99% of people who have the disease, but it gives a false positive for 3% of people who do not have the disease. If a person tests positive, what is the probability that the person has the disease?

12. **Advertising** The table below shows the number of men and the number of women who responded either positively or negatively to a new commercial. If 1 person is chosen from this group, find the probability that the person is a woman, given that the person responded negatively. Round to the nearest thousandth.

	Positive	Negative	Total
Men	684	736	1420
Women	753	642	1395
Total	1437	1378	2815

13. **Genotypes** Straight or curly hair for a hamster is determined by a dominant allele S, corresponding to straight hair, and a recessive allele s, which gives curly hair. If 1 parent is of genotype Ss and the other is of genotype ss, compute the probability that the offspring of the parents will have curly hair.

14. **Inventories** A software company is preparing a bid to create a new inventory program for an auto parts company. The software company estimates the probabilities of potential profit (or loss) as shown in the table below. What is the profit expectation for the company?

Profit/loss	Probability
$75,000	0.18
$50,000	0.36
$25,000	0.31
$0	0.08
−$10,000	0.05
−$20,000	0.02

The United States, as a nation, is dependent on world trade. And world trade is dependent on internationally standardized units of measurement. Almost all countries use the metric system as their sole system of measurement. The United States is one of only a few countries that has not converted to the metric system as its official system of measurement.

In this Appendix, we will present the metric system of measurement and explain how to convert between different units.

The basic unit of *length,* or distance, in the metric system is the **meter** (m). One meter is approximately the distance from a doorknob to the floor. All units of length in the metric system are derived from the meter. Prefixes added to the basic unit denote the length of the unit. For example, the prefix *centi-* means "one hundredth;" therefore, 1 centimeter is 1 one hundredth of a meter (0.01 m).

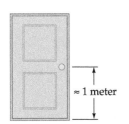

≈ 1 meter

kilo- = 1 000	1 kilometer (km)	= 1 000 meters (m)
hecto- = 100	1 hectometer (hm)	= 100 m
deca- = 10	1 decameter (dam)	= 10 m
	1 meter (m)	= 1 m
deci- = 0.1	1 decimeter (dm)	= 0.1 m
centi- = 0.01	1 centimeter (cm)	= 0.01 m
milli- = 0.001	1 millimeter (mm)	= 0.001 m

Notice that in this list 1000 is written as 1 000, with a space between the 1 and the zeros. When writing numbers using metric units, each group of three numbers is separated by a space instead of a comma. A space is also used after each group of three numbers to the right of a decimal. For example, 31,245.2976 is written 31 245.297 6 in metric notation.

question Which unit in the metric system is one thousandth of a meter?

Mass and weight are closely related. *Weight* is a measure of how strongly gravity is pulling on an object. Therefore, an object's weight is less in space than on Earth's surface. However, the amount of material in the object, its *mass,* remains the same. On the surface of Earth, the terms *mass* and *weight* can be used interchangeably.

The basic unit of mass in the metric system is the **gram** (g). If a box 1 centimeter long on each side is filled with pure water, the mass of that water is 1 gram.

1 cm

1 cm

1 cm

1 gram = the mass of water in a box that is 1 centimeter long on each side

answer The millimeter is one thousandth of a meter.

The units of mass in the metric system have the same prefixes as the units of length.

1 kilogram (kg)   = 1 000 grams (g)

1 hectogram (hg) = 100 g

1 decagram (dag) = 10 g

1 gram (g)   = 1 g

1 decigram (dg) = 0.1 g

1 centigram (cg) = 0.01 g

1 milligram (mg) = 0.001 g

Weight ≈ 1 gram

The gram is a small unit of mass. A paperclip weighs about 1 gram. In many applications, the kilogram (1 000 grams) is a more useful unit of mass. This textbook weighs about 2.5 kilograms.

question Which unit in the metric system is equal to 1 000 grams?

Liquid substances are measured in units of *capacity*.

The basic unit of capacity in the metric system is the **liter** (L). One liter is defined as the capacity of a box that is 10 centimeters long on each side.

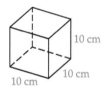

10 cm

10 cm

10 cm

1 liter = the capacity of a box that is 10 centimeters long on each side

The units of capacity in the metric system have the same prefixes as the units of length.

1 kiloliter (kl)   = 1 000 liters (L)

1 hectoliter (hl) = 100 L

1 decaliter (dal) = 10 L

1 liter (L)   = 1 L

1 deciliter (dl) = 0.1 L

1 centiliter (cl) = 0.01 L

1 milliliter (ml) = 0.001 L

POINT OF INTEREST

The definition of 1 in. has been changed as a consequence of the wide acceptance of the metric system. One inch is now exactly 25.4 mm.

Converting between units in the metric system involves moving the decimal point to the right or to the left. Listing the units in order from largest to smallest will indicate how many places to move the decimal point and in which direction.

To convert 3 800 cm to meters, write the units of length in order from largest to smallest.

km   hm   dam   m   dm   cm   mm

2 positions

• Converting from cm to m requires moving 2 positions to the left.

3 800 cm = 38.00 m

2 places

• Move the decimal point the same number of places and in the same direction.

answer The kilogram is equal to 1 000 grams.

Convert 27 kg to grams.

kg   hg   dag   g   dg   cg   mg

         3 positions

- Write the units of mass in order from largest to smallest.
- Converting kg to g requires moving 3 positions to the right.

27 kg = 27 000 g

        3 places

- Move the decimal point the same number of places and in the same direction.

▼ example 1   **Convert Units in the Metric System of Measurement**

Convert.

a. 4.08 m to centimeters   **b.** 5.93 g to milligrams

c. 82 ml to liters   **d.** 9 kl to liters

Solution

a. Write the units of length from largest to smallest.

    km   hm   dam   (m)   dm   (cm)   mm

Converting m to cm requires moving 2 positions to the right.

    4.08 m = 408 cm

b. Write the units of mass from largest to smallest.

    kg   hg   dag   (g)   dg   cg   (mg)

Converting g to mg requires moving 3 positions to the right.

    5.93 g = 5 930 mg

c. Write the units of capacity from largest to smallest.

    kl   hl   dal   (L)   dl   cl   (ml)

Converting ml to L requires moving 3 positions to the left.

    82 ml = 0.082 L

d. Write the units of capacity from largest to smallest.

    (kl)   hl   dal   (L)   dl   cl   ml

Converting kl to L requires moving 3 positions to the right.

    9 kl = 9 000 L

▼ check your progress 1   Convert.

a. 1 295 m to kilometers   **b.** 7 543 g to kilograms

c. 6.3 L to milliliters   **d.** 2 kl to liters

Solution *See page S50.*

Other prefixes in the metric system are becoming more common as a result of technological advances in the computer industry. For example:

    tera-   = 1 000 000 000 000

    giga-   = 1 000 000 000

    mega-   = 1 000 000

    micro- = 0.000 001

    nano-   = 0.000 000 001

    pico-   = 0.000 000 000 001

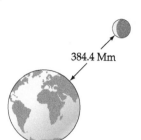

384.4 Mm

The amount of memory in a computer hard drive is generally measured in gigabytes. The speed of a computer is measured in picoseconds.

Here are a few more examples of how these prefixes are used.

The mass of Earth gains 40 Gg (gigagrams) each year from captured meteorites and cosmic dust.

The average distance from Earth to the moon is 384.4 Mm (megameters) and the average distance from Earth to the sun is 149.5 Gm (gigameters).

The wavelength of yellow light is 590 nm (nanometers).

The diameter of a hydrogen atom is about 70 pm (picometers).

There are additional prefixes in the metric system, representing both larger and smaller units. We may hear them more and more often as computer chips hold more and more information, as computers get faster and faster, and as we learn more and more about objects in our universe that are great distances away.

EXERCISES   (Suggested Assignment: Exercises 1–69, odds)

1. In the metric system, what is the basic unit of length? Of liquid measure? Of weight?

2. a. 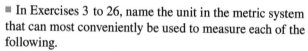 Explain how to convert meters to centimeters.

   b. Explain how to convert milliliters to liters.

In Exercises 3 to 26, name the unit in the metric system that can most conveniently be used to measure each of the following.

3. The distance from New York to London

4. The weight of a truck

5. A person's waist

6. The amount of coffee in a mug

7. The weight of a thumbtack

8. The amount of water in a swimming pool

9. The distance a baseball player hits a baseball

10. A person's hat size

11. The amount of fat in a slice of cheddar cheese

12. A person's weight

13. The maple syrup served with pancakes

14. The amount of water in a water cooler

15. The amount of vitamin C in a vitamin tablet

16. A serving of cereal

17. The width of a hair

18. A person's height

19. The amount of medication in an aspirin

20. The weight of a lawnmower

21. The weight of a slice of bread

22. The contents of a bottle of salad dressing

23. The amount of water a family uses monthly

24. The newspapers collected at a recycling center

25. The amount of liquid in a bowl of soup

26. The distance to the bank

27. a. Complete the table.

Metric system prefix	Symbol	Magnitude	Means multiply the basic unit by:
tera-	T	10^{12}	1 000 000 000 000
giga-	G	?	1 000 000 000
mega-	M	10^6	?
kilo-	?	?	1 000
hecto-	h	?	100
deca-	da	10^1	?
deci-	d	$\dfrac{1}{10}$	?
centi-	?	$\dfrac{1}{10^2}$	?
milli-	?	?	0.001
micro-	μ (mu)	$\dfrac{1}{10^6}$	?
nano-	n	$\dfrac{1}{10^9}$	?
pico-	p	?	0.000 000 000 001

b. How can the magnitude column in the table above be used to determine how many places to move the decimal point when converting to the basic unit in the metric system?

■ In Exercises 28 to 57, convert the given measure.

28. 42 cm = ———— mm

29. 91 cm = ———— mm

30. 360 g = ———— kg

31. 1 856 g = ———— kg

32. 5 194 ml = ———— L

33. 7 285 ml = ———— L

34. 2 m = ———— mm

35. 8 m = ———— mm

36. 217 mg = ———— g

37. 34 mg = ———— g

38. 4.52 L = ———— ml

39. 0.029 7 L = ———— ml

40. 8 406 m = ———— km

41. 7 530 m = ———— km

42. 2.4 kg = ———— g

43. 9.2 kg = ———— g

44. 6.18 kl = ———— L

45. 0.036 kl = ———— L

46. 9.612 km = ———— m

47. 2.35 km = ———— m

48. 0.24 g = ———— mg

49. 0.083 g = ———— mg

50. 298 cm = ———— m

51. 71.6 cm = ———— m

52. 2 431 L = ———— kl

53. 6 302 L = ———— kl

54. 0.66 m = ———— cm

55. 4.58 m = ———— cm

56. 243 mm = ———— cm

57. 92 mm = ———— cm

58. The Olympics   **a.** One of the events in the summer Olympics is the 50 000-meter walk. How many kilometers do the entrants in this event walk?

   b. One of the events in the winter Olympic games is the 10 000-meter speed skating event. How many kilometers do the entrants in this event skate?

59. Gemology   A carat is a unit of weight equal to 200 mg. Find the weight in grams of a 10-carat precious stone.

60. Sewing   How many pieces of material, each 75 cm long, can be cut from a bolt of fabric that is 6 m long?

61. Water Treatment   An athletic club uses 800 ml of chlorine each day for its swimming pool. How many liters of chlorine are used in a month of 30 days?

62. Carpentry   Each of the four shelves in a bookcase measures 175 cm. Find the cost of the shelves when the price of lumber is $15.75 per meter.

63.   **Beverages**   See the news clipping below. How many 140-milliter servings are in one carafe of Pure Premium?

IN THE NEWS
Tropicana Goes Metric

Tropicana has introduced metric-sized packaging. The first was the Pure Premium 1.75-liter carafe. Other sizes of packaging include 200 ml, 400 ml, 450 ml, and 1 L.

SOURCE: www.geometric.us

---

64. Consumerism   A 1.19-kilogram container of Quaker Oats contains 30 servings. Find the number of grams in one serving of the oatmeal. Round to the nearest gram.

65. Nutrition   A patient is advised to supplement her diet with 2 g of calcium per day. The calcium tablets she purchases contain 500 mg of calcium per tablet. How many tablets per day should the patient take?

66. Education   A laboratory assistant is in charge of ordering acid for three chemistry classes of 30 students each. Each student requires 80 ml of acid. How many liters of acid should be ordered? The assistant must order by the whole liter.

67. Consumerism   A case of 12 one-liter bottles of apple juice costs $39.60. A case of 24 cans, each can containing 340 ml of apple juice, costs $29.50. Which case of apple juice costs less per milliliter?

68.   **Adopt-A-Highway**   See the news clipping below. Find the average number of meters adopted by one of the groups in the Missouri Adopt-a-Highway program. Round to the nearest meter.

IN THE NEWS
Highway Adoption Proves Popular

The Missouri Adopt-a-Highway Program, which has been in existence for 20 years, currently has 3772 groups in the program. The groups have adopted 8 502 km along the roadways.

SOURCE: www.modot.org

---

69. Light   The distance between Earth and the Sun is 150 000 000 km. Light travels 300 000 000 m in 1 s. How many seconds does it take for light to reach Earth from the Sun?

70.   Explain why is it advantageous to have internationally standardized units of measure.

E X T E N S I O N S

Critical Thinking

71. Business   A service station operator bought 85 kl of gasoline for $51,500. The gasoline was sold for $0.859 per liter. Find the profit on the 85 kl of gasoline.

72. Business   For $299, a cosmetician buys 5 L of moisturizer and repackages it in 125-milliliter jars. Each jar costs the cosmetician $1.10. Each jar of moisturizer is sold for $17.90. Find the profit on the 5 L of moisturizer.

73. Business   A health food store buys nuts in 10-kilogram containers and repackages the nuts for resale. The store packages the nuts in 200-gram bags, costing $0.12 each, and sells them for $5.78 per bag. Find the profit on a 10-kilogram container of nuts costing $150.

Cooperative Learning

74.   Form two debating teams. One team should argue in favor of changing to the metric system in the United States, and the other should argue against it.

U.S. Customary System

Length		**Capacity**		**Weight**		**Area**	
in.	inches	oz	fluid ounces	oz	ounces	in^2	square inches
ft	feet	c	cups	lb	pounds	sq ft	square feet
yd	yards	qt	quarts				
mi	miles	gal	gallons				

Metric System

Length		**Capacity**		**Weight**		**Area**	
mm	millimeter (0.001 m)	ml	milliliter (0.001 L)	mg	milligram (0.001 g)	cm^2	square centimeters
cm	centimeter (0.01 m)	cl	centiliter (0.01 L)	cg	centigram (0.01 g)	m^2	square meters
dm	decimeter (0.1 m)	dl	deciliter (0.1 L)	dg	decigram (0.1 g)		
m	meter	L	liter	g	gram		
dam	decameter (10 m)	dal	decaliter (10 L)	dag	decagram (10 g)		
hm	hectometer (100 m)	hl	hectoliter (100 L)	hg	hectogram (100 g)		
km	kilometers (1 000 m)	kl	kiloliter (1 000 L)	kg	kilogram (1 000 g)		

Time

h	hours	min	minutes	s	seconds

CHAPTER 2

SECTION 2.1

▼ **check your progress 1**, *page 52*

The only months that start with the letter A are April and August. When we use the roster method, the set is given by {April, August}.

▼ **check your progress 2**, *page 53*

The set {March, May} is the set of all months that start with the letter M.

▼ **check your progress 3**, *page 54*

a. {0, 1, 2, 3} **b.** {12, 13, 14, 15, 16, 17, 18, 19}

c. {−4, −3, −2, −1}

▼ **check your progress 4**, *page 54*

a. False **b.** True **c.** True **d.** True

▼ **check your progress 5**, *page 55*

a. $\{x \,|\, x \in I \text{ and } x < 9\}$ **b.** $\{x \,|\, x \in N \text{ and } x > 4\}$

▼ **check your progress 6**, *page 55*

a. $n(C) = 5$ **b.** $n(D) = 1$ **c.** $n(E) = 0$

▼ **check your progress 7**, *page 56*

a. The sets are not equal but they both contain six elements. Thus the sets are equivalent.

b. The sets are not equal but they both contain 16 elements. Thus the sets are equivalent.

SECTION 2.2

▼ **check your progress 1**, *page 63*

a. $M = \{0, 4, 6, 17\}$. The set of elements in $U = \{0, 2, 3, 4, 6, 7, 17\}$ but not in M is $M' = \{2, 3, 7\}$.

b. $P = \{2, 4, 6\}$. The set of elements in $U = \{0, 2, 3, 4, 6, 7, 17\}$ but not in P is $P' = \{0, 3, 7, 17\}$.

▼ **check your progress 2**, *page 64*

a. False. The number 3 is an element of the first set but not an element of the second set. Therefore, the first set is not a subset of the second set.

b. True. The set of counting numbers is the same set as the set of natural numbers, and every set is a subset of itself.

c. True. The empty set is a subset of every set.

d. True. Each element of the first set is an integer.

▼ **check your progress 3**, *page 65*

a. Yes, because every natural number is a whole number, and the whole numbers include 0, which is not a natural number.

b. The first set is not a proper subset of the second set because the sets are equal.

▼ **check your progress 4**, *page 65*

Subsets with zero elements: { }

Subsets with one element: {a}, {b}, {c}, {d}, {e}

Subsets with two elements: {a, b}, {a, c}, {a, d}, {a, e}, {b, c}, {b, d}, {b, e}, {c, d}, {c, e}, {d, e}

Subsets with three elements: {a, b, c}, {a, b, d}, {a, b, e}, {a, c, d}, {a, c, e}, {a, d, e}, {b, c, d}, {b, c, e}, {b, d, e}, {c, d, e}

Subsets with four elements: {a, b, c, d}, {a, b, c, e}, {a, b, d, e}, {a, c, d, e}, {b, c, d, e}

Subsets with five elements: {a, b, c, d, e}

▼ **check your progress 5**, *page 66*

a. The company offers 11 upgrade options. Each option is independent of the other options. Thus the company can produce

$$2^{11} = 2048$$

different versions of the car.

b. Use the method of guessing and checking to find the smallest natural number n for which $2^n > 8000$.

$$2^{11} = 2048$$
$$2^{12} = 4096$$
$$2^{13} = 8192$$

The company must provide a minimum of 13 upgrade options if it wishes to offer at least 8000 different versions of the car.

SECTION 2.3

▼ **check your progress 1**, *page 72*

a. $D \cap E = \{0, 3, 8, 9\} \cap \{3, 4, 8, 9, 11\}$
$\quad = \{3, 8, 9\}$

b. $D \cap F = \{0, 3, 8, 9\} \cap \{0, 2, 6, 8\}$
$\quad = \{0, 8\}$

▼ **check your progress 2**, *page 72*

a. $D \cup E = \{0, 4, 8, 9\} \cup \{1, 4, 5, 7\}$
$\quad = \{0, 1, 4, 5, 7, 8, 9\}$

b. $D \cup F = \{0, 4, 8, 9\} \cup \{2, 6, 8\}$
$\quad = \{0, 2, 4, 6, 8, 9\}$

▼ **check your progress 3**, *page 73*

a. The set $D \cap (E' \cup F)$ can be described as "the set of all elements that are in D, and in F or not E."

b. The set $L' \cup M$ can be described as "the set of all elements that are in M or are not in L."

▼ **check your progress 4**, *page 74*

To determine the region(s) represented by $(A \cap B)'$, first determine the region(s) in Figure 2.1 that are represented by $A \cap B$.

Set	Region or regions	Venn diagram
$A \cap B$	i The region common to A and B	
$(A \cap B)'$	ii, iii, iv The regions in U that are not in $A \cap B$	

Now determine the region(s) in Figure 2.1 that are represented by $A' \cup B'$.

Set	Region or regions	Venn diagram
A'	iii, iv The regions outside of A	
B'	ii, iv The regions outside of B	
$A' \cup B'$	ii, iii, iv The regions formed by joining the regions in A' and the regions in B'	

The expressions $(A \cap B)'$ and $A' \cup B'$ are both represented by regions ii, iii, and iv. Thus $(A \cap B)' = A' \cup B'$ for all sets A and B.

▼ **check your progress 5**, *page 75*

The following solutions reference the regions shown in Figure 2.2, which is displayed below for convenience.

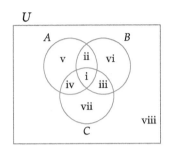

a. $A \cap B$ is represented by the regions i and ii. C is represented by regions i, iii, iv, and vii. $(A \cap B) \cap C$ is represented by the regions that are common to $A \cap B$ and C.

Thus $(A \cap B) \cap C$ is represented by region i.

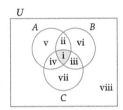

b. A is represented by the regions i, ii, iv, and v. B' is represented by the regions outside of B: iv, v, vii, and viii.

$A \cup B'$ is represented by the regions formed by joining the regions in A and the regions in B'.

Thus $A \cup B'$ is represented by the regions i, ii, iv, v, vii, and viii.

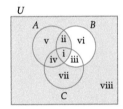

c. C' is represented by the regions outside of circle C: ii, v, vi, and viii.

B is represented by all the regions inside circle B: i, ii, iii, and vi.

$C' \cap B$ is represented by the regions that are common to C' and B.

Thus $C' \cap B$ is represented by regions ii and vi.

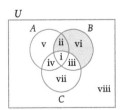

▼ **check your progress 6**, *page 77*

Determine the regions represented by $A \cap (B \cup C)$.

Set	Region or regions	Venn diagram
A	i, ii, iv, v The regions in A	*(Venn diagram with regions v, ii, vi, iv, i, iii, vii, viii; A region shaded)*
$B \cup C$	i, ii, iii, iv, vi, vii The regions in B joined with the regions in C	*(Venn diagram with B and C regions shaded)*
$A \cap (B \cup C)$	i, ii, iv The regions common to A and $B \cup C$	*(Venn diagram with regions i, ii, iv shaded)*

Now determine the regions represented by $(A \cap B) \cup (A \cap C)$.

Set	Region or regions	Venn diagram
$A \cap B$	i, ii The regions common to A and B	*(Venn diagram with regions i, ii shaded)*
$A \cap C$	i, iv The regions common to A and C	*(Venn diagram with regions i, iv shaded)*
$(A \cap B) \cup (A \cap C)$	i, ii, iv The regions in $A \cap B$ joined with the regions in $A \cap C$	*(Venn diagram with regions i, ii, iv shaded)*

The expressions $A \cap (B \cup C)$ and $(A \cap B) \cup (A \cap C)$ are both represented by the regions i, ii, and iv. Thus $A \cap (B \cup C) = (A \cap B) \cup (A \cap C)$ for all sets A and B.

▼ **check your progress 7**, *page 78*

a. Because Alex is in blood group A, not in blood group B, and is Rh+, his blood type is A+.

b. Roberto is in both blood group A and blood group B. Roberto is not Rh+. Thus Roberto's blood type is AB−.

▼ **check your progress 8**, *page 79*

a. Alex's blood type is A+. The blood transfusion table shows that a person with blood type A+ can safely receive type A− blood.

b. The blood transfusion table shows that a person with type AB+ blood can safely receive each of the eight different types of blood. Thus a person with AB+ blood is classified as a universal recipient.

SECTION 2.4

▼ **check your progress 1**, *page 86*

The intersection of the two sets includes the 85 students who like both volleyball and basketball.

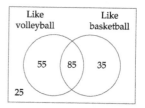

a. Because 140 students like volleyball and 85 like both sports, there must be $140 - 85 = 55$ students who like only volleyball.

b. Because 120 students like basketball and 85 like both sports, there must be $120 - 85 = 35$ students who like only basketball.

c. The Venn diagram shows that the number of students who like only volleyball plus the number who like only basketball plus the number who like both sports is $55 + 35 + 85 = 175$. Thus of the 200 students surveyed, only $200 - 175 = 25$ do not like either of the sports.

▼ **check your progress 2**, *page 87*

The intersection of the three sets includes the 15 people who like all three activities.

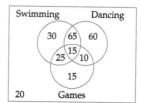

a. There are 25 people who like dancing and games. This includes the 15 people who like all three activities. Thus there must be another $25 - 15 = 10$ people who like only dancing and games. There are 40 people who like swimming and games. Thus there must be another $40 - 15 = 25$ people who like only swimming and games. There are 80 people who like swimming and dancing. Thus there must be another $80 - 15 = 65$ people who like only swimming and dancing. Hence $10 + 25 + 65 = 100$ people who like exactly two of the three activities.

b. There are 135 people who like swimming. We have determined that 15 people like all three activities, 25 like only swimming and games, and 65 like only swimming and dancing. This means that $135 - (15 + 25 + 65) = 30$ people like only swimming.

c. There are a total of 240 passengers surveyed. The Venn diagram shows that $15 + 25 + 10 + 15 + 30 + 65 + 60 = 220$ passengers like at least one of the activities. Thus $240 - 220 = 20$ passengers like none of the activities.

▼ **check your progress 3**, *page 89*

Let $B = \{$the set of students who play basketball$\}$.
Let $S = \{$the set of students who play soccer$\}$.

$$
\begin{aligned}
n(B \cup S) &= n(B) + n(S) - n(B \cap S) \\
&= 80 + 60 - 24 \\
&= 116
\end{aligned}
$$

Using the inclusion-exclusion principle, we see that 116 students play either basketball or soccer.

▼ **check your progress 4**, *page 89*

$$
\begin{aligned}
n(A \cup B) &= n(A) + n(B) - n(A \cap B) \\
852 &= 785 + 162 - n(A \cap B) \\
852 &= 947 - n(A \cap B) \\
n(A \cap B) &= 947 - 852 \\
n(A \cap B) &= 95
\end{aligned}
$$

▼ **check your progress 5**, *page 90*

$$
\begin{aligned}
p(A \cup \text{Rh}+) &= p(A) + p(\text{Rh}+) - p(A \cap \text{Rh}+) \\
91\% &= 44\% + 84\% - p(A \cap \text{Rh}+) \\
91\% &= 128\% - p(A \cap \text{Rh}+) \\
p(A \cap \text{Rh}+) &= 128\% - 91\% \\
p(A \cap \text{Rh}+) &= 37\%
\end{aligned}
$$

About 37% of the U.S. population has the A antigen and is Rh+.

▼ **check your progress 6**, *page 91*

a. The number 410 appears in both the column labeled "Yahoo!" and the row labeled "children." The table shows that 410 children surveyed use Yahoo! as a search engine. Thus

$$n(Y \cap C) = 410$$

b. The set $B \cap M'$ is the set of surveyed Bing users who are women or children. Thus

$$
\begin{aligned}
n(B \cap M') &= 325 + 40 \\
&= 365
\end{aligned}
$$

c. The set $G \cap M$ represents the set of surveyed Google users who are men. The table shows that this set includes 440 people. The set $G \cap W$ represents the set of surveyed Google users who are women. The table shows that this set includes 390 people. Thus $n((G \cap M) \cup (G \cap W)) = 440 + 390 = 830$.

SECTION 2.5

▼ **check your progress 1**, *page 97*

Write the sets so that one is aligned below the other. Draw arrows to show how you wish to pair the elements of each set. One possible method is shown in the following figure.

$$N = \{1, 2, 3, 4, \ldots, \quad n, \quad \ldots\}$$
$$\updownarrow \updownarrow \updownarrow \updownarrow \qquad \updownarrow$$
$$D = \{1, 3, 5, 7, \ldots, 2n - 1, \ldots\}$$

In the preceding correspondence, each natural number $n \in N$ is paired with the odd number $(2n - 1) \in D$. The *general correspondence*

$n \leftrightarrow (2n - 1)$ enables us to determine exactly which element of D will be paired with any given element of N, and vice versa. For instance, under this correspondence, $8 \in N$ is paired with the odd number $2 \cdot 8 - 1 = 15$, and $21 \in D$ is paired with the natural number $\frac{21 + 1}{2} = 11$. The general correspondence $n \leftrightarrow (2n - 1)$ establishes a one-to-one correspondence between the sets.

▼ **check your progress 2**, *page 98*

One proper subset of V is $P = \{41, 42, 43, 44, \ldots, 40 + n \ldots\}$, which was produced by deleting 40 from set V. To establish a one-to-one correspondence between V and P, consider the following diagram.

$$V = \{40, 41, 42, 43, \ldots, 39 + n, \ldots\}$$
$$\updownarrow \updownarrow \updownarrow \updownarrow \qquad \updownarrow$$
$$P = \{41, 42, 43, 44, \ldots, 40 + n, \ldots\}$$

In the above correspondence, each element of the form $39 + n$ from set V is paired with an element of the form $40 + n$ from set P. The general correspondence $(39 + n) \leftrightarrow (40 + n)$ establishes a one-to-one correspondence between V and P. Because V can be placed in a one-to-one correspondence with a proper subset of itself, V is an infinite set.

▼ **check your progress 3**, *page 99*

The following figure shows that we can establish a one-to-one correspondence between M and the set of natural numbers N by pairing $\frac{1}{n + 1}$ of set M with n of set N.

$$M = \left\{ \frac{1}{2}, \frac{1}{3}, \frac{1}{4}, \frac{1}{5}, \ldots, \frac{1}{n + 1}, \ldots \right\}$$
$$\updownarrow \updownarrow \updownarrow \updownarrow \qquad \updownarrow$$
$$N = \{ 1, \quad 2, \quad 3, \quad 4, \ldots, \quad n, \quad \ldots \}$$

Thus the cardinality of M must be the same as the cardinality of N, which is $\aleph_0$.

CHAPTER 3

SECTION 3.1

▼ **check your progress 1**, *page 113*

a. The sentence "Open the door" is a command. It is not a statement.

b. The word *large* is not a precise term. It is not possible to determine whether the sentence "7055 is a large number" is true or false, and thus the sentence is not a statement.

c. You may not know whether the given sentence is true or false, but you know that the sentence is either true or false and that it is not both true and false. Thus the sentence is a statement.

d. The sentence $x > 3$ is a statement because for any given value of x, the inequality $x > 3$ is true or false, but not both.

▼ **check your progress 2**, *page 114*

a. The *Queen Mary* 2 is not the world's largest cruise ship.

b. The fire engine is red.

▼ **check your progress 3**, *page 115*

a. $\sim p \wedge r$

b. $\sim s \wedge \sim r$

c. $r \leftrightarrow q$

d. $p \rightarrow \sim r$

▼ **check your progress 4**, *page 115*

$e \wedge \sim t$: All men are created equal and I am not trading places.

$a \vee \sim t$: I get Abe's place or I am not trading places.

$e \rightarrow t$: If all men are created equal, then I am trading places.

$t \leftrightarrow g$: I am trading places if and only if I get George's place.

▼ **check your progress 5**, *page 117*

a. If Kesha's singing style is similar to Uffie's and she has messy hair, then she is a rapper.

b. $\sim r \rightarrow (\sim q \wedge \sim p)$

▼ **check your progress 6**, *page 118*

a. True. A conjunction of two statements is true provided that both statements are true.

b. True. A disjunction of two statements is true provided that at least one statement is true.

c. False. If both statements of a disjunction are false, then the disjunction is false.

▼ **check your progress 7**, *page 119*

a. Some bears are not brown.

b. Some smartphones are expensive.

c. All vegetables are green.

SECTION 3.2

▼ **check your progress 1**, *page 124*

a.

p	q	$\sim p$	$\sim q$	$p \wedge \sim q$	$\sim p \vee q$	$(p \wedge \sim q) \vee (\sim p \vee q)$	
T	T	F	F	F	T	T	row 1
T	F	F	T	T	F	T	row 2
F	T	T	F	F	T	T	row 3
F	F	T	T	F	T	T	row 4
		1	2	3	4	5	

b. p is true and q is false in row 2 of the above truth table. The truth value of $(p \wedge \sim q) \vee (\sim p \vee q)$ in row 2 is T (true).

▼ **check your progress 2**, *page 125*

a.

p	q	r	$\sim p$	$\sim r$	$\sim p \wedge r$	$q \wedge \sim r$	$(\sim p \wedge r) \vee (q \wedge \sim r)$	
T	T	T	F	F	F	F	F	row 1
T	T	F	F	T	F	T	T	row 2
T	F	T	F	F	F	F	F	row 3
T	F	F	F	T	F	F	F	row 4
F	T	T	T	F	T	F	T	row 5
F	T	F	T	T	F	T	T	row 6
F	F	T	T	F	T	F	T	row 7
F	F	F	T	T	F	F	F	row 8
			1	2	3	4	5	

b. p is false, q is true, and r is false in row 6 of the above truth table. The truth value of $(\sim p \wedge r) \vee (q \wedge \sim r)$ in row 6 is T (true).

▼ **check your progress 3**, *page 126*

The given statement has two simple statements. Thus you should use a standard form that has $2^2 = 4$ rows.

Step 1 Enter the truth values for each simple statement and their negations. See columns 1, 2, and 3 in the table below.

Step 2 Use the truth values in columns 2 and 3 to determine the truth values to enter under the "and" connective. See column 4 in the table below.

Step 3 Use the truth values in columns 1 and 4 to determine the truth values to enter under the "or" connective. See column 5 in the table below.

p	q	$\sim p$	$\vee$	$(p$	$\wedge$	$q)$
T	T	F	T	T	T	T
T	F	F	F	T	F	F
F	T	T	T	F	F	T
F	F	T	T	F	F	F
		1	5	2	4	3

The truth table for $\sim p \vee (p \wedge q)$ is displayed in column 5.

▼ **check your progress 4**, *page 127*

p	q	p	$\vee$	$(p$	$\wedge$	$\sim q)$
T	T	T	T	T	F	F
T	F	T	T	T	T	T
F	T	F	F	F	F	F
F	F	F	F	F	F	T
		1	5	2	4	3

The above truth table shows that $p \equiv p \vee (p \wedge \sim q)$.

▼ **check your progress 5**, *page 128*

Let d represent "I am going to the dance." Let g represent "I am going to the game." The original sentence in symbolic form is $\sim(d \wedge g)$. Applying one of De Morgan's laws, we find that $\sim(d \wedge g) \equiv \sim d \vee \sim g$. Thus an equivalent form of "It is not true that, I am going to the dance and I am going to the game" is "I am not going to the dance or I am not going to the game."

▼ **check your progress 6**, *page 128*

The following truth table shows that $p \wedge (\sim p \wedge q)$ is always false. Thus $p \wedge (\sim p \wedge q)$ is a self-contradiction.

p	q	p	$\wedge$	$(\sim p$	$\wedge$	$q)$
T	T	T	F	F	F	T
T	F	T	F	F	F	F
F	T	F	F	T	T	T
F	F	F	F	T	F	F
		1	5	2	4	3

▼ **check your progress 1**, *page 133*

a. *Antecedent:* I study for at least 6 hours
Consequent: I will get an A on the test

b. *Antecedent:* I get the job
Consequent: I will buy a new car

c. *Antecedent:* you can dream it
Consequent: you can do it

▼ **check your progress 2**, *page 135*

a. Because the antecedent is true and the consequent is false, the statement is a false statement.

b. Because the antecedent is false, the statement is a true statement.

c. Because the consequent is true, the statement is a true statement.

▼ **check your progress 3**, *page 135*

p	q	$[p$	$\wedge$	$(p$	$\rightarrow$	$q)]$	$\rightarrow$	q
T	T	T	T	T	T	T	T	T
T	F	T	F	T	F	F	T	F
F	T	F	F	F	T	T	T	T
F	F	F	F	F	T	F	T	F
		1	6	2	5	3	7	4

▼ **check your progress 4**, *page 136*

a. I will move to Georgia or I will live in Houston.

b. The number is not divisible by 2 or the number is even.

▼ **check your progress 5**, *page 137*

a. I finished the report and I did not go to the concert.

b. The square of n is 25 and n is not 5 or −5.

▼ **check your progress 6**, *page 137*

a. Let $x = 6.5$. Then the first inequality of the biconditional is false, and the second inequality of the biconditional is true. Thus the given biconditional statement is false.

b. Both inequalities of the biconditional are true for $x > 2$, and both inequalities are false for $x \leq 2$. Because both inequalities have the same truth value for any real number x, the given biconditional is true.

▼ **check your progress 1**, *page 141*

a. If a geometric figure is a square, then it is a rectangle.

b. If I am older than 30, then I am at least 21.

▼ check your progress **2**, *page 142*

Converse: If we are not going to have a quiz tomorrow, then we will have a quiz today.

Inverse: If we don't have a quiz today, then we will have a quiz tomorrow.

Contrapositive: If we have a quiz tomorrow, then we will not have a quiz today.

▼ check your progress **3**, *page 143*

a. The second statement is the inverse of the first statement. Thus the statements are not equivalent. This can also be demonstrated by the fact that the first statement is true for $c = 0$ and the second statement is false for $c = 0$.

b. The second statement is the contrapositive of the first statement. Thus the statements are equivalent.

▼ check your progress **4**, *page 143*

a. *Contrapositive:* If x is an odd integer, then $3 + x$ is an even integer. The contrapositive is true and so the original statement is also true.

b. *Contrapositive:* If two triangles are congruent triangles, then the two triangles are similar triangles. The contrapositive is true and so the original statement is also true.

c. *Contrapositive:* If tomorrow is Thursday, then today is Wednesday. The contrapositive is true and so the original statement is also true.

SECTION 3.5

▼ check your progress **1**, *page 148*

Let p represent the statement "She got on the plane." Let r represent the statement "She will regret it." Then the symbolic form of the argument is

$$\sim p \rightarrow r$$
$$\underline{\sim r}$$
$$\therefore p$$

▼ check your progress **2**, *page 149*

Let r represent the statement "The stock market rises." Let f represent the statement "The bond market will fall." Then the symbolic form of the argument is

$$r \rightarrow f$$
$$\underline{\sim f}$$
$$\therefore \sim r$$

The truth table for this argument is as follows:

		First premise	Second premise	Conclusion	
r	f	$r \rightarrow f$	$\sim f$	$\sim r$	
T	T	T	F	F	row 1
T	F	F	T	F	row 2
F	T	T	F	T	row 3
F	F	T	T	T	row 4

Row 4 is the only row in which all the premises are true, so it is the only row that we examine. Because the conclusion is true in row 4, the argument is valid.

▼ check your progress **3**, *page 150*

Let a represent the statement "I arrive before 8 A.M." Let f represent the statement "I will make the flight." Let p represent the statement "I will give the presentation." Then the symbolic form of the argument is

$$a \rightarrow f$$
$$\underline{f \rightarrow p}$$
$$\therefore a \rightarrow p$$

The truth table for this argument is as follows:

			First premise	Second premise	Conclusion	
a	*f*	*p*	$a \rightarrow f$	$f \rightarrow p$	$a \rightarrow p$	
T	T	T	T	T	T	row 1
T	T	F	T	F	F	row 2
T	F	T	F	T	T	row 3
T	F	F	F	T	F	row 4
F	T	T	T	T	T	row 5
F	T	F	T	F	T	row 6
F	F	T	T	T	T	row 7
F	F	F	T	T	T	row 8

The only rows in which all the premises are true are rows 1, 5, 7, and 8. In each of these rows the conclusion is also true. Thus the argument is a valid argument.

▼ check your progress **4**, *page 151*

Let *f* represent "I go to Florida for spring break." Let $\sim s$ represent "I will not study." Then the symbolic form of the argument is

$$f \rightarrow \sim s$$
$$\underline{\sim f}$$
$$\therefore s$$

This argument has the form of the fallacy of the inverse. Thus the argument is invalid.

▼ check your progress **5**, *page 153*

Let *r* represent "I read a math book." Let *f* represent "I start to fall asleep." Let *d* represent "I drink a soda." Let *e* represent "I eat a candy bar." Then the symbolic form of the argument is

$$r \rightarrow f$$
$$f \rightarrow d$$
$$\underline{d \rightarrow e}$$
$$\therefore r \rightarrow e$$

The argument has the form of the extended law of syllogism. Thus the argument is valid.

▼ check your progress **6**, *page 154*

We are given the following premises:

$$\sim m \vee t$$
$$t \rightarrow \sim d$$
$$e \vee g$$
$$\underline{e \rightarrow d}$$
$$\therefore ?$$

The first premise can be written as $m \rightarrow t$, the third premise can be written as $\sim e \rightarrow g$, and the fourth premise can be written as $\sim d \rightarrow \sim e$. Thus the argument can be expressed in the following equivalent form.

$$m \rightarrow t$$
$$t \rightarrow \sim d$$
$$\sim e \rightarrow g$$
$$\underline{\sim d \rightarrow \sim e}$$
$$\therefore ?$$

If we switch the order of the third and fourth premises, then we have the following equivalent form.

$$m \rightarrow t$$
$$t \rightarrow \sim d$$
$$\sim d \rightarrow \sim e$$
$$\underline{\sim e \rightarrow g}$$
$$\therefore ?$$

An application of the extended law of syllogism produces $m \rightarrow g$ as a valid conclusion for the argument. *Note:* Although $m \rightarrow \sim e$ is also a valid conclusion for the argument, we do not list it as our answer because it can be obtained without using all of the given premises.

SECTION 3.6

▼ check your progress **1**, *page 159*

The following Euler diagram shows that the argument is valid.

▼ **check your progress 2**, *page 160*

From the given premises we can conclude that 7 may or may not be a prime number. Thus the argument is invalid.

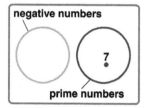
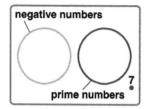

▼ **check your progress 3**, *page 161*

From the given premises we can construct two possible Euler diagrams.

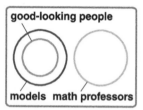
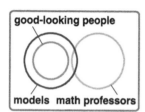

From the rightmost Euler diagram we can determine that the argument is invalid.

▼ **check your progress 4**, *page 161*

The following Euler diagram illustrates that all squares are quadrilaterals, so the argument is a valid argument.

▼ **check your progress 5**, *page 162*

The following Euler diagrams illustrate two possible cases. In both cases we see that all white rabbits like tomatoes.

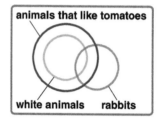
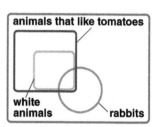

CHAPTER 6

SECTION 6.1

▼ check your progress **1**, *page 301*

𓏲𓏲𓎡𓎡𓎡𓎡𓏲𓏲𓎡𓎡𓎡|||

▼ check your progress **2**, *page 301*
$(1 \times 1{,}000{,}000) + (3 \times 100{,}000) + (1 \times 10{,}000) + (4 \times 1000)$
$+ (3 \times 100) + (2 \times 10) + (1 \times 6) = 1{,}314{,}326$

▼ check your progress **3**, *page 301*

23,341
+10,562

(Egyptian numeral symbols)

Replace 10 heel bones with one scroll to produce:

(Egyptian numeral symbols)

which is
33,903.

▼ check your progress **4**, *page 302*

61,432
−45,121

(Egyptian numeral symbols)

Replace one pointing finger with 10 lotus flowers
to produce:

(Egyptian numeral symbols)

which is
16,311.

▼ check your progress **5**, *page 303*
$\text{MCDXLV} = M + (CD) + (XL) + V$
$= 1000 + 400 + 40 + 5 = 1445$

▼ check your progress **6**, *page 304*
$473 = 400 + 70 + 3 = CD + LXX + III = CDLXXIII$

▼ check your progress **7**, *page 304*
a. $\overline{\text{VII}}\text{CCLIV} = \overline{\text{VII}} + CCLIV = 7000 + 254 = 7254$
b. $8070 = 8000 + 70 = \overline{\text{VIII}} + LXX = \overline{\text{VIII}}LXX$

SECTION 6.2

▼ check your progress **1**, *page 308*
$17{,}325 = 10{,}000 + 7000 + 300 + 20 + 5$
$= (1 \times 10{,}000) + (7 \times 1000) + (3 \times 100) + (2 \times 10) + 5$
$= (1 \times 10^4) + (7 \times 10^3) + (3 \times 10^2) + (2 \times 10^1)$
$+ (5 \times 10^0)$

▼ check your progress **2**, *page 308*
$(5 \times 10^4) + (9 \times 10^3) + (2 \times 10^2) + (7 \times 10^1) + (4 \times 10^0)$
$= (5 \times 10{,}000) + (9 \times 1000) + (2 \times 100)$
$+ (7 \times 10) + (4 \times 1)$
$= 50{,}000 + 9000 + 200 + 70 + 4$
$= 59{,}274$

▼ check your progress **3**, *page 309*
$152 = (1 \times 100) + (5 \times 10) + 2$
$+ 234 = (2 \times 100) + (3 \times 10) + 4$
$(3 \times 100) + (8 \times 10) + 6 = 386$

▼ check your progress **4**, *page 309*
$147 = (1 \times 100) + (4 \times 10) + 7$
$+ 329 = (3 \times 100) + (2 \times 10) + 9$
$(4 \times 100) + (6 \times 10) + 16$

Replace 16 with $(1 \times 10) + 6$

$= (4 \times 100) + (6 \times 10) + (1 \times 10) + 6$
$= (4 \times 100) + (7 \times 10) + 6$
$= 476$

▼ check your progress **5**, *page 310*
$382 = (3 \times 100) + (8 \times 10) + 2$
$- 157 = (1 \times 100) + (5 \times 10) + 7$

Because $7 > 2$, it is necessary to borrow by rewriting (8×10) as
$(7 \times 10) + 10$.

$382 = (3 \times 100) + (7 \times 10) + 12$
$- 157 = (1 \times 100) + (5 \times 10) + 7$
$(2 \times 100) + (2 \times 10) + 5 = 225$

▼ check your progress **6**, *page 311*

(Babylonian numeral symbols)

$= (21 \times 60^2) + (5 \times 60) + (34 \times 1)$
$= 75{,}600 + 300 + 34 = 75{,}934$

▼ check your progress **7**, *page 311*

$$\begin{array}{r} 3 \\ 3600\overline{)12578} \\ 10800 \\ \hline 1778 \end{array} \qquad \begin{array}{r} 29 \\ 60\overline{)1778} \\ 120 \\ \hline 578 \\ 540 \\ \hline 38 \end{array}$$

Thus $12{,}578 = (3 \times 60^2) + (29 \times 60) + (38 \times 1) =$

(Babylonian numeral symbols)

▼ check your progress **8**, *page 312*
Combine the symbols for each place value.

(Babylonian numeral symbols)

Replace ten ❙s in the ones place with a ❬.

(Babylonian numeral symbols)

Take away 60 from the ones place and add 1 to the 60s place.

(Babylonian numeral symbols)

Take away 60 from the 60s place and add 1 to the 60^2 place.

(Babylonian numeral symbols)

Thus

(Babylonian numeral symbols)

▼ **check your progress 9**, *page 313*

a. $(16 \times 360) + (0 \times 20) + (1 \times 1) = 5761$

b. $(9 \times 7200) + (1 \times 360) + (10 \times 20) + (4 \times 1) = 65,364$

▼ **check your progress 10**, *page 314*

$$
\begin{array}{r} 1 \\ \hline 7200)\overline{11480} \\ 7200 \\ \hline 4280 \end{array}
\qquad
\begin{array}{r} 11 \\ \hline 360)\overline{4280} \\ 360 \\ \hline 680 \\ 360 \\ \hline 320 \end{array}
\qquad
\begin{array}{r} 16 \\ \hline 20)\overline{320} \\ 20 \\ \hline 120 \\ 120 \\ \hline 0 \end{array}
$$

Thus $11,480 = (1 \times 7200) + (11 \times 360) + (16 \times 20) + (0 \times 1)$.
In Mayan numerals this is

SECTION 6.3

▼ **check your progress 1**, *page 318*

$3156_{seven} = (3 \times 7^3) + (1 \times 7^2) + (5 \times 7^1) + (6 \times 7^0)$
$= (3 \times 343) + (1 \times 49) + (5 \times 7) + (6 \times 1)$
$= 1029 + 49 + 35 + 6$
$= 1119$

▼ **check your progress 2**, *page 319*

$111000101_{two} = (1 \times 2^8) + (1 \times 2^7) + (1 \times 2^6) + (0 \times 2^5)$
$+ (0 \times 2^4) + (0 \times 2^3) + (1 \times 2^2)$
$+ (0 \times 2^1) + (1 \times 2^0)$
$= (1 \times 256) + (1 \times 128) + (1 \times 64) + (0 \times 32)$
$+ (0 \times 16) + (0 \times 8) + (1 \times 4)$
$+ (0 \times 2) + (1 \times 1)$
$= 256 + 128 + 64 + 0 + 0 + 0 + 4 + 0 + 1$
$= 453$

▼ **check your progress 3**, *page 319*

$A5B_{twelve} = (10 \times 12^2) + (5 \times 12^1) + (11 \times 12^0)$
$= 1440 + 60 + 11$
$= 1511$

▼ **check your progress 4**, *page 320*

$C24F_{sixteen} = (12 \times 16^3) + (2 \times 16^2) + (4 \times 16^1) + (15 \times 16^0)$
$= 49,152 + 512 + 64 + 15$
$= 49,743$

▼ **check your progress 5**, *page 321*

a.
$$
\begin{array}{r|r}
5 & 1952 \\
\hline
5 & 390 \quad 2 \\
5 & 78 \quad 0 \\
5 & 15 \quad 3 \\
\hline
& 3 \quad 0
\end{array}
$$
$1952 = 30302_{five}$

b.
$$
\begin{array}{r|r}
12 & 1952 \\
\hline
12 & 162 \quad 8 \\
12 & 13 \quad 6 \\
\hline
& 1 \quad 1
\end{array}
$$
$1952 = 1168_{twelve}$

▼ **check your progress 6**, *page 321*

$$
\begin{array}{ccccc}
6 & 3 & 2 & 1 & 0_{eight} \\
\| & \| & \| & \| & \| \\
110 & 011 & 010 & 001 & 000_{two}
\end{array}
$$

$63210_{eight} = 110011010001000_{two}$

▼ **check your progress 7**, *page 322*

$$
\begin{array}{cccc}
111 & 010 & 011 & 100_{two} \\
\| & \| & \| & \| \\
7 & 2 & 3 & 4_{eight}
\end{array}
$$

$111010011100_{two} = 7234_{eight}$

▼ **check your progress 8**, *page 322*

$$
\begin{array}{ccc}
C & 5 & A_{sixteen} \\
\| & \| & \| \\
1100 & 0101 & 1010_{two}
\end{array}
$$

$C5A_{sixteen} = 110001011010_{two}$

▼ **check your progress 9**, *page 323*

Insert a zero to
make a group of four.

$$
\begin{array}{cccc}
0101 & 0001 & 1101 & 0010_{two} \\
\| & \| & \| & \| \\
5 & 1 & D & 2_{sixteen}
\end{array}
$$

$101000111010010_{two} = 51D2_{sixteen}$

▼ **check your progress 10**, *page 323*

$1110010_{two} = 114$

SECTION 6.4

▼ **check your progress 1**, *page 328*

$$
\begin{array}{r}
1\,1\quad 1 \\
1\,1\,0\,0\,1_{two} \\
+ \quad 1\,1\,0\,1_{two} \\
\hline
1\,0\,0\,1\,1\,0_{two}
\end{array}
$$

▼ **check your progress 2**, *page 328*

$$
\begin{array}{r}
1\,1 \\
3\,2_{four} \\
+ \quad 1\,2_{four} \\
\hline
1\,1\,0_{four}
\end{array}
$$

▼ **check your progress 3**, *page 329*

$$
\begin{array}{r}
1\,2 \\
3\,5_{seven} \\
4\,6_{seven} \\
+ \quad 2\,4_{seven} \\
\hline
1\,4\,1_{seven}
\end{array}
$$

▼ **check your progress 4**, *page 330*

$$
\begin{array}{r}
1\,1 \\
A\,C\,4_{sixteen} \\
+ \quad 6\,E\,8_{sixteen} \\
\hline
1\,1\,A\,C_{sixteen}
\end{array}
$$

▼ check your progress 5, *page 330*

$$2 + 1$$

$$\begin{array}{r} \not3\,6\,5_{nine} \\ -\,1\,8\,3_{nine} \\ \hline \end{array}$$

$$10$$

$$\begin{array}{r} 2\,\not6\,5_{nine} \\ -\,1\,8\,3_{nine} \\ \hline \end{array}$$

$$16$$

$$\begin{array}{r} 2\,\not6\,5_{nine} \\ -\,1\,8\,3_{nine} \\ \hline 1\,7\,2_{nine} \end{array}$$

Because $8_{nine} > 6_{nine}$, it is necessary to borrow from the 3 in the first column at the left.

Borrow 1 nine from the first column and add 9 $= 10_{nine}$ to the 6 in the middle column.

$16_{nine} - 8_{nine} = 15 - 8$
$= 7$
$= 7_{nine}$

▼ check your progress 6, *page 331*

$$\begin{array}{r} 7 \quad 10 \\ \not8 \quad \not3 \quad A_{twelve} \\ -\,4 \quad 6 \quad 7_{twelve} \\ \hline 3 \quad 9 \quad 3_{twelve} \end{array}$$

- $A_{twelve} - 7_{twelve} = 10 - 7 = 3 = 3_{twelve}$
- $10_{twelve} + 3_{twelve} = 13_{twelve} = 15$ $15 - 6 = 9 = 9_{twelve}$
- $7_{twelve} - 4_{twelve} = 3_{twelve}$

▼ check your progress 7, *page 332*

$$\begin{array}{r} 1 \\ 2 \quad 1 \quad 3_{four} \\ \times \qquad\qquad 2_{four} \\ \hline 1 \quad 0 \quad 3 \quad 2_{four} \end{array}$$

- $2_{four} \times 3_{four} = 12_{four}$
- $2_{four} \times 1_{four} + 1_{four} = 3_{four}$
- $2_{four} \times 2_{four} = 10_{four}$

▼ check your progress 8, *page 333*

$$\begin{array}{r} 2 \\ 3 \quad 4_{eight} \\ \times \quad 2 \quad 5_{eight} \\ \hline 2 \quad 1 \quad 4_{eight} \end{array}$$

- $5_{eight} \times 4_{eight} = 20 = 24_{eight}$
- $5_{eight} \times 3_{eight} + 2_{eight} = 15 + 2 = 17 = 21_{eight}$

$$\begin{array}{r} 1 \\ 3 \quad 4_{eight} \\ \times \quad 2 \quad 5_{eight} \\ \hline 2 \quad 1 \quad 4_{eight} \\ 7 \quad 0_{eight} \\ \hline 1 \quad 1 \quad 1 \quad 4_{eight} \end{array}$$

- $2_{eight} \times 4_{eight} = 8 = 10_{eight}$
- $2_{eight} \times 3_{eight} + 1_{eight} = 6 + 1 = 7 = 7_{eight}$

▼ check your progress 9, *page 334*

First list a few multiples of 3_{five}.

$3_{five} \times 0_{five} = 0_{five}$
$3_{five} \times 1_{five} = 3_{five}$
$3_{five} \times 2_{five} = 11_{five}$
$3_{five} \times 3_{five} = 14_{five}$
$3_{five} \times 4_{five} = 22_{five}$

$$\begin{array}{r} 1 \\ 3_{five}\overline{)3\,2\,4_{five}} \\ \underline{3} \\ 2 \end{array}$$

$$\begin{array}{r} 1\,0 \\ 3_{five}\overline{)3\,2\,4_{five}} \\ \underline{3} \\ 2 \\ \underline{0} \\ 2\,4 \end{array}$$

$$\begin{array}{r} 1\,0\,4 \\ 3_{five}\overline{)3\,2\,4_{five}} \\ \underline{3} \\ 2 \\ \underline{0} \\ 2\,4 \\ \underline{2\,2} \\ 2 \end{array}$$

Thus $324_{five} \div 3_{five} = 104_{five}$ with a remainder of 2_{five}.

▼ check your progress 10, *page 335*

The divisor is 10_{two}. The multiples of the divisor are $10_{two} \times 0_{two} = 0_{two}$ and $10_{two} \times 1_{two} = 10_{two}$.

$$\begin{array}{r} 1\,1\,1\,0\,0\,1_{two} \\ 10_{two}\overline{)1\,1\,1\,0\,0\,1\,1_{two}} \\ \underline{1\,0} \\ 1\,1 \\ \underline{1\,0} \\ 1\,0 \\ \underline{1\,0} \\ 0\,0 \\ \underline{0} \\ 0\,1 \\ \underline{0} \\ 1\,1 \\ \underline{1\,0} \\ 1 \end{array}$$

Thus $1110011_{two} \div 10_{two} = 111001_{two}$ with a remainder of 1_{two}.

SECTION 6.5

▼ check your progress 1, *page 338*

a. Divide 9 by 1, 2, 3, ... , 9 to determine that the only natural number divisors of 9 are 1, 3, and 9.

b. Divide 11 by 1, 2, 3, ... , 11 to determine that the only natural number divisors of 11 are 1 and 11.

c. Divide 24 by 1, 2, 3, ... , 24 to determine that the only natural number divisors of 24 are 1, 2, 3, 4, 6, 8, 12, and 24.

▼ check your progress 2, *page 339*

a. The only divisors of 47 are 1 and 47. Thus 47 is a prime number.

b. 171 is divisible by 3, 9, 19, and 57. Thus 171 is a composite number.

c. The divisors of 91 are 1, 7, 13, and 91. Thus 91 is a composite number.

▼ check your progress 3, *page 340*

a. The sum of the digits of 341,565 is 24; therefore, 341,565 is divisible by 3.

b. The number 341,565 is not divisible by 4 because the number formed by the last two digits, 65, is not divisible by 4.

c. The number 341,565 is not divisible by 10 because it does not end in 0.

d. The sum of the digits with even place-value powers is 14. The sum of the digits with odd place-value powers is 10. The difference of these sums is 4. Thus 341,565 is not divisible by 11.

▼ check your progress **4**, *page 341*

a.

$315 = 3^2 \cdot 5 \cdot 7$

b.

$273 = 3 \cdot 7 \cdot 13$

c.

$1309 = 7 \cdot 11 \cdot 17$

SECTION 6.6

▼ check your progress **1**, *page 348*

a. The proper factors of 24 are 1, 2, 3, 4, 6, 8, and 12. The sum of these proper factors is 36. Because 24 is less than the sum of its proper factors, 24 is an abundant number.

b. The proper factors of 28 are 1, 2, 4, 7, and 14. The sum of these proper factors is 28. Because 28 equals the sum of its proper factors, 28 is a perfect number.

c. The proper factors of 35 are 1, 5, and 7. The sum of these proper factors is 13. Because 35 is larger than the sum of its proper factors, 35 is a deficient number.

▼ check your progress **2**, *page 349*

$2^7 - 1 = 127$, which is a prime number.

▼ check your progress **3**, *page 349*

The exponent $n = 61$ is a prime number, and we are given that $2^{61} - 1$ is a prime number, so the perfect number we seek is $2^{60}(2^{61} - 1)$.

▼ check your progress **4**, *page 352*

First consider $2^{2976221}$. The base b is 2. The exponent x is 2,976,221.

$$(x \log b) + 1 = (2{,}976{,}221 \log 2) + 1$$
$$\approx 895{,}931.8 + 1$$
$$= 895{,}932.8$$

The greatest integer of 895,932.8 is 895,932. Thus $2^{2976221}$ has 895,932 digits. The number $2^{2976221}$ is not a power of 10, so the Mersenne number $2^{2976221} - 1$ also has 895,932 digits.

▼ check your progress **5**, *page 353*

Substituting 9 for x, 11 for y, and 4 for n in $x^n + y^n = z^n$ yields

$$9^4 + 11^4 = z^4$$
$$6561 + 14{,}641 = z^4$$
$$21{,}202 = z^4$$

The real solution of $z^4 = 21{,}202$ is $\sqrt[4]{21{,}202} \approx 12.066858$, which is not a natural number. Thus $x = 9$, $y = 11$, and $n = 4$ do not satisfy the equation $x^n + y^n = z^n$ where z is a natural number.

▼ **check your progress 3**, *page 476*

The years 2008, 2012, and 2016 are leap years, so there are 3 years between the two dates with 366 days and 6 years with 365 days. The total number of days between the dates is $3 \cdot 366 + 6 \cdot 365 = 3288$. $3288 \div 7 = 469$ remainder 5, so $3288 \equiv 5 \bmod 7$. The day of the week 3288 days after Tuesday, February 12, 2008, will be the same as the day 5 days later, a Sunday.

▼ **check your progress 4**, *page 477*

$51 + 72 = 123$, and $123 \div 3 = 41$ remainder 0, so $(51 + 72) \bmod 3 \equiv 0$.

▼ **check your progress 5**, *page 478*

$21 - 43 = -22$, a negative number, so we must find x so that $-22 \equiv x \bmod 7$. Thus we must find x so that $\dfrac{-22 - x}{7} = \dfrac{-(22 + x)}{7}$ is an integer. Trying the whole number values of x less than 7, we find that when $x = 6$, $\dfrac{-(22 + 6)}{7} = \dfrac{-28}{7} = -4$. $(21 - 43) \bmod 7 \equiv 6$.

▼ **check your progress 6**, *page 478*

Tuesday corresponds to 2 (see the chart on page 474), so the day of the week 93 days from now is represented by $(2 + 93) \bmod 7$. Because $95 \div 7 = 13$ remainder 4, $(2 + 93) \bmod 7 \equiv 4$, which corresponds to Thursday.

▼ **check your progress 7**, *page 479*

$33 \cdot 41 = 1353$ and $1353 \div 17 = 79$ remainder 10, so $(33 \cdot 41) \bmod 17 \equiv 10$.

▼ **check your progress 8**, *page 480*

Substitute each whole number from 0 to 11 into the congruence.

$4(0) + 1 \not\equiv 5 \bmod 12$	Not a solution
$4(1) + 1 \equiv 5 \bmod 12$	1 is a solution.
$4(2) + 1 \not\equiv 5 \bmod 12$	Not a solution
$4(3) + 1 \not\equiv 5 \bmod 12$	Not a solution
$4(4) + 1 \equiv 5 \bmod 12$	4 is a solution.
$4(5) + 1 \not\equiv 5 \bmod 12$	Not a solution
$4(6) + 1 \not\equiv 5 \bmod 12$	Not a solution
$4(7) + 1 \equiv 5 \bmod 12$	7 is a solution.
$4(8) + 1 \not\equiv 5 \bmod 12$	Not a solution
$4(9) + 1 \not\equiv 5 \bmod 12$	Not a solution
$4(10) + 1 \equiv 5 \bmod 12$	10 is a solution.
$4(11) + 1 \not\equiv 5 \bmod 12$	Not a solution

The solutions from 0 to 11 are 1, 4, 7, and 10. The remaining solutions are obtained by repeatedly adding the modulus 12 to these solutions. So the solutions are 1, 4, 7, 10, 13, 16, 19, 22, ...

▼ **check your progress 9**, *page 480*

In mod 12 arithmetic, $6 + 6 = 12$, so the additive inverse of 6 is 6.

CHAPTER 8

SECTION 8.1

▼ **check your progress 1**, *page 474*

a. $6 \oplus 10 = 4$

b. $5 \oplus 9 = 2$

c. $7 \ominus 11 = 8$

d. $5 \ominus 10 = 7$

▼ **check your progress 2**, *page 475*

a. Find $\dfrac{7 - 12}{5} = \dfrac{-5}{5} = -1$. Because -1 is an integer, $7 \equiv 12 \bmod 5$ is a true congruence.

b. Find $\dfrac{15 - 1}{8} = \dfrac{14}{8} = \dfrac{7}{4}$. Because $\dfrac{7}{4}$ is not an integer, $15 \equiv 1 \bmod 8$ is not a true congruence.

▼ **check your progress 10**, *page 481*

Solve the congruence equation $5x \equiv 1 \bmod 11$ by substituting whole number values of x less than the modulus.

$5(1) \not\equiv 1 \bmod 11$

$5(2) \not\equiv 1 \bmod 11$

$5(3) \not\equiv 1 \bmod 11$

$5(4) \not\equiv 1 \bmod 11$

$5(5) \not\equiv 1 \bmod 11$

$5(6) \not\equiv 1 \bmod 11$

$5(7) \not\equiv 1 \bmod 11$

$5(8) \not\equiv 1 \bmod 11$

$5(9) \equiv 1 \bmod 11$

In mod 11 arithmetic, the multiplicative inverse of 5 is 9.

SECTION 8.2

▼ **check your progress 1**, *page 485*

Check the ISBN congruence equation.

$$d_{13} \equiv 10 - [9 + 3(7) + 8 + 3(0) + 7 + 3(6) + 0 + 3(7) + 3 + 3(2) + 6 + 3(1)] \bmod 10$$
$$\equiv 10 - 102 \bmod 10$$
$$\equiv 10 - 2 = 8$$

Because the check digit does not match, the ISBN is invalid.

▼ **check your progress 2**, *page 486*

Check the UPC congruence equation.

$$1(3) + 3(1) + 2(3) + 3(1) + 4(3) + 2(1) + 6(3) + 5(1) + 9(3) + 3(1) + 3(3) + 9 \equiv ? \bmod 10$$
$$100 \equiv 0 \bmod 10$$

Because $100 \equiv 0 \bmod 10$, the UPC is valid.

▼ **check your progress 3**, *page 487*

Highlight every other digit, reading from right to left:

6 0 1 1 0 1 2 3 9 1 4 5 2 3 1 7

Double the highlighted digits:

12 0 2 1 0 1 4 3 18 1 8 5 4 3 2 7

Add all the digits, treating two-digit numbers as two single digits:

$$(1 + 2) + 0 + 2 + 1 + 0 + 1 + 4 + 3 + (1 + 8) + 1 + 8 + 5 + 4 + 3 + 2 + 7 = 53$$

Because $53 \not\equiv 0 \bmod 10$, this is not a valid credit card number.

▼ **check your progress 4**, *page 490*

a. The encrypting congruence is $c \equiv (p + 17) \bmod 26$.

A	$c \equiv (1 + 17) \bmod 26 \equiv 18 \bmod 26 \equiv 18$	Code A as R.
L	$c \equiv (12 + 17) \bmod 26 \equiv 29 \bmod 26 \equiv 3$	Code L as C.
P	$c \equiv (16 + 17) \bmod 26 \equiv 33 \bmod 26 \equiv 7$	Code P as G.
I	$c \equiv (9 + 17) \bmod 26 \equiv 26 \bmod 26 \equiv 0$	Code I as Z.
N	$c \equiv (14 + 17) \bmod 26 \equiv 31 \bmod 26 \equiv 5$	Code N as E.
E	$c \equiv (5 + 17) \bmod 26 \equiv 22 \bmod 26 \equiv 22$	Code E as V.
S	$c \equiv (19 + 17) \bmod 26 \equiv 36 \bmod 26 \equiv 10$	Code S as J.

K $c \equiv (11 + 17) \bmod 26 \equiv 28 \bmod 26 \equiv 2$ Code K as B.

G $c \equiv (7 + 17) \bmod 26 \equiv 24 \bmod 26 \equiv 24$ Code G as X.

Thus the plaintext ALPINE SKIING is coded as RCGZEV JBZZEX.

b. To decode, because $m = 17, n = 26 - 17 = 9$, and the decoding congruence is $p \equiv (c + 9) \bmod 26$.

T $c \equiv (20 + 9) \bmod 26 \equiv 29 \bmod 26 \equiv 3$ Decode T as C.

I $c \equiv (9 + 9) \bmod 26 \equiv 18 \bmod 26 \equiv 18$ Decode I as R.

F $c \equiv (6 + 9) \bmod 26 \equiv 15 \bmod 26 \equiv 15$ Decode F as O.

J $c \equiv (10 + 9) \bmod 26 \equiv 19 \bmod 26 \equiv 19$ Decode J as S.

Continuing, the ciphertext TIFJJ TFLEKIP JBZZEX decodes as CROSS COUNTRY SKIING.

▼ check your progress 5, *page 490*

The encrypting congruence is $c \equiv (3p + 1) \bmod 26$.

C $c \equiv (3 \cdot 3 + 1) \bmod 26 \equiv 10 \bmod 26 = 10$ Code C as J.

O $c \equiv (3 \cdot 15 + 1) \bmod 26 \equiv 46 \bmod 26 = 20$ Code O as T.

L $c \equiv (3 \cdot 12 + 1) \bmod 26 \equiv 37 \bmod 26 = 11$ Code L as K.

R $c \equiv (3 \cdot 18 + 1) \bmod 26 \equiv 55 \bmod 26 = 3$ Code R as C.

Continuing, the plaintext COLOR MONITOR is coded as JTKTC NTQBITC.

▼ check your progress 6, *page 491*

Solve the congruence equation $c \equiv (7p + 1) \bmod 26$ for p.

$$c = 7p + 1$$
$$c - 1 = 7p \qquad \text{• Subtract 1 from each side of the equation.}$$
$$15(c - 1) = 15(7p) \qquad \text{• Multiply each side of the equation by the multiplicative inverse of 7.}$$
$$\text{Because } 7 \cdot 15 \equiv 1 \bmod 26, \text{ multiply each side by 15.}$$

$$[15(c - 1)] \bmod 26 \equiv p$$

The decoding congruence is $p \equiv [15(c - 1)] \bmod 26$.

I $p \equiv [15(9 - 1)] \bmod 26 \equiv 120 \bmod 26 \equiv 16$ Decode I as P.

G $p \equiv [15(7 - 1)] \bmod 26 \equiv 90 \bmod 26 \equiv 12$ Decode G as L.

H $p \equiv [15(8 - 1)] \bmod 26 \equiv 105 \bmod 26 \equiv 1$ Decode H as A.

T $p \equiv [15(20 - 1)] \bmod 26 \equiv 285 \bmod 26 \equiv 25$ Decode T as Y.

Continuing, the ciphertext IGHT OHGG decodes as PLAY BALL.

SECTION 8.3

▼ check your progress 1, *page 497*

1. Determine whether the operation is closed by finding all possible products.

   $(1 \cdot 1) \bmod 4 \equiv 1 \bmod 4 = 1$

   $(1 \cdot 2) \bmod 4 \equiv 2 \bmod 4 = 2$

   $(1 \cdot 3) \bmod 4 \equiv 3 \bmod 4 = 3$

   $(2 \cdot 2) \bmod 4 \equiv 4 \bmod 4 = 0$

   $(2 \cdot 3) \bmod 4 \equiv 6 \bmod 4 = 2$

   $(3 \cdot 3) \bmod 4 \equiv 9 \bmod 4 = 1$

   The result of each product is in the set. The operation is closed.

2. Modulo operations satisfy the associative property.

3. 1 is the identity element.

4. Determine whether each element has an inverse. From the above calculation, 1 and 3 have an inverse. However,

   $(2 \cdot 2) \bmod 4 \equiv 0 \bmod 4 = 0 \neq 1$. Therefore, 2 does not have an inverse.

   The set and operation do not form a group.

▼ check your progress 2, *page 500*

Rotate the original triangle, *I*, about the line of symmetry through the bottom right vertex, followed by a clockwise rotation of 240°.

    followed by R_{240}

Therefore, $R_r \Delta R_{240} = R_l$.

▼ check your progress **3**, *page 501*

$$R_r \Delta R_{240} = \begin{pmatrix} 1 & 2 & 3 \\ 2 & 1 & 3 \end{pmatrix} \Delta \begin{pmatrix} 1 & 2 & 3 \\ 3 & 1 & 2 \end{pmatrix}$$

- $1 \rightarrow 2 \rightarrow 1$. Thus $1 \rightarrow 1$.
- $2 \rightarrow 1 \rightarrow 3$. Thus $2 \rightarrow 3$.
- $3 \rightarrow 3 \rightarrow 2$. Thus $3 \rightarrow 2$.

$$= \begin{pmatrix} 1 & 2 & 3 \\ 1 & 3 & 2 \end{pmatrix} = R_l$$

▼ check your progress **4**, *page 502*

$E\Delta B = \begin{pmatrix} 1 & 2 & 3 \\ 2 & 1 & 3 \end{pmatrix} \Delta \begin{pmatrix} 1 & 2 & 3 \\ 3 & 1 & 2 \end{pmatrix}$. 1 is replaced by 2, which is then
replaced by 1 in the second permutation. Thus 1 remains as 1. 2 is replaced
by 1, which is then replaced by 3, so ultimately, 2 is replaced by 3. Finally,
3 remains as 3 in the first permutation but is replaced by 2 in the second, so
ultimately 3 is replaced by 2.

The result is $\begin{pmatrix} 1 & 2 & 3 \\ 1 & 3 & 2 \end{pmatrix}$, which is C. Thus $E\Delta B = C$.

▼ check your progress **5**, *page 502*

$D = \begin{pmatrix} 1 & 2 & 3 \\ 3 & 2 & 1 \end{pmatrix}$ replaces 1 with 3, 2 with 2, and 3 with 1.

Reversing these, we need to replace 3 with 1, leave 2 alone, and

replace 1 with 3. This is the element $\begin{pmatrix} 1 & 2 & 3 \\ 3 & 2 & 1 \end{pmatrix}$, which is D again.

Thus D is its own inverse.

▼ check your progress **2**, *page 735*

a. {1, 3, 5, 7, 9} **b.** {0, 3, 6, 9} **c.** {8, 9}

▼ check your progress **3**, *page 736*

	H	**T**
1	1H	1T
2	2H	2T
3	3H	3T
4	4H	4T
5	5H	5T
6	6H	6T

The sample space has 12 elements:
{1H, 1T, 2H, 2T, 3H, 3T, 4H, 4T, 5H, 5T, 6H, 6T}

▼ check your progress **4**, *page 737*

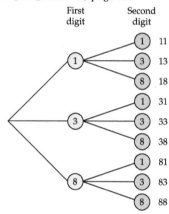

▼ check your progress **5**, *page 738*

Any of the 9 runners could win the gold medal, so $n_1 = 9$. That leaves $n_2 = 8$ runners that could win silver, and $n_3 = 7$ possibilities for bronze. By the counting principle, there are $9 \cdot 8 \cdot 7 = 504$ possible ways the medals can be awarded.

▼ check your progress **6**, *page 739*

a. Because a student can receive more than 1 award and each award has 5 possible destinations, there are $5 \cdot 5 \cdot 5 = 125$ ways for the awards to be given.

b. Once a student has an award, that student cannot receive another award, so there are $5 \cdot 4 \cdot 3 = 60$ different ways for the awards to be given.

S E C T I O N 12.2

▼ check your progress **1**, *page 745*

a. $7! + 4! = (7 \cdot 6 \cdot 5 \cdot 4 \cdot 3 \cdot 2 \cdot 1) + (4 \cdot 3 \cdot 2 \cdot 1)$
     $= 5040 + 24 = 5064$

b. $\dfrac{8!}{4!} = \dfrac{8 \cdot 7 \cdot 6 \cdot 5 \cdot \cancel{4!}}{\cancel{4!}} = 8 \cdot 7 \cdot 6 \cdot 5 = 1680$

CHAPTER 12

S E C T I O N 12.1

▼ check your progress **1**, *page 734*

The possible outcomes are {M, i, s, p}. There are 4 possible outcomes.

▼ check your progress **2**, *page 746*

Because the players are ranked, the number of different golf teams possible is the number of permutations of 8 players selected 5 at a time.

$$P(8, 5) = \frac{8!}{(8-5)!} = \frac{8!}{3!} = \frac{8 \cdot 7 \cdot 6 \cdot 5 \cdot 4 \cdot 3!}{3!}$$
$$= 8 \cdot 7 \cdot 6 \cdot 5 \cdot 4 = 6720$$

There are 6720 possible golf teams.

▼ check your progress **3**, *page 747*

The order in which the cars finish is important, so the number of ways to place first, second, and third is

$$P(43, 3) = \frac{43!}{(43-3)!} = \frac{43 \cdot 42 \cdot 41 \cdot 40!}{40!} = 74{,}046$$

There are 74,046 different ways to award the first, second, and third place prizes.

▼ check your progress **4**, *page 748*

a. With no restrictions, there are 7 tutors available for 7 hours, so the number of schedules is

$$P(7, 7) = \frac{7!}{(7-7)!} = \frac{7!}{0!} = 7! = 5040$$

There are 5040 possible schedules.

b. This is a multistage experiment; there are 3! ways to schedule the juniors and 4! ways to schedule the seniors. By the counting principle, the number of different tutoring schedules is $3! \cdot 4! = 6 \cdot 24 = 144$.

There are 144 tutor schedules.

▼ check your progress **5**, *page 749*

a. With $n = 8$ coins and $k_1 = 3$ (number of pennies), $k_2 = 2$ (number of nickels), and $k_3 = 3$ (number of dimes), the number of different possible stacks is

$$\frac{8!}{3! \cdot 2! \cdot 3!} = \frac{8 \cdot 7 \cdot 6 \cdot 5 \cdot 4 \cdot 3!}{3! \cdot 2! \cdot 3!} = \frac{8 \cdot 7 \cdot 6 \cdot 5 \cdot 4}{6 \cdot 2} = 560$$

There are 560 possible stacks.

b. Not including the dimes, there are $\frac{5!}{3! \cdot 2!} = 10$ ways to stack the pennies and nickels. The dimes are identical, so there is only 1 way to arrange the dimes together, but there are 6 different locations in the stack of pennies and nickels into which the dimes could be placed. By the counting principle, the total number of ways in which the stack of coins can be arranged if the dimes are together is $10 \cdot 6 = 60$.

▼ check your progress **6**, *page 751*

The order in which the waiters and waitresses are chosen is not important, so the number of ways to choose 9 people from 16 is

$$C(16, 9) = \frac{16!}{9! \cdot (16-9)!} = \frac{16!}{9! \cdot 7!}$$
$$= \frac{16 \cdot 15 \cdot 14 \cdot 13 \cdot 12 \cdot 11 \cdot 10 \cdot 9!}{9! \cdot 7!}$$
$$= \frac{16 \cdot 15 \cdot 14 \cdot 13 \cdot 12 \cdot 11 \cdot 10}{7 \cdot 6 \cdot 5 \cdot 4 \cdot 3 \cdot 2 \cdot 1} = 11{,}440$$

There are 11,440 possible groups of 9 waiters and waitresses.

▼ check your progress **7**, *page 751*

There are $C(4, 3)$ ways for the auditor to choose 3 corporate tax returns and $C(6, 2)$ ways to choose 2 individual tax returns. By the counting principle, the total number of ways in which the auditor can choose the returns is

$$C(4, 3) \cdot C(6, 2) = \frac{4!}{3! \cdot 1!} \cdot \frac{6!}{2! \cdot 4!} = 4 \cdot 15 = 60$$

There are 60 ways to choose the tax returns.

▼ check your progress **8**, *page 752*

For any single suit, there are $C(13, 4)$ ways of choosing 4 cards. That leaves $52 - 13 = 39$ cards of other suits from which to choose the fifth card. In addition, there are 4 different suits we could start with. By the counting principle, the number of 5-card combinations containing 4 cards of the same suit is

$$4 \cdot C(13, 4) \cdot 39 = 4 \cdot \frac{13!}{4! \cdot 9!} \cdot 39 = 4 \cdot 715 \cdot 39 = 111{,}540$$

There are 111,540 five-card combinations containing 4 cards of the same suit.

SECTION 12.3

▼ check your progress **1**, *page 757*

$S = \{\text{HH, HT, TH, TT}\}$

▼ check your progress **2**, *page 759*

The sample space for rolling a single die is $S = \{1, 2, 3, 4, 5, 6\}$. The elements in the event that an odd number will be rolled are $E = \{1, 3, 5\}$. Then

$$P(E) = \frac{n(E)}{n(S)} = \frac{3}{6} = \frac{1}{2}$$

The probability that an odd number will be rolled is $\frac{1}{2}$.

▼ check your progress **3**, *page 759*

The sample space is shown in Figure 12.6 on page 759. Let E be the event that the sum of the pips on the upward faces is 7; the elements of this event are

$$E = \{\boxed{\cdot}\ \boxed{\vdots\vdots}, \boxed{\because}\ \boxed{\vdots}, \boxed{\therefore}\ \boxed{\cdot\cdot}, \boxed{\cdot\cdot}\ \boxed{\therefore}, \boxed{\vdots}\ \boxed{\because}, \boxed{\vdots\vdots}\ \boxed{\cdot}\}.$$

Then the probability of rolling a 7 is

$$P(E) = \frac{n(E)}{n(S)} = \frac{6}{36} = \frac{1}{6}$$

The probability that the sum is 7 is $\frac{1}{6}$.

▼ check your progress **4**, *page 760*

Let E be the event that a person between the ages of 39 and 49 is selected. Then

$$P(E) = \frac{773}{3228} \approx 0.24$$

The probability the selected person is between the ages of 39 and 49 is approximately 0.24.

▼ **check your progress 5**, *page 761*

Make a Punnett square.

Parents	C	c
C	CC	Cc
c	Cc	cc

To be white, the offspring must be cc. From the table, only 1 of the 4 possible genotypes is cc, so the probability that an offspring will be white is $\frac{1}{4}$.

▼ **check your progress 6**, *page 762*

Let E be the event of selecting a blue ball. Because there are 5 blue balls, there are 5 favorable outcomes, leaving 7 unfavorable outcomes.

$$\text{Odds against } E = \frac{\text{number of unfavorable outcomes}}{\text{number of favorable outcomes}} = \frac{7}{5}$$

The odds against selecting a blue ball from the box are 7 to 5.

▼ **check your progress 7**, *page 763*

Let E represent the event of an earthquake of magnitude 6.7 or greater in the Bay Area in the next 30 years. Then the probability of E, $P(E)$, is 0.6. The odds in favor of this event are given by

$$\text{Odds in favor} = \frac{P(E)}{1 - P(E)}$$

$$= \frac{0.6}{1 - 0.6} = \frac{0.6}{0.4} = \frac{3}{2}$$

The odds in favor of this event are 3 to 2.

SECTION 12.4

▼ **check your progress 1**, *page 769*

Let A be the event of rolling a 7, and let B be the event of rolling an 11. From the sample space on page 721, $P(A) = \frac{6}{36} = \frac{1}{6}$ and $P(B) = \frac{2}{36} = \frac{1}{18}$. Because A and B are mutually exclusive events,

$$P(A \text{ or } B) = P(A) + P(B) = \frac{1}{6} + \frac{1}{18} = \frac{4}{18} = \frac{2}{9}$$

The probability of rolling a 7 or an 11 is $\frac{2}{9}$.

▼ **check your progress 2**, *page 771*

Let $A = \{\text{people with a degree in business}\}$ and $B = \{\text{people with a starting salary between \$20,000 and \$24,999}\}$. Then, from the table, $n(A) = 4 + 16 + 21 + 35 + 22 = 98$, $n(B) = 4 + 16 + 3 + 16 = 39$, and $n(A \text{ and } B) = 16$. The total number of people represented in the table is 206.

$$P(A \text{ or } B) = P(A) + P(B) - P(A \text{ and } B)$$

$$= \frac{98}{206} + \frac{39}{206} - \frac{16}{206} = \frac{121}{206} \approx 0.587$$

The probability of choosing a person who has a degree in business or a starting salary between \$20,000 and \$24,999 is about 58.7%.

▼ **check your progress 3**, *page 772*

If E is the event that a person has type A blood, then E^C is the event that the person does not have type A blood, and

$$P(E^c) = 1 - P(E) = 1 - 0.34 = 0.66$$

The probability that a person does not have type A blood is 66%.

▼ **check your progress 4**, *page 773*

Let $E = \{\text{at least 1 roll of sum 7}\}$; then $E^c = \{\text{no sum of 7 is rolled}\}$. Using the table on page 759, there are 36 possibilities for each toss of the dice. Thus, $n(S) = 36 \cdot 36 \cdot 36 = 46,656$. For each roll of the dice, there are 30 numbers that do not total 7, so $n(E^c) = 30 \cdot 30 \cdot 30 = 27,000$. Then

$$P(E) = 1 - P(E^c) = 1 - \frac{27,000}{46,656} = \frac{19,656}{46,656} \approx 0.421$$

There is about a 42.1% chance of rolling a sum of 7 at least once.

▼ **check your progress 5**, *page 774*

Let $E = \{\text{at least one \$100 bill}\}$; then $E^c = \{\text{no \$100 bills}\}$. The number of elements in the sample space is the number of ways to choose 4 bills from 35:

$$n(S) = C(35, 4) = \frac{35!}{4!\,(35 - 4)!} = \frac{35!}{4!\,31!} = 52,360$$

To count the number of ways not to choose any \$100 bills, we need to compute the number of ways to choose 4 \$1 bills from the 31 \$1 bills available.

$$n(E^c) = C(31, 4) = \frac{31!}{4!\,(31 - 4)!} = \frac{31!}{4!\,27!} = 31,465$$

$$P(E) = 1 - P(E^c) = 1 - \frac{n(E^c)}{n(S)}$$

$$= 1 - \frac{31,465}{52,360} = \frac{20,895}{52,360} \approx 0.399$$

The probability of pulling out at least one \$100 bill is about 39.9%.

SECTION 12.5

▼ **check your progress 1**, *page 779*

Let $B = \{\text{the sum is 6}\}$ and $A = \{\text{the first die is not a 3}\}$. From the table on page 759, there are 4 possible rolls of the dice for which the first die is not a 3 and the sum is 6. So $P(A \text{ and } B) = \frac{4}{36} = \frac{1}{9}$. There are 30 possibilities for which the first die is not a 3, so $P(A) = \frac{30}{36} = \frac{5}{6}$. Then

$$P(B\,|\,A) = \frac{P(A \text{ and } B)}{P(A)} = \frac{\frac{1}{9}}{\frac{5}{6}} = \frac{2}{15}$$

The probability of rolling a 6 given that the first die is not a 3 is $\frac{2}{15}$.

▼ **check your progress 2**, *page 782*

Let $A = \{\text{a spade is dealt first}\}$, $B = \{\text{a heart is dealt second}\}$, and $C = \{\text{a spade is dealt third}\}$. Then

$$P(A \text{ and } B \text{ and } C) = P(A) \cdot P(B\,|\,A) \cdot P(C\,|\,A \text{ and } B)$$

$$= \frac{13}{52} \cdot \frac{13}{51} \cdot \frac{12}{50} = \frac{13}{850}$$

The probability is $\frac{13}{850}$, or about 0.015.

▼ check your progress **3**, *page 783*

Each coin toss is independent of the others, because the probability of getting heads on any toss is not affected by the results of the other coin tosses. Let $E_1 = \{$heads on the first toss$\}$, $E_2 = \{$heads on the second toss$\}$, and $E_3 = \{$heads on the third toss$\}$. The events are independent, and the probability of flipping heads is $\frac{1}{2}$, so

$$P(E_1 \text{ and } E_2 \text{ and } E_3) = P(E_1) \cdot P(E_2) \cdot P(E_3) = \frac{1}{2} \cdot \frac{1}{2} \cdot \frac{1}{2} = \frac{1}{8}$$

▼ check your progress **4**, *page 784*

Let D be the event that a person has the genetic defect, and let T be the event that the test for the defect is positive. We are asked for $P(D\,|\,T)$, which can be calculated by

$$P(D\,|\,T) = \frac{P(D \text{ and } T)}{P(T)}$$

A tree diagram will help us compute the needed probabilities.

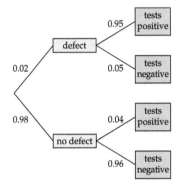

From the diagram, $P(D \text{ and } T) = P(D) \cdot P(T\,|\,D) = (0.02)(0.95)$. To compute $P(T)$, we need to combine two branches from the diagram, one corresponding to a correct positive test result when the person has the defect, and one corresponding to a false positive result when the person does not have the defect: $P(T) = (0.02)(0.95) + (0.98)(0.04)$. Then

$$P(D\,|\,T) = \frac{P(D \text{ and } T)}{P(T)} = \frac{(0.02)(0.95)}{(0.02)(0.95) + (0.98)(0.04)} \approx 0.326$$

There is only a 32.6% chance that a person who tests positive actually has the defect.

SECTION 12.6

▼ check your progress **1**, *page 789*

Let S_1 be the event that the roulette ball lands on a number from 1 to 12, in which case the player wins $10. There are 38 possible numbers, so $P(S_1) = \frac{12}{38}$. Let S_2 be the event that a number from 1 to 12 does not come up, in which case the player loses $5. Then $P(S_2) = 1 - \frac{12}{38} = \frac{26}{38}$.

$$\text{Expectation} = P(S_1) \cdot S_1 + P(S_2) \cdot S_2$$

$$= \frac{12}{38}(10) + \frac{26}{38}(-5) = -\frac{5}{19} \approx -0.263$$

The player's expectation is about −$0.263.

▼ check your progress **2**, *page 791*

Let S_1 be the event that the person will die within 1 year. Then $P(S_1) = 0.000753$, and the company must pay out $10,000. Because the company received a premium of $45, the actual loss is $9955. Let S_2 be the event that the policy holder does not die during the year of the policy. Then $P(S_2) = 0.999247$ and the company keeps the premium. The expectation is

$$\text{Expectation} = P(S_1) \cdot S_1 + P(S_2) \cdot S_2$$

$$= 0.000753(-9955) + 0.999247(45)$$

$$= 37.47$$

The company's expectation is $37.47.

▼ check your progress **3**, *page 791*

$$\text{Expectation} = 0.05(500,000) + 0.30(250,000) + 0.35(150,000)$$

$$+ 0.20(-100,000) + 0.10(-350,000)$$

$$= 97,500$$

The company's profit expectation is $97,500.

CHAPTER 2

EXERCISE SET 2.1   *page 59*

1. {penny, nickel, dime, quarter}   **3.** {Mercury, Mars}   **5.** The answer in the year 2011: {George W. Bush, Barack Obama}
7. {−5, −4, −3, −2, −1}   **9.** {7}   **11.** { }
In Exercises 13 to 19, only one possible answer is given.
13. The set of days of the week that begin with the letter T   **15.** The set consisting of the two planets in our solar system that are closest to the
sun   **17.** The set of single-digit natural numbers   **19.** The set of natural numbers less than or equal to 7   **21.** True   **23.** False;
$b \in \{a, b, c\}$, but {b} is not an element of {a, b, c}.   **25.** False; {0} has one element, whereas $\varnothing$ has no elements.   **27.** False; the word "good" is
subjective.   **29.** False; 0 is an element of the first set, but 0 is not an element of the second set.
In Exercises 31 to 39, only one possible answer is given.
31. {$x \mid x \in N$ and $x < 13$}   **33.** {$x \mid x$ is a multiple of 5 and $4 < x < 16$}   **35.** {$x \mid x$ is the name of a month that has 31 days}
37. {$x \mid x$ is the name of a U.S. state that begins with the letter A}   **39.** {$x \mid x$ is a season that starts with the letter s}
41. {Michigan, Wisconsin}   **43.** {Texas, Florida, Ohio}   **45.** {2005, 2006, 2007}   **47.** {2009}
49. {2005, 2006, 2007, 2009}   **51.** {2004, 2005}   **53.** {1994, 1995, 1996, 1997, 1998}   **55.** {2004, 2005, 2006}
57. 11   **59.** 0   **61.** 4   **63.** 16   **65.** 121   **67.** Neither   **69.** Both   **71.** Equivalent   **73.** Equivalent
75. Not well defined   **77.** Not well defined   **79.** Well defined   **81.** Well defined   **83.** Not well defined
85. Not well defined   **87.** $A = B$ **89.** $A \neq B$ **91.** Answers will vary; however, the set of all real numbers between 0 and 1 is one example
of a set that cannot be written using the roster method.

EXERCISE SET 2.2   *page 69*

1. {0, 1, 3, 5, 8}   **3.** $U = \{0, 1, 2, 3, 4, 5, 6, 7, 8\}$ **5.** {0, 7, 8}   **7.** {0, 2, 4, 6, 8}   **9.** $\subseteq$ **11.** $\not\subseteq$ **13.** $\subseteq$
15. $\subseteq$ **17.** $\subseteq$ **19.** True   **21.** True   **23.** True   **25.** False   **27.** True   **29.** True   **31.** False
33. False   **35.** False   **37.** 18 hours   **39.** $\varnothing$, {α}, {β}, {α, β}   **41.** $\varnothing$, {I}, {II}, {III}, {I, II}, {I, III}, {II, III}, {I, II, III}
43. 4 subsets   **45.** 128 subsets   **47.** 2048 subsets   **49.** 1 subset   **51. a.** 15 different sets   **b.** 10 different sums
c. Two different sets of coins can have the same value.   **53. a.** 6 elements   **b.** 2 elements   **c.** 4 elements   **d.** 1 element
55. a. 1024 omelets   **b.** 12 ingredients   **57. a.** {1, 2, 3} has only three elements, namely 1, 2, and 3. Because {2} is not equal to 1, 2, or 3,
{2} $\notin$ {1, 2, 3}.   **b.** 1 is not a set, so it cannot be a subset.   **c.** The given set has the elements 1 and {1}. Because $1 \neq \{1\}$, there are exactly two elements in {1, {1}}.   **59. a.** {A, B, C}, {A, B, D}, {A, B, E}, {A, C, D}, {A, C, E}, {A, D, E}, {B, C, D}, {B, C, E}, {B, D, E}, {C, D, E}, {A, B, C, D},
{A, B, C, E}, {A, B, D, E}, {A, C, D, E}, {B, C, D, E}, {A, B, C, D, E}   **b.** {A}, {B}, {C}, {D}, {E}, {A, B}, {A, C}, {A, D}, {A, E}, {B, C}, {B, D}, {B, E},
{C, D}, {C, E}, {D, E}

EXERCISE SET 2.3   *page 82*

1. {1, 2, 4, 5, 6, 8}   **3.** {4, 6}   **5.** {3, 7}   **7.** $U = \{1, 2, 3, 4, 5, 6, 7, 8\}$ **9.** $\varnothing$ **11.** $B = \{1, 2, 5, 8\}$
13. $U = \{1, 2, 3, 4, 5, 6, 7, 8\}$ **15.** {2, 5, 8}   **17.** {1, 3, 4, 6, 7}   **19.** {2, 5, 8}
In Exercises 21 to 27, one possible answer is given. Your answers may vary from the given answers.
21. The set of all elements that are not in L or are in T **23.** The set of all elements that are in A, or are in C but not in B
25. The set of all elements that are in T, and are also in J or not in K **27.** The set of all elements that are in both W and V, or are in both W and Z

29.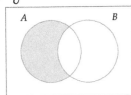

31.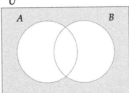

33.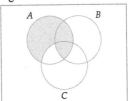

35.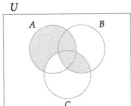

37. Not equal   **39.** Equal   **41.** Not equal   **43.** Not equal

45. Equal   **47. a.** and **b.** Answers will vary depending upon the time the Twitter Venn was produced.   **49.** Yellow   **51.** Cyan
53. Red

In Exercises 55 to 63, one possible answer is given. Your answers may vary from the given answers.

55. $A \cap B'$ **57.** $(A \cup B)'$ **59.** $B \cup C$ **61.** $C \cap (A \cup B)'$ **63.** $(A \cup B)' \cup (A \cap B \cap C)$

65. a.
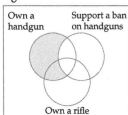
b.
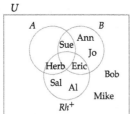
c.
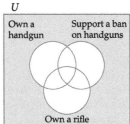

67.
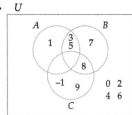
69.
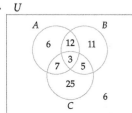

See the *Student Solutions Manual* for the verification for Exercise 71.   **73.** $\{3, 9\}$ **75.** $\{2, 8\}$ **77.** $\{3, 9\}$
79. Responses will vary.   **81.**
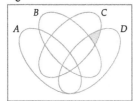

EXERCISE SET 2.4   *page 92*

1. 7   **3.** 8   **5.** 8   **7.** 12   **9.** $n(A \cup B) = 7$; $n(A) + n(B) - n(A \cap B) = 4 + 5 - 2 = 7$ **11.** 113   **13.** 1060
15.

17. a. 180 investors   **b.** 200 investors   **19. a.** 15%   **b.** 13%

21. a. 450 customers   **b.** 140 customers   **c.** 130 customers   **23. a.** 109 households   **b.** 328 households
c. 104 households   **25. a.** 101 people   **b.** 370 people   **c.** 380 people   **d.** 373 people   **e.** 225 people
f. 530 people   **27. a.** 72 elements   **b.** 47 elements   **c.** 25 elements   **d.** 0 elements   **29. a.** 200 users
b. 271 users   **c.** 16 users

EXERCISE SET 2.5   *page 104*

1. a. One possible one-to-one correspondence   **b.** 6   **3.** Pair $(2n - 1)$ of D with $(3n)$ of M.   **5.** $\aleph_0$ **7.** c
between V and M is given by

$$V = \{a, e, i\}$$
↕ ↕ ↕
$$M = \{3, 6, 9\}$$

9. c   **11.** Equivalent   **13.** Equivalent   **15.** Let $S = \{10, 20, 30, \dots, 10n, \dots\}$. Then S is a proper subset of A. A rule for a one-to-one correspondence between A and S is $(5n) \leftrightarrow (10n)$. Because A can be placed in a one-to-one correspondence with a proper subset of itself, A is an infinite set.

17. Let $R = \left\{\dfrac{3}{4}, \dfrac{5}{6}, \dfrac{7}{8}, ..., \dfrac{2n+1}{2n+2}, ...\right\}$. Then R is a proper subset of C. A rule for a one-to-one correspondence between C and R is

$\left(\dfrac{2n-1}{2n}\right) \leftrightarrow \left(\dfrac{2n+1}{2n+2}\right)$. Because C can be placed in a one-to-one correspondence with a proper subset of itself, C is an infinite set.

In Exercises 19 to 25, let $N = \{1, 2, 3, 4, ..., n, ...\}$. Then a one-to-one correspondence between the given sets and the set of natural numbers N is given by the following general correspondences.

19. $(n + 49) \leftrightarrow n$ **21.** $\left(\dfrac{1}{3^{n-1}}\right) \leftrightarrow n$ **23.** $(10^n) \leftrightarrow n$ **25.** $(n^3) \leftrightarrow n$

27. a. For any natural number n, the two natural numbers preceding $3n$ are not multiples of 3. Pair these two numbers, $3n - 2$ and $3n - 1$, with the multiples of 3 given by $6n - 3$ and $6n$, respectively. Using the two general correspondences $(6n - 3) \leftrightarrow (3n - 2)$ and $(6n) \leftrightarrow (3n - 1)$ (as shown below), we can establish a one-to-one correspondence between the multiples of 3 (set M) and the set K of all natural numbers that are not multiples of 3.

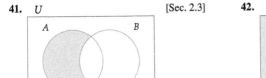

$M = \{3, 6, 9, 12, 15, 18, ..., 6n - 3, \quad 6n, \quad ...\}$

$K = \{1, 2, 4, 5, \ 7, \ 8, \ ..., \ 3n - 2, 3n - 1, ...\}$

The following answers in parts b and c were produced by using the correspondences established in part a. **b.** 302 **c.** 1800

29. The set of real numbers x such that $0 < x < \pi$ is equivalent to the set of all real numbers.

CHAPTER 2 REVIEW EXERCISES   *page 108*

1. {January, June, July} [Sec. 2.1]   **2.** {Alaska, Hawaii} [Sec. 2.1]   **3.** {0, 1, 2, 3, 4, 5, 6, 7} [Sec. 2.1]   **4.** $\{-8, 8\}$ [Sec. 2.1]
5. {1, 2, 3, 4} [Sec. 2.1]   **6.** {1, 2, 3, 4, 5, 6} [Sec. 2.1]   **7.** $\{x \mid x \in I \text{ and } x > -6\}$ [Sec. 2.1]   **8.** $\{x \mid x$ is the name of a month with exactly 30 days} [Sec. 2.1]   **9.** $\{x \mid x$ is the name of a U.S. state that begins with the letter K} [Sec. 2.1]   **10.** $\{x^3 \mid x = 1, 2, 3, 4, \text{ or } 5\}$ [Sec. 2.1]   **11.** Equivalent [Sec. 2.1]   **12.** Both equal and equivalent [Sec. 2.1]   **13.** False [Sec. 2.1]   **14.** True [Sec. 2.1]
15. True [Sec. 2.1]   **16.** False [Sec. 2.1]   **17.** {6, 10} [Sec. 2.3]   **18.** {2, 6, 10, 16, 18} [Sec. 2.3]
19. $C = \{14, 16\}$ [Sec. 2.3]   **20.** {2, 6, 8, 10, 12, 16, 18} [Sec. 2.3]   **21.** {2, 6, 10, 16} [Sec. 2.3]   **22.** {8, 12} [Sec. 2.3]
23. {6, 8, 10, 12, 14, 16, 18} [Sec. 2.3]   **24.** {8, 12} [Sec. 2.3]   **25.** No [Sec. 2.2]   **26.** No [Sec. 2.2]
27. Proper subset [Sec. 2.2]   **28.** Proper subset [Sec. 2.2]   **29.** Not a proper subset [Sec. 2.2]   **30.** Not a proper subset [Sec. 2.2]
31. $\varnothing, \{I\}, \{II\}, \{I, II\}$ [Sec. 2.2]   **32.** $\varnothing, \{s\}, \{u\}, \{n\}, \{s, u\}, \{s, n\}, \{u, n\}, \{s, u, n\}$ [Sec. 2.2]   **33.** $\varnothing$, {penny}, {nickel}, {dime}, {quarter}, {penny, nickel}, {penny, dime}, {penny, quarter}, {nickel, dime}, {nickel, quarter}, {dime, quarter}, {penny, nickel, dime}, {penny, nickel, quarter}, {penny, dime, quarter}, {nickel, dime, quarter}, {penny, nickel, dime, quarter} [Sec. 2.2]   **34.** $\varnothing$, {A}, {B}, {C}, {D}, {E}, {A, B}, {A, C}, {A, D}, {A, E}, {B, C}, {B, D}, {B, E}, {C, D}, {C, E}, {D, E}, {A, B, C}, {A, B, D}, {A, B, E}, {A, C, D}, {A, C, E}, {A, D, E}, {B, C, D}, {B, C, E}, {B, D, E}, {C, D, E}, {A, B, C, D}, {A, B, C, E}, {A, B, D, E}, {A, C, D, E}, {B, C, D, E}, {A, B, C, D, E}
[Sec. 2.2]   **35.** $2^4 = 16$ subsets [Sec. 2.2]   **36.** $2^{26} = 67{,}108{,}864$ subsets [Sec. 2.2]   **37.** $2^{15} = 32{,}768$ subsets [Sec. 2.2]
38. $2^7 = 128$ subsets [Sec. 2.2]   **39.** True [Sec. 2.3]   **40.** True [Sec. 2.3]

41. [Sec. 2.3]

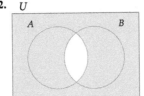

42. [Sec. 2.3]

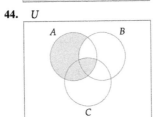

43. [Sec. 2.3]

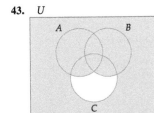

44. [Sec. 2.3]

45. Equal [Sec. 2.3]

46. Not equal [Sec. 2.3]   **47.** Not equal [Sec. 2.3]   **48.** Not equal [Sec. 2.3]
49. $(A \cup B)' \cap C$ or $C \cap (A' \cap B')$ [Sec. 2.3]   **50.** $(A \cap B) \cup (B \cap C')$ [Sec. 2.3]
51. [Sec. 2.3]

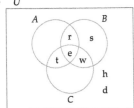

52. [Sec. 2.3]

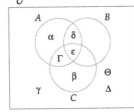

53. 391 members [Sec. 2.4]

54. a. 42 customers   **b.** 31 customers   **c.** 20 customers   **d.** 142 customers   [Sec. 2.4]   **55.** 6 athletes   [Sec. 2.4]
56. 874 students   [Sec. 2.4]

57. One possible one-to-one correspondence between $\{1, 3, 6, 10\}$ and $\{1, 2, 3, 4\}$ is given by

$\{1, 3, 6, 10\}$
↕ ↕ ↕ ↕
$\{1, 2, 3, 4\}$ [Sec. 2.5]

58. $\{x \mid x > 10 \text{ and } x \in N\} = \{11, 12, 13, 14, \ldots, n + 10, \ldots\}$
Thus a one-to-one correspondence between the sets is given by

$\{11, 12, 13, 14, \ldots, n + 10, \ldots\}$
↕ ↕ ↕ ↕ ↕
$\{2, \quad 4, \quad 6, \quad 8, \quad \ldots, \quad 2n, \quad \ldots\}$ [Sec. 2.5]

59. One possible one-to-one correspondence between the sets is given by

$\{3, \quad 6, \quad 9, \quad \ldots, 3n, \ldots\}$
↕ ↕ ↕ ↕
$\{10, 100, 1000, \ldots, 10^n, \ldots\}$ [Sec. 2.5]

60. In the following figure, the line from E that passes through $\overline{AB}$ and $\overline{CD}$ illustrates a method of establishing a one-to-one correspondence between $\{x \mid 0 \le x \le 1\}$ and $\{x \mid 0 \le x \le 4\}$.

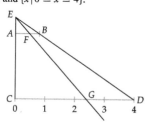

[Sec. 2.5]

61. A proper subset of A is $S = \{10, 14, 18, \ldots, 4n + 6, \ldots\}$. A one-to-one correspondence between A and S is given by

$A = \{6, 10, 14, 18, \ldots, 4n + 2, \ldots\}$
↕ ↕ ↕ ↕ ↕
$S = \{10, 14, 18, 22, \ldots, 4n + 6, \ldots\}$

Because A can be placed in a one-to-one correspondence with a proper subset of itself, A is an infinite set.   [Sec. 2.5]

62. A proper subset of B is $T = \left\{\dfrac{1}{2}, \dfrac{1}{4}, \dfrac{1}{8}, \dfrac{1}{16}, \ldots, \dfrac{1}{2^n}, \ldots\right\}$. A one-to-one correspondence between B and T is given by

$$B = \left\{1, \dfrac{1}{2}, \dfrac{1}{4}, \dfrac{1}{8}, \ldots, \dfrac{1}{2^{n-1}}, \ldots\right\}$$
↕ ↕ ↕ ↕ ↕
$$T = \left\{\dfrac{1}{2}, \dfrac{1}{4}, \dfrac{1}{8}, \dfrac{1}{16}, \ldots, \dfrac{1}{2^n}, \ldots\right\}$$

Because B can be placed in a one-to-one correspondence with a proper subset of itself, B is an infinite set.   [Sec. 2.5]

63. 5   [Sec. 2.1]   **64.** 10   [Sec. 2.1]   **65.** 2   [Sec. 2.1]   **66.** 5   [Sec. 2.1]   **67.** $\aleph_0$ [Sec. 2.5]
68. $\aleph_0$ [Sec. 2.5]   **69.** c [Sec. 2.5]   **70.** c [Sec. 2.5]   **71.** $\aleph_0$ [Sec. 2.5]   **72.** c [Sec. 2.5]

CHAPTER 2 TEST   *page 110*

1. $\{1, 2, 4, 5, 6, 7, 9, 10\}$ [Sec. 2.2, Example 1; Sec. 2.3, Example 1]   **2.** $\{2, 9, 10\}$ [Sec. 2.2, Example 1; Sec. 2.3, Example 1]
3. $\{1, 2, 3, 4, 6, 9, 10\}$ [Sec. 2.2, Example 1; Sec. 2.3, Examples 1 and 2]   **4.** $\{5, 7, 8\}$ [Sec. 2.2, Example 1; Sec. 2.3, Examples 1 and 2]
5. $\{x \mid x \in W \text{ and } x < 7\}$ [Sec. 2.1, Example 5]   **6.** $\{x \mid x \in I \text{ and } -3 \le x \le 2\}$ [Sec. 2.1, Example 5]
7. a. 4   **b.** $\aleph_0$ [Sec. 2.1, Example 6; Sec. 2.5, Example 3]   **8. a.** Neither   **b.** Equivalent   [Sec. 2.1, Example 7]
9. a. Equivalent   **b.** Equivalent   [Sec. 2.5, Example 3]
10. $\{a\}, \{b\}, \{c\}, \{d\}$
$\{a, b\}, \{a, c\}, \{a, d\}, \{b, c\}, \{b, d\}, \{c, d\}$
$\{a, b, c\}, \{a, b, d\}, \{a, c, d\}, \{b, c, d\}$
$\{a, b, c, d\}$ [Sec. 2.2, Example 4]

11. $2^{21} = 2,097,152$ subsets   [Sec. 2.2, Example 5]

12. [Sec. 2.3, Example 6]

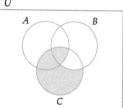

13. [Sec. 2.3, Example 6]

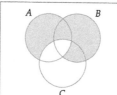

14. $A' \cap B'$ [Sec. 2.3, Example 4 and Check Your Progress 4]   **15. a.** $2^9 = 512$ different versions   **b.** 12 options   [Sec. 2.2, Example 5]
16. 1164   [Sec. 2.4, Example 3]   **17.** 541 families   [Sec. 2.4, Example 1]   **18. a.** 232 households   **b.** 102 households
c. 857 households   **d.** 79 households   [Sec. 2.4, Example 2]
19. $\{5, 10, 15, 20, 25, ..., \ 5n, \ ...\}$ **20.** $\{3, \ 6, \ 9, \ 12, ..., 3n, ...\}$
$\{0, \ 1, \ 2, \ 3, \ 4, \ ..., n - 1, ...\}$ $\{6, 12, 18, 24, ..., 6n, ...\}$
$(5n) \leftrightarrow (n - 1)$ [Sec. 2.5, Example 1] $(3n) \leftrightarrow (6n)$ [Sec. 2.5, Example 2]

CHAPTER 3

EXERCISE SET 3.1   *page 121*

1. Not a statement   **3.** Statement   **5.** Not a statement   **7.** Statement   **9.** One simple statement is "The principal will attend the class on Tuesday." The other simple statement is "The principal will attend the class on Wednesday."   **11.** One simple statement is "A triangle is an acute triangle." The other simple statement is "It has three acute angles."   **13.** The Giants did not lose the game.   **15.** The game went into over-time.   **17.** $w \rightarrow t$, conditional   **19.** $l \leftrightarrow a$, biconditional   **21.** $d \rightarrow f$, conditional   **23.** $m \vee c$, disjunction   **25.** The tour goes to Italy and the tour does not go to Spain.   **27.** If we go to Venice, then we will not go to Florence.   **29.** We will go to Florence if and only if we do not go to Venice.   **31.** Taylor Swift is a singer or she is an actress, and she is not a songwriter.   **33.** If Taylor Swift is a singer, then she is not a songwriter and she is not an actress.   **35.** Taylor Swift is an actress and a singer, if and only if she is not a songwriter.
37. $(p \vee q) \wedge \sim r$ **39.** $(q \wedge r) \rightarrow \sim p$ **41.** $s \rightarrow (q \wedge \sim p)$ **43.** True   **45** True   **47.** True   **49.** True   **51.** No lions are playful.   **53.** Some classic movies were not first produced in black and white.   **55.** Some even numbers are odd numbers.   **57.** Some cars do not run on gasoline.   **59.** $p \rightarrow q$, where p represents "you can count your money" and q represents "you don't have a billion dollars."   **61.** $p \rightarrow q$, where p represents "people concentrated on the really important things in life" and q represents "there'd be a shortage of fishing poles."   **63.** $p \leftrightarrow q$, where p represents "an angle is a right angle" and q represents "its measure is 90°."   **65.** $p \rightarrow q$, where p represents "two sides of a triangle are equal in length" and q represents "the angles opposite those sides are congruent."   **67.** $p \rightarrow q$, where p represents "it is a square" and q represents "it is a rectangle."   **69.** 6 cups. In any teapot the tea level cannot rise above its spout opening, because any extra tea will flow out the spout. Because both spout openings are at the same height, the maximum number of cups they can hold is the same.

EXERCISE SET 3.2   *page 131*

1. True   **3.** False   **5.** False   **7.** False   **9.** False   **11. a.** If p is false, then $p \wedge (q \vee r)$ must be a false statement.   **b.** For a conjunctive statement to be true, it is necessary that all components of the statement be true. Because it is given that one of the components (p) is false, $p \wedge (q \vee r)$ must be a false statement.

p	q	**13.**	**15.**	**17.**
T	T	T	F	F
T	F	F	T	T
F	T	T	F	F
F	F	T	F	T

p	q	r	**19.**	**21.**	**23.**	**25.**	**27.**
T	T	T	F	T	T	T	F
T	T	F	F	T	F	T	T
T	F	T	F	T	F	T	T
T	F	F	F	T	F	T	F
F	T	T	F	F	T	T	F
F	T	F	F	F	F	T	F
F	F	T	F	T	T	F	T
F	F	F	T	T	F	T	F

See the *Student Solutions Manual* for the solutions to Exercises 29 to 35.   **37.** It did not rain and it did not snow.   **39.** She did not visit either France or Italy.   **41.** She did not get a promotion and she did not receive a raise.   **43.** Tautology   **45.** Tautology   **47.** Tautology
49. Self-contradiction   **51.** Self-contradiction   **53.** Not a self-contradiction   **55.** The symbol $\leq$ means "less than or equal to."
57. $2^5 = 32$ **59.** F F F T T T F T F F F T F F F F   **61.** Circle the 1 and the three 9s. Then invert the paper so that the digits are upside down and hand it back to your friend.

1. *Antecedent:* I had the money
Consequent: I would buy the painting

3. *Antecedent:* they had a guard dog
Consequent: no one would trespass on their property

5. *Antecedent:* I change my major
Consequent: I must reapply for admission

7. True **9.** True **11.** True **13.** False

p	q	15.	17.
T	T	T	T
T	F	T	T
F	T	T	T
F	F	T	T

p	q	r	19.	21.	23.
T	T	T	T	T	T
T	T	F	T	F	T
T	F	T	T	T	T
T	F	F	T	T	T
F	T	T	T	T	T
F	T	F	T	F	T
F	F	T	T	T	T
F	F	F	T	T	T

25. She cannot sing or she would be perfect for the part. **27.** Either x is not an irrational number or x is not a terminating decimal.
29. The fog must lift or our flight will be cancelled. **31.** They offered me the contract and I didn't accept.
33. Pigs have wings and they still can't fly. **35.** She traveled to Italy and she didn't visit her relatives. **37.** False
39. True **41.** False **43.** True **45.** True **47.** $p \rightarrow v$ **49.** $t \rightarrow \sim v$ **51.** $(\sim t \wedge p) \rightarrow v$
53. Not equivalent **55.** Not equivalent **57.** Equivalent **59.** If a number is a rational number, then it is a real number.
61. If an animal is a sauropod, then it is herbivorous. **63.** Turn two of the valves on. After one minute, turn off one of these valves. When you get up to the field, the sprinklers will be running on the region that is controlled by the valve that is still in the on position. The region that is wet but not receiving any water is controlled by the valve you turned off. The region that is completely dry is the region that is controlled by the valve that you left in the off position.

1. If we take the aerobics class, then we will be in good shape for the ski trip. **3.** If the number is an odd prime number, then it is greater than 2.
5. If he has the talent to play a keyboard, then he can join the band. **7.** If I was able to prepare for the test, then I had the textbook.
9. If you ran the Boston marathon, then you are in excellent shape. **11. a.** If I quit this job, then I am rich. **b.** If I were not rich, then I would not quit this job. **c.** If I would not quit this job, then I would not be rich. **13. a.** If we are not able to attend the party, then she did not return soon. **b.** If she returns soon, then we will be able to attend the party. **c.** If we are able to attend the party, then she returned soon. **15. a.** If a figure is a quadrilateral, then it is a parallelogram. **b.** If a figure is not a parallelogram, then it is not a quadrilateral.
c. If a figure is not a quadrilateral, then it is not a parallelogram. **17. a.** If I am able to get current information about astronomy, then I have access to the Internet. **b.** If I do not have access to the Internet, then I will not be able to get current information about astronomy.
c. If I am not able to get current information about astronomy, then I don't have access to the Internet. **19. a.** If we don't have enough money for dinner, then we took a taxi. **b.** If we did not take a taxi, then we will have enough money for dinner. **c.** If we have enough money for dinner, then we did not take a taxi. **21. a.** If she can extend her vacation for at least two days, then she will visit Kauai. **b.** If she does not visit Kauai, then she could not extend her vacation for at least two days. **c.** If she cannot extend her vacation for at least two days, then she will not visit Kauai.
23. a. If two lines are parallel, then the two lines are perpendicular to a given line. **b.** If two lines are not perpendicular to a given line, then the two lines are not parallel. **c.** If two lines are not parallel, then the two lines are not both perpendicular to a given line. **25.** Not equivalent
27. Equivalent **29.** Not equivalent **31.** If $x = 7$, then $3x - 7 \neq 11$. The original statement is true. **33.** If $|a| = 3$, then $a = 3$. The original statement is false. **35.** If $a + b = 25$, then $\sqrt{a + b} = 5$. The original statement is true. **37.** $p \rightarrow q$ **39. a.** and **b.** Answers will vary. **41.** If you can dream it, then you can do it. **43.** If I were a dancer, then I would not be a singer. **45.** A conditional statement and its contrapositive are equivalent. They always have the same truth values. **47.** The Hatter is telling the truth.

Solution to the *Where is the Missing Dollar?* puzzle in the Math Matters on page 154:
There is no missing dollar. The desk clerk has $25. The men received a total of $3 from the refund and the bellhop has $2 of the refund.
$25 + $3 + $2 = $30, which is the original total the men paid. *Note:* In the puzzle it states that each man paid $9 for the room, but this is not correct. Each man paid $9 for his share of the room and his share of the tip, if you can call it a tip, that the bellhop pocketed. Thus the men spent a total of $3 \times $9 = 27 for the room and the tip to the bellhop. The remaining $3 was given back to them by the bellhop.

1. $r \rightarrow c$
$\underline{\quad r \quad}$
$\therefore c$

3. $g \rightarrow s$
$\underline{\quad \sim g \quad}$
$\therefore \sim s$

5. $s \rightarrow i$
$\underline{\quad s \quad}$
$\therefore i$

7. $\sim p \rightarrow \sim a$
$\underline{\quad a \quad}$
$\therefore p$

9. Invalid **11.** Invalid **13.** Valid

15. Invalid   **17.** Invalid   **19.** Invalid   **21.** Valid   **23.** Valid   **25.**

$$h \to r$$
$$\underline{\sim h}$$
$$\therefore \sim r$$
Invalid

27.

$$\sim b \to d$$
$$\underline{b \lor d}$$
$$\therefore b$$
Invalid

29.

$$c \to t$$
$$\underline{t}$$
$$\therefore c$$
Invalid

31. Valid argument, modus tollens   **33.** Invalid argument, fallacy of the inverse   **35.** Valid argument, law of syllogism

37. Valid argument, modus ponens   **39.** Valid argument, modus tollens

See the *Student Solutions Manual* for the solutions to Exercises 41 to 45.   **47.** q **49.** it is not a theropod.   **51.** 12. Any number multiplied by 0 produces 0. Thus the only arithemtic you need to perform are the last two operations.

EXERCISE SET 3.6   *page 163*

1. Valid   **3.** Valid   **5.** Valid   **7.** Valid   **9.** Invalid   **11.** Invalid   **13.** Invalid   **15.** Valid   **17.** Valid
19. Invalid   **21.** All reuben sandwiches need mustard.   **23.** 1001 ends with a 5.   **25.** Some horses are grey.
27. a. Invalid   **b.** Invalid   **c.** Invalid   **d.** Invalid   **e.** Valid   **f.** Valid   **29.** c

CHAPTER 3 REVIEW EXERCISES   *page 168*

1. Not a statement   [Sec. 3.1]   **2.** Statement   [Sec. 3.1]   **3.** Statement   [Sec. 3.1]   **4.** Statement   [Sec. 3.1]
5. Not a statement   [Sec. 3.1]   **6.** Statement   [Sec. 3.1]   **7.** $m \land b$, conjunction   [Sec. 3.1]   **8.** $d \to e$, conditional   [Sec. 3.1]
9. $g \leftrightarrow d$, biconditional   [Sec. 3.1]   **10.** $t \to s$, conditional   [Sec. 3.1]   **11.** No dogs bite. [Sec. 3.1]   **12.** Some desserts at the Cove restaurant are not good. [Sec. 3.1]   **13.** Some winners do not receive a prize. [Sec. 3.1]   **14.** All cameras use film. [Sec. 3.1]
15. Some students finished the assignment. [Sec. 3.1]   **16.** Nobody enjoyed the story. [Sec. 3.1]   **17.** True   [Sec. 3.1]
18. True   [Sec. 3.1]   **19.** True   [Sec. 3.1]   **20.** True   [Sec. 3.1]   **21.** False   [Sec. 3.2]   **22.** False   [Sec. 3.2/3.3]
23. True   [Sec. 3.2]   **24.** True   [Sec. 3.2/3.3]   **25.** True   [Sec. 3.2/3.3]   **26.** False   [Sec. 3.2/3.3]

p	q	**27.** [Sec. 3.2/3.3]	**28.** [Sec. 3.2/3.3]	**29.** [Sec. 3.2/3.3]	**30.** [Sec. 3.2/3.3]
T	T	T	F	F	T
T	F	T	F	F	T
F	T	T	T	F	F
F	F	F	F	F	T

p	q	r	**31.** [Sec. 3.2/3.3]	**32.** [Sec. 3.2/3.3]	**33.** [Sec. 3.2/3.3]	**34.** [Sec. 3.2/3.3]
T	T	T	T	F	F	T
T	T	F	T	T	F	T
T	F	T	T	T	T	T
T	F	F	F	T	T	T
F	T	T	T	F	F	F
F	T	F	T	F	F	T
F	F	T	T	T	F	T
F	F	F	T	T	F	T

35. Bob passed the English proficiency test or he did not register for a speech course. [Sec. 3.2]   **36.** It is not true that, Ellen went to work this morning or she took her medication. [Sec. 3.2]   **37.** It is not the case that, Wendy will not go to the store this afternoon and she will be able to prepare her fettuccine al pesto recipe. [Sec. 3.2]   **38.** It is not the case that, Gina did not enjoy the movie or she enjoyed the party. [Sec. 3.2/3.3]

See the *Student Solutions Manual* for solutions to Exercises 39 to 42.

43. Self-contradiction [Sec. 3.2] **44.** Tautology [Sec. 3.2/3.3] **45.** Tautology [Sec. 3.2/3.3] **46.** Tautology [Sec. 3.2/3.3]
47. *Antecedent:* he has talent **48.** *Antecedent:* I had a credential
Consequent: he will succeed [Sec. 3.3] *Consequent:* I could get the job [Sec. 3.3]
49. *Antecedent:* I join the fitness club **50.** *Antecedent:* I will attend
Consequent: I will follow the exercise program [Sec. 3.3] *Consequent:* it is free [Sec. 3.3]
51. She is not tall or she would be on the volleyball team. [Sec. 3.3] **52.** He cannot stay awake or he would finish the report. [Sec. 3.3]
53. Rob is ill or he would start. [Sec. 3.3] **54.** Sharon will not be promoted or she closes the deal. [Sec. 3.3]
55. I get my paycheck and I do not purchase a ticket. [Sec. 3.3] **56.** The tomatoes will get big and you did not provide them with plenty of
water. [Sec. 3.3] **57.** You entered Cleggmore University and you did not have a high score on the SAT exam. [Sec. 3.3] **58.** Ryan enrolled at a
university and he did not enroll at Yale. [Sec. 3.3] **59.** False [Sec. 3.3] **60.** True [Sec. 3.3] **61.** False [Sec. 3.3]
62. False [Sec. 3.3] **63.** If a real number has a nonrepeating, nonterminating decimal form, then the real number is irrational. [Sec. 3.4]
64. If you are a politician, then you are well known. [Sec. 3.4] **65.** If I can sell my condominium, then I can buy the house. [Sec. 3.4]
66. If a number is divisible by 9, then the number is divisible by 3. [Sec. 3.4] **67. a.** *Converse:* If $x > 3$, then $x + 4 > 7$.
b. *Inverse:* If $x + 4 \le 7$, then $x \le 3$. **c.** *Contrapositive:* If $x \le 3$, then $x + 4 \le 7$. [Sec. 3.4] **68. a.** *Converse:* If a recipe can be prepared
in less than 20 minutes, then the recipe is in this book. **b.** *Inverse:* If a recipe is not in this book, then the recipe cannot be prepared in less than 20
minutes. **c.** *Contrapositive:* If a recipe cannot be prepared in less than 20 minutes, then the recipe is not in this book. [Sec. 3.4]
69. a. *Converse:* If $(a + b)$ is divisible by 3, then a and b are both divisible by 3. **b.** *Inverse:* If a and b are not both divisible by 3, then $(a + b)$ is not
divisible by 3. **c.** *Contrapositive:* If $(a + b)$ is not divisible by 3, then a and b are not both divisible by 3. [Sec. 3.4] **70. a.** *Converse:*
If they come, then you built it. **b.** *Inverse:* If you do not build it, then they will not come. **c.** *Contrapositive:* If they do not come, then you did not
build it. [Sec. 3.4] **71. a.** *Converse:* If it has exactly two parallel sides, then it is a trapezoid. **b.** *Inverse:* If it is not a trapezoid, then it does not
have exactly two parallel sides. **c.** *Contrapositive:* If it does not have exactly two parallel sides, then it is not a trapezoid. [Sec. 3.4]
72. a. *Converse:* If they returned, then they liked it. **b.** *Inverse:* If they do not like it, then they will not return. **c.** *Contrapositive:* If they do
not return, then they did not like it. [Sec. 3.4] **73.** $q \to p$, the converse of the original statement [Sec. 3.4] **74.** True **75.** If x is an odd
prime number, then $x > 2$. [Sec. 3.4] **76.** If the senator attends the meeting, then she will vote on the motion. [Sec. 3.4] **77.** If their manager
contacts me, then I will purchase some of their products. [Sec. 3.4] **78.** If I can rollerblade, then Ginny can rollerblade. [Sec. 3.4]
79. Valid [Sec. 3.5] **80.** Valid [Sec. 3.5] **81.** Invalid [Sec. 3.5] **82.** Valid [Sec. 3.5] **83.** Valid argument, disjunctive syllogism [Sec. 3.5] **84.** Valid argument, law of syllogism [Sec. 3.5] **85.** Invalid argument, fallacy of the inverse [Sec. 3.5]
86. Valid argument, disjunctive syllogism [Sec. 3.5] **87.** Valid argument, modus tollens [Sec. 3.5] **88.** Invalid argument, fallacy of the
inverse [Sec. 3.5] **89.** Valid [Sec. 3.6] **90.** Invalid [Sec. 3.6] **91.** Invalid [Sec. 3.6] **92.** Valid [Sec. 3.6]

CHAPTER 3 TEST *page 170*

1. a. Not a statement **b.** Statement [Sec. 3.1, Example 1] **2. a.** All trees are green. **b.** Some apartments are available. [Sec. 3.1, Example 2] **3. a.** False **b.** True [Sec. 3.1, Example 6] **4. a.** False **b.** True [Sec. 3.3, Example 3]

p	q	5. [Sec. 3.3, Example 3]
T	T	T
T	F	T
F	T	T
F	F	T

7. It is not true that, Elle ate breakfast or took a lunch break. [Sec. 3.2, Example 5]

p	q	r	6. [Sec. 3.3, Example 3]
T	T	T	F
T	T	F	T
T	F	T	F
T	F	F	F
F	T	T	F
F	T	F	T
F	F	T	T
F	F	F	F

8. A tautology is a statement that is always true. [Sec. 3.2, Example 6]   **9.** $\sim p \vee q$ [Sec. 3.3, Example 4]   **10. a.** False
b. False   [Sec. 3.3, Example 2]   **11. a.** *Converse:* If $x > 4$, then $x + 7 > 11$.   **b.** *Inverse:* If $x + 7 \leq 11$, then $x \leq 4$.
c. *Contrapositive:* If $x \leq 4$, then $x + 7 \leq 11$.   [Sec. 3.4, Example 2]

12. $p \to q$
$\underline{p}$
$\therefore \; q$ [Sec. 3.5, Example 4]

13. $p \to q$
$\underline{q \to r}$
$\therefore \; p \to r$ [Sec. 3.5, Example 4]

14. Valid   [Sec. 3.5, Example 2]

15. Invalid   [Sec. 3.5, Example 3]   **16.** Invalid argument; the argument is a fallacy of the inverse. [Sec. 3.5, Example 4]
17. Valid argument; the argument is a disjunctive syllogism. [Sec. 3.5, Example 4]   **18.** Invalid argument, as shown by an Euler diagram.
[Sec. 3.6, Example 2]   **19.** Invalid argument, as shown by an Euler diagram. [Sec. 3.6, Example 2]   **20.** Invalid argument; the argument is a
fallacy of the converse. [Sec. 3.5, Example 4]

CHAPTER 6

EXERCISE SET 6.1   *page 306*

1. ∩∩∩∩|||||| **3.** ⌐||| **5.** ꜩꜩ99999∩∩∩∩∩|||||||||| **7.** //ꜩꜩꜩ99991||

9. //////ꜩꜩꜩꜩꜩ999999999   **11.** ⚘⚭⚭ꜩꜩꜩꜩꜩ991||| **13.** 2134   **15.** 845   **17.** 1232   **19.** 221,011

21. 65,769   **23.** 5,122,406   **25.** 94   **27.** 666   **29.** 32   **31.** 56   **33.** 650   **35.** 1409   **37.** 1240   **39.** 840

41. 9044   **43.** 11,461   **45.** CLVII   **47.** DXLII   **49.** MCXCVII   **51.** DCCLXXXVII   **53.** DCLXXXIII

55. V̄DCCCXCVIII   **57.** 504   **59.** 203   **61.** 595   **63.** 2484   **65. a.** and **b.** Answers will vary.

EXERCISE SET 6.2   *page 316*

1. $(4 \times 10^1) + (8 \times 10^0)$ **3.** $(4 \times 10^2) + (2 \times 10^1) + (0 \times 10^0)$ **5.** $(6 \times 10^3) + (8 \times 10^2) + (0 \times 10^1) + (3 \times 10^0)$
7. $(1 \times 10^4) + (0 \times 10^3) + (2 \times 10^2) + (0 \times 10^1) + (8 \times 10^0)$ **9.** 456   **11.** 5076   **13.** 35,407   **15.** 683,040   **17.** 76
19. 395   **21.** 2481   **23.** 27   **25.** 3363   **27.** 10,311   **29.** 23   **31.** 97   **33.** 72,133   **35.** 2,171,466
37. 〈〈〈𝖸𝖸   **39.** 𝖸𝖸 𝖸𝖸𝖸𝖸𝖸𝖸𝖸   **41.** 𝖸 〈〈〈𝖸𝖸𝖸𝖸 〈〈〈𝖸𝖸𝖸𝖸𝖸𝖸   **43.** 𝖸𝖸 〈〈〈〈〈𝖸𝖸𝖸𝖸𝖸 〈〈𝖸𝖸𝖸𝖸
45. 𝖸𝖸𝖸𝖸𝖸 〈〈〈〈〈𝖸𝖸𝖸𝖸 〈〈〈〈𝖸𝖸𝖸𝖸𝖸   **47.** 𝖸 𝖸𝖸𝖸𝖸𝖸𝖸𝖸𝖸   **49.** 𝖸 𝖸𝖸𝖸𝖸𝖸𝖸 𝖸𝖸𝖸   **51.** 〈𝖸 〈〈〈〈〈𝖸𝖸𝖸 〈𝖸𝖸𝖸𝖸
53. 194   **55.** 1803   **57.** 14,492   **59.** 36,103   **61.** ——   **63.** ··   **65.** ····   **67.** ·
69. a. and b. Answers will vary.

EXERCISE SET 6.3   *page 325*

1. 73   **3.** 61   **5.** 718   **7.** 485   **9.** 181   **11.** 2032_{five} **13.** 12540_{six} **15.** 22886_{nine} **17.** $111111011100_{\text{two}}$
19. $1B7_{\text{twelve}}$ **21.** 13   **23.** 27   **25.** 100   **27.** 139   **29.** 41   **31.** 90   **33.** 1338   **35.** 26_{eight} **37.** 23033_{four}
39. 24_{five} **41.** 2446_{nine} **43.** 126_{eight} **45.** $7C_{\text{sixteen}}$ **47.** 11101010_{two} **49.** 312_{eight} **51.** 151_{sixteen}
53. $1011111011110011_{\text{two}}$ **55.** $10111010010111001111_{\text{two}}$ **57.** Answers will vary.   **59.** 54   **61.**   **63.**
65. 256   **67. a. and b.** Answers will vary.

EXERCISE SET 6.4   *page 336*

1. 332_{five} **3.** 6562_{seven} **5.** 1001000_{two} **7.** 1271_{twelve} **9.** $D036_{\text{sixteen}}$ **11.** 1124_{six} **13.** 241_{five} **15.** 6542_{eight}
17. 1111_{two} **19.** 1111001_{two} **21.** $411A_{\text{twelve}}$ **23.** 384_{nine} **25.** 523_{six} **27.** 1201_{three} **29.** 45234_{eight} **31.** 1010100_{two}
33. 14207_{eight} **35.** 321222_{four} **37.** $3A61_{\text{sixteen}}$ **39.** 33_{four}; remainder 0_{four} **41.** Quotient 33_{four}; remainder 0_{four}
43. Quotient 1223_{six}; remainder 1_{six} **45.** Quotient 1110_{two}; remainder 0_{two} **47.** Quotient $A8_{\text{twelve}}$; remainder 3_{twelve}
49. Quotient 14_{five}; remainder 11_{five} **51.** Eight   **53. a.** 629   **b.** $384 = 110000000_{\text{two}}$; $245 = 11110101_{\text{two}}$ **c.** 1001110101_{two}
d. 629   **e.** Same   **55. a.** 6422   **b.** $247 = 11110111_{\text{two}}$; $26 = 11010_{\text{two}}$ **c.** $1100100010110_{\text{two}}$ **d.** 6422   **e.** Same
57. Base seven   **59.** In a base one numeration system, 0 would be the only numeral, and the place values would be $1^0, 1^1, 1^2, 1^3, \ldots$, each of which
equals 1. Thus 0 is the only number you could write using a base one numeration system.   **61.** M = 1, A = 4, S = 3, and O = 0

EXERCISE SET 6.5   *page 345*

1. 1, 2, 4, 5, 10, 20   **3.** 1, 5, 13, 65   **5.** 1, 41   **7.** 1, 2, 5, 10, 11, 22, 55, 110   **9.** 1, 5, 7, 11, 35, 55, 77, 385   **11.** Composite
13. Prime   **15.** Prime   **17.** Prime   **19.** Composite   **21.** 2, 3, 5, 6, and 10   **23.** 3   **25.** 2, 3, 4, 6, and 8
27. 2, 5, and 10   **29.** $2 \cdot 3^2$ **31.** $2^3 \cdot 3 \cdot 5$ **33.** $5^2 \cdot 17$ **35.** 2^{10} **37.** $2^3 \cdot 3 \cdot 263$ **39.** $2 \cdot 3^2 \cdot 1013$
41. 2, 3, 5, 7, 11, 13, 17, 19, 23, 29, 31, 37, 41, 43, 47, 53, 59, 61, 67, 71, 73, 79, 83, 89, 97, 101, 103, 107, 109, 113, 127, 131, 137, 139, 149, 151, 157, 163, 167,
173, 179, 181, 191, 193, 197, 199   **43.** 3 and 5, 5 and 7, 11 and 13, 17 and 19, 29 and 31, 41 and 43, 59 and 61, 71 and 73, 101 and 103, 107 and 109, 137
and 139, 149 and 151, 179 and 181, 191 and 193, 197 and 199   **45.** 311 and 313, or 347 and 349   In Exercise 47, parts a to f, only one possible sum
is given.   **47. a.** $24 = 5 + 19$ **b.** $50 = 3 + 47$ **c.** $86 = 3 + 83$ **d.** $144 = 5 + 139$ **e.** $210 = 11 + 199$
f. $264 = 7 + 257$ **49.** Yes   **51.** Yes   **53.** No   **55.** Yes   **57.** Yes   **59.** Yes   **61.** No   **63.** Yes
65. a. $n = 3$ **b.** $n = 4$ **67.** To determine whether a given number is divisible by 17, multiply the ones digit of the given number by 5. Find the
difference between this result and the number formed by omitting the ones digit from the given number. Keep repeating this procedure until you obtain a
small final difference. If the final difference is divisible by 17, then the given number is divisible by 17. If the final difference is not divisible by 17, then the
given number is not divisible by 17.   **69.** 12   **71.** 8   **73.** 24

EXERCISE SET 6.6   *page 355*

1. Abundant   **3.** Deficient   **5.** Deficient   **7.** Abundant   **9.** Deficient   **11.** Deficient   **13.** Abundant   **15.** Abundant
17. Prime   **19.** Prime   **21.** $2^{126}(2^{127} - 1)$ **23.** 6   **25.** 420,921   **27.** 2,098,960   **29.** 11,185,272
31. $9^5 + 15^5 = 818,424$. Because $15^5 < 818,424$ and $16^5 > 818,424$, we know there is no natural number z such that $z^5 = 9^5 + 15^5$.
33. a. False. For instance, if $n = 11$, then $2^{11} - 1 = 2047 = 23 \cdot 89$.   **b.** False. Fermat's last theorem was the last of Fermat's theories (conjectures)
that other mathematicians were able to establish.   **c.** True   **d.** Conjecture.   **35. a.** $12^7 - 12 = 35,831,796$, which is divisible by 7.
b. $8^{11} - 8 = 8,589,934,584$, which is divisible by 11.   **37.** $8128 = 1^3 + 3^3 + 5^3 + \cdots + 13^3 + 15^3$.

See the *Student Solutions Manual* for the verifications in Exercise 39. **41.** The first five Fermat numbers formed using $m = 0, 1, 2, 3$, and 4 are all prime numbers. In 1732, Euler discovered that the sixth Fermat number, 4,294,967,297, formed using $m = 5$, is not a prime number because it is divisible by 641.

CHAPTER 6 REVIEW EXERCISES *page 359*

1. ⵡⵡⵡⵡ⋈⋈~ƒƒƒƒƒ999∩∩||||| [Sec. 6.1] **2.** ⵡⵡⵡ~//ƒƒƒƒ∩∩∩||| [Sec. 6.1]

3. 223,013 [Sec. 6.1] **4.** 221,354 [Sec. 6.1] **5.** 349 [Sec. 6.1] **6.** 774 [Sec. 6.1] **7.** 9640 [Sec. 6.1] **8.** 92,444 [Sec. 6.1]
9. DLXVII [Sec. 6.1] **10.** DCCCXXIII [Sec. 6.1] **11.** MMCDLXXXIX [Sec. 6.1] **12.** MCCCXXXV [Sec. 6.1]
13. $(4 \times 10^2) + (3 \times 10^1) + (2 \times 10^0)$ [Sec. 6.2] **14.** $(4 \times 10^5) + (5 \times 10^4) + (6 \times 10^3) + (3 \times 10^2) + (2 \times 10^1) + (7 \times 10^0)$ [Sec. 6.2]
15. 5,038,204 [Sec. 6.2] **16.** 387,960 [Sec. 6.2] **17.** 801 [Sec. 6.2] **18.** 1603 [Sec. 6.2] **19.** 76,441 [Sec. 6.2]
20. 87,393 [Sec. 6.2] **21.** ◄ⵟⵟ ⵟ [Sec. 6.2] **22.** ◄ⵟⵟⵟⵟⵟⵟⵟ ▲ [Sec. 6.2] **23.** ⵟⵟⵟ ◄◄ⵟⵟⵟⵟⵟⵟⵟⵟ ⵟⵟⵟ [Sec. 6.2]

24. ⵟⵟⵟⵟⵟ ◄◄ⵟ ◄◄ⵟ [Sec. 6.2] **25.** 194 [Sec. 6.2] **26.** 267 [Sec. 6.2] **27.** 2178 [Sec. 6.2] **28.** 6580 [Sec. 6.2]

29. • [Sec. 6.2] **30.** ≡ [Sec. 6.2] **31.** ⎯ [Sec. 6.2] **32.** ⎯ [Sec. 6.2] **33.** 29 [Sec. 6.3]
  •••   •   ••••
• •   • •   • •

34. 146 [Sec. 6.3] **35.** 227 [Sec. 6.3] **36.** 286 [Sec. 6.3] **37.** 1200_{three} [Sec. 6.3] **38.** 234_{seven} [Sec. 6.3] **39.** 714_{eleven} [Sec. 6.3]
40. $18B9_{twelve}$ [Sec. 6.3] **41.** 1153_{six} [Sec. 6.3] **42.** 640_{eight} [Sec. 6.3] **43.** 458_{nine} [Sec. 6.3] **44.** $B62_{twelve}$ [Sec. 6.3]
45. 34_{eight} [Sec. 6.3] **46.** 124_{eight} [Sec. 6.3] **47.** $38D_{sixteen}$ [Sec. 6.3] **48.** $754_{sixteen}$ [Sec. 6.3] **49.** 10101_{two} [Sec. 6.3]
50. 1100111010_{two} [Sec. 6.3] **51.** 1001010_{two} [Sec. 6.3] **52.** 110001110010_{two} [Sec. 6.3] **53.** 410 [Sec. 6.3] **54.** 277 [Sec. 6.3]
55. 1041 [Sec. 6.3] **56.** 1616 [Sec. 6.3] **57.** 423_{six} [Sec. 6.4] **58.** 1240_{eight} [Sec. 6.4] **59.** 536_{nine} [Sec. 6.4]
60. 1113_{four} [Sec. 6.4] **61.** 16412_{eight} [Sec. 6.4] **62.** 324203_{five} [Sec. 6.4] **63.** Quotient 11100_{two}; remainder 1_{two} [Sec. 6.4]
64. Quotient 21_{four}; remainder 3_{four} [Sec. 6.4] **65.** 3, 5, 9, and 11 [Sec. 6.5] **66.** 2, 4, and 11 [Sec. 6.5] **67.** Composite [Sec. 6.5]
68. Composite [Sec. 6.5] **69.** Composite [Sec. 6.5] **70.** Composite [Sec. 6.5] **71.** $3^2 \cdot 5$ [Sec. 6.5] **72.** $2 \cdot 3^3$ [Sec. 6.5]
73. $3^2 \cdot 17$ [Sec. 6.5] **74.** $3 \cdot 5 \cdot 19$ [Sec. 6.5] **75.** Perfect [Sec. 6.6] **76.** Deficient [Sec. 6.6] **77.** Abundant [Sec. 6.6]
78. Abundant [Sec. 6.6] **79.** $2^{60}(2^{61} - 1)$ [Sec. 6.6] **80.** $2^{1278}(2^{1279} - 1)$ [Sec. 6.6] **81.** 368 [Sec. 6.1] **82.** 513 [Sec. 6.1]
83. 1162 [Sec. 6.1] **84.** 3003 [Sec. 6.1] **85.** 39,751 [Sec. 6.6] **86.** 895,932 [Sec. 6.6] **87.** Zero [Sec. 6.6] **88.** No [Sec. 6.6]

CHAPTER 6 TEST *page 361*

1. ƒƒƒ9∩∩|||| [Sec. 6.1, Example 1] **2.** 4263 [Sec. 6.1, Example 2] **3.** 1447 [Sec. 6.1, Example 5] **4.** MMDCIX [Sec. 6.1, Example 6]

5. $(6 \times 10^4) + (7 \times 10^3) + (4 \times 10^2) + (8 \times 10^1) + (5 \times 10^0)$ [Sec. 6.2, Example 1] **6.** 530,284 [Sec. 6.2, Example 2]
7. 37,274 [Sec. 6.2, Example 6] **8.** ⵟⵟ ◄◄◄◄ⵟ ◄ⵟⵟⵟⵟⵟ [Sec. 6.2, Example 7] **9.** 1305 [Sec. 6.2, Example 9]

10. • [Sec. 6.2, Example 10] **11.** 854 [Sec. 6.3, Example 1] **12. a.** 4144_{eight} **b.** $12B0_{twelve}$ [Sec. 6.3, Example 5]
  • •
• •

13. 100101110111_{two} [Sec. 6.3, Example 6] **14.** $AB7_{sixteen}$ [Sec. 6.3, Example 9] **15.** 112_{five} [Sec. 6.4, Example 2]
16. 313_{eight} [Sec. 6.4, Example 5] **17.** 11100110_{two} [Sec. 6.4, Example 7] **18.** Quotient 61_{seven}; remainder 3_{seven} [Sec. 6.4, Example 9]
19. $2 \cdot 5 \cdot 23$ [Sec. 6.5, Example 4] **20.** Composite [Sec. 6.5, Example 2] **21. a.** No **b.** Yes **c.** No [Sec. 6.5, Example 3]
22. a. Yes **b.** No **c.** No [Sec. 6.5, Example 3] **23.** Abundant [Sec. 6.6, Example 1] **24.** $2^{16}(2^{17} - 1)$ [Sec. 6.6, Example 3]

CHAPTER 8

EXERCISE SET 8.1   *page 482*

1. 8   **3.** 12   **5.** 2   **7.** 4   **9.** 4   **11.** 7   **13.** 11   **15.** 7   **17.** 0300   **19.** 0400   **21.** 2000
23. 2100   **25.** 3   **27.** 6   **29.** True   **31.** False   **33.** True   **35.** False   **37.** True
39. Possible answers are 2, 8, 14, 20, 26, 32, 38, ...   **41.** 3   **43.** 3   **45.** 2   **47.** 10   **49.** 3   **51.** 3
53. 5   **55.** 4   **57.** 3   **59.** 7   **61.** 2   **63. a.** 6 o'clock   **b.** 5 o'clock   **65. a.** Tuesday   **b.** Monday
67. Saturday   **69.** Friday   **71.** 1, 4, 7, 10, 13, 16, ...   **73.** 1, 6, 11, 16, 21, 26, ...   **75.** 0, 2, 4, 6, 8, 10, 12, ...
77. No solutions   **79.** 0, 2, 4, 6, 8, 10, 12, ...   **81.** No solutions   **83.** 5, 7   **85.** 3, 3   **87.** 5, 3   **89.** 6
91. 6   **93.** 2   **97.** 11:00   **99.** 4

EXERCISE SET 8.2   *page 493*

1. No   **3.** Yes   **5.** Yes   **7.** 1   **9.** 0   **11.** 5   **13.** 6   **15.** 5   **17.** 7   **19.** 2   **21.** 0
23. 7   **25.** 3   **27.** Yes   **29.** Yes   **31.** Yes   **33.** No   **35.** No   **37.** Yes
39. BPZMM UCASMBMMZA   **41.** UF'E M SUDX   **43.** VWLFNV DQG VWRQHV   **45.** AGE OF ENLIGHTENMENT
47. FRIEND IN NEED   **49.** DANGER WILL ROBINSON   **51.** FORTUNE COOKIE
53. PHQ ZLOOLQJOB EHOLHYH ZKDW WKHB ZLVK   **55.** JUSQD UT LURNUR   **57.** PODONNQN NSBQK
59. TURN BACK THE CLOCK   **61.** BARREL OF MONKEYS   **63.** Because the check digit is simply the sum of the first 10 digits mod 9, the same digits in a different order will give the same sum and hence the same check digit.

EXERCISE SET 8.3   *page 505*

1. a. Yes   **b.** No   **3. a.** Yes   **b.** No   **5.** Yes   **7.** Yes   **9.** No; property 4 fails.   **11.** Yes
13. Yes   **15.** No; property 4 fails.   **17.** Yes   **19.** R_l **21.** R_{120} **23.** R_{120} **25.** R_r

27. R_l **29.** $I = \begin{pmatrix} 1 & 2 & 3 & 4 \\ 1 & 2 & 3 & 4 \end{pmatrix}, R_{90} = \begin{pmatrix} 1 & 2 & 3 & 4 \\ 2 & 3 & 4 & 1 \end{pmatrix}, R_{180} = \begin{pmatrix} 1 & 2 & 3 & 4 \\ 3 & 4 & 1 & 2 \end{pmatrix}, R_{270} = \begin{pmatrix} 1 & 2 & 3 & 4 \\ 4 & 1 & 2 & 3 \end{pmatrix}, R_v = \begin{pmatrix} 1 & 2 & 3 & 4 \\ 4 & 3 & 2 & 1 \end{pmatrix},$

$R_h = \begin{pmatrix} 1 & 2 & 3 & 4 \\ 2 & 1 & 4 & 3 \end{pmatrix}, R_r = \begin{pmatrix} 1 & 2 & 3 & 4 \\ 3 & 2 & 1 & 4 \end{pmatrix}, R_l = \begin{pmatrix} 1 & 2 & 3 & 4 \\ 1 & 4 & 3 & 2 \end{pmatrix}$ **31.** R_r **33.** R_{90} **35.** I **37.** D

39. E **41.** B **43.** $\begin{pmatrix} 1 & 2 & 3 & 4 \\ 1 & 2 & 3 & 4 \end{pmatrix}, \begin{pmatrix} 1 & 2 & 3 & 4 \\ 1 & 2 & 4 & 3 \end{pmatrix}, \begin{pmatrix} 1 & 2 & 3 & 4 \\ 1 & 3 & 2 & 4 \end{pmatrix}, \begin{pmatrix} 1 & 2 & 3 & 4 \\ 1 & 3 & 4 & 2 \end{pmatrix}, \begin{pmatrix} 1 & 2 & 3 & 4 \\ 1 & 4 & 2 & 3 \end{pmatrix}, \begin{pmatrix} 1 & 2 & 3 & 4 \\ 1 & 4 & 3 & 2 \end{pmatrix},$

$\begin{pmatrix} 1 & 2 & 3 & 4 \\ 2 & 1 & 3 & 4 \end{pmatrix}, \begin{pmatrix} 1 & 2 & 3 & 4 \\ 2 & 1 & 4 & 3 \end{pmatrix}, \begin{pmatrix} 1 & 2 & 3 & 4 \\ 2 & 3 & 1 & 4 \end{pmatrix}, \begin{pmatrix} 1 & 2 & 3 & 4 \\ 2 & 3 & 4 & 1 \end{pmatrix}, \begin{pmatrix} 1 & 2 & 3 & 4 \\ 2 & 4 & 1 & 3 \end{pmatrix}, \begin{pmatrix} 1 & 2 & 3 & 4 \\ 2 & 4 & 3 & 1 \end{pmatrix}, \begin{pmatrix} 1 & 2 & 3 & 4 \\ 3 & 1 & 2 & 4 \end{pmatrix}, \begin{pmatrix} 1 & 2 & 3 & 4 \\ 3 & 1 & 4 & 2 \end{pmatrix},$

$\begin{pmatrix} 1 & 2 & 3 & 4 \\ 3 & 2 & 1 & 4 \end{pmatrix}, \begin{pmatrix} 1 & 2 & 3 & 4 \\ 3 & 2 & 4 & 1 \end{pmatrix}, \begin{pmatrix} 1 & 2 & 3 & 4 \\ 3 & 4 & 1 & 2 \end{pmatrix}, \begin{pmatrix} 1 & 2 & 3 & 4 \\ 3 & 4 & 2 & 1 \end{pmatrix}, \begin{pmatrix} 1 & 2 & 3 & 4 \\ 4 & 1 & 2 & 3 \end{pmatrix}, \begin{pmatrix} 1 & 2 & 3 & 4 \\ 4 & 1 & 3 & 2 \end{pmatrix}, \begin{pmatrix} 1 & 2 & 3 & 4 \\ 4 & 2 & 1 & 3 \end{pmatrix}, \begin{pmatrix} 1 & 2 & 3 & 4 \\ 4 & 2 & 3 & 1 \end{pmatrix},$

$\begin{pmatrix} 1 & 2 & 3 & 4 \\ 4 & 3 & 1 & 2 \end{pmatrix}, \begin{pmatrix} 1 & 2 & 3 & 4 \\ 4 & 3 & 2 & 1 \end{pmatrix}$ **45.** $\begin{pmatrix} 1 & 2 & 3 & 4 \\ 2 & 1 & 3 & 4 \end{pmatrix}$ **47.** $\begin{pmatrix} 1 & 2 & 3 & 4 \\ 3 & 1 & 4 & 2 \end{pmatrix}$ **49.** $\begin{pmatrix} 1 & 2 & 3 & 4 \\ 4 & 3 & 2 & 1 \end{pmatrix}$ **51.** d

53. c **55.** Answers will vary.   **57.** Yes   **59. a.** and **b.** Answers will vary.   **c.** Values of n that are prime

61. a.

⊕	1	2	3	4	5
1	1	2	3	4	5
2	2	3	4	5	1
3	3	4	5	1	2
4	4	5	1	2	3
5	5	1	2	3	4

b. Yes   **c.** 1   **d.** 1 is its own inverse, 2 and 5 are inverses; 3 and 4 are inverses.

CHAPTER 8 REVIEW EXERCISES   *page 510*

1. 2 [Sec. 8.1]   **2.** 2 [Sec. 8.1]   **3.** 5 [Sec. 8.1]   **4.** 6 [Sec. 8.1]   **5.** 9 [Sec. 8.1]
6. 4 [Sec. 8.1]   **7.** 11 [Sec. 8.1]   **8.** 7 [Sec. 8.1]   **9.** 3 [Sec. 8.1]   **10.** 4 [Sec. 8.1]
11. True [Sec. 8.1]   **12.** False [Sec. 8.1]   **13.** False [Sec. 8.1]   **14.** True [Sec. 8.1]   **15.** 2 [Sec. 8.1]
16. 4 [Sec. 8.1]   **17.** 0 [Sec. 8.1]   **18.** 3 [Sec. 8.1]   **19.** 8 [Sec. 8.1]   **20.** 3 [Sec. 8.1]
21. 7 [Sec. 8.1]   **22.** 5 [Sec. 8.1]   **23. a.** 2 o'clock   **b.** 6 o'clock [Sec. 8.1]   **24.** Monday [Sec. 8.1]
25. 3, 7, 11, 15, 19, 23, … [Sec. 8.1]   **26.** 7, 16, 25, 34, 43, 52, … [Sec. 8.1]   **27.** 0, 5, 10, 15, 20, 25, 30, … [Sec. 8.1]
28. 4, 15, 26, 37, 48, 59, 70, … [Sec. 8.1]   **29.** 2, 3 [Sec. 8.1]   **30.** 5, 7 [Sec. 8.1]   **31.** 6 [Sec. 8.1]
32. 2 [Sec. 8.1]   **33.** 8 [Sec. 8.2]   **34.** 5 [Sec. 8.2]   **35.** 2 [Sec. 8.2]   **36.** 1 [Sec. 8.2]
37. No [Sec. 8.2]   **38.** Yes [Sec. 8.2]   **39.** No [Sec. 8.2]   **40.** No [Sec. 8.2]   **41.** THF AOL MVYJL IL DPAO FVB
[Sec. 8.2]   **42.** NLYNPW LWW AWLYD [Sec. 8.2]   **43.** GOOD LUCK TOMORROW [Sec. 8.2]
44. THE DAY HAS ARRIVED [Sec. 8.2]   **45.** UVR YX NDU PGVU [Sec. 8.2]   **46.** YOU PASSED THE TEST [Sec. 8.2]
47. Yes [Sec. 8.3]   **48.** Yes [Sec. 8.3]   **49.** No, properties 1, 3, and 4 fail. [Sec. 8.3]   **50.** Yes [Sec. 8.3]
51. R_{240} [Sec. 8.3]   **52.** R_l [Sec. 8.3]   **53.** R_r [Sec. 8.3]   **54.** R_{240} [Sec. 8.3]   **55.** A [Sec. 8.3]
56. I [Sec. 8.3]   **57.** D [Sec. 8.3]   **58.** B [Sec. 8.3]

CHAPTER 8 TEST   *page 511*

1. a. 3   **b.** 5 [Sec. 8.1, Example 1]   **2.** Monday [Sec. 8.1, Example 3]   **3. a.** True   **b.** False [Sec. 8.1, Example 2]
4. 4 [Sec. 8.1, Example 4]   **5.** 6 [Sec. 8.1, Example 5]   **6.** 8 [Sec. 8.1, Example 7]   **7. a.** 6 o'clock
b. 5 o'clock [Sec. 8.1, Example 6]   **8.** 5, 14, 23, 32, 41, 50, … [Sec. 8.1, Example 8]   **9.** 1, 3, 5, 7, 9, 11, … [Sec. 8.1, Example 8]
10. 4, 2 [Sec. 8.1, Example 9 and Example 10]   **11.** 0 [Sec. 8.2, Example 1]   **12.** 0 [Sec. 8.2, Example 2]
13. Yes [Sec. 8.2, Example 3]   **14.** BOZYBD LKMU [Sec. 8.2, Example 4]   **15.** NEVER QUIT [Sec. 8.2, Example 4]
16. Yes [Sec. 8.3, Check Your Progress 1]   **17.** No, property 4 fails; many elements do not have an inverse. [Sec. 8.3, Example 1]
18. a. R_t **b.** R_r [Sec. 8.3, Example 2]   **19.** $\begin{pmatrix} 1 & 2 & 3 \\ 1 & 3 & 2 \end{pmatrix}$ [Sec. 8.3, Example 4]   **20.** $\begin{pmatrix} 1 & 2 & 3 \\ 2 & 3 & 1 \end{pmatrix}$ [Sec. 8.3, Example 5]

CHAPTER 12

EXERCISE SET 12.1 *page 741*

1. {0, 2, 4, 6, 8}   **3.** {Monday, Tuesday, Wednesday, Thursday, Friday, Saturday, Sunday}
5. {HH, TT, HT, TH}   **7.** {1H, 2H, 3H, 4H, 5H, 6H, 1T, 2T, 3T, 4T, 5T, 6T}
9. {$S_1E_1D_1$, $S_1E_1D_2$, $S_1E_2D_1$, $S_1E_2D_2$, $S_1E_3D_1$, $S_1E_3D_2$, $S_2E_1D_1$, $S_2E_1D_2$, $S_2E_2D_1$, $S_2E_2D_2$, $S_2E_3D_1$, $S_2E_3D_2$}
11. {ABCD, ABDC, ACBD, ACDB, ADBC, ADCB}   **13.** 12   **15.** 4^{20} **17.** 7000   **19.** 47,916,000   **21.** 90   **23.** 18
25. 62   **27.** 24   **29.** 15   **31.** 13^4 **33.** 1   **35.** Answers will vary.   **37.** 150; 25   **39.** Answers will vary.

EXERCISE SET 12.2 *page 753*

1. 40,320   **3.** 362,760   **5.** 120   **7.** 6720   **9.** 181,440   **11.** 1   **13.** 40,320   **15.** 3360   **17.** $\dfrac{1}{720}$ **19.** 36
21. 1   **23.** 525   **25.** $\dfrac{60}{143}$ **27.** $\dfrac{360}{1001}$ **29.** 2880   **31.** 21   **33.** 792   **35.** No   **37.** 120   **39.** 43,680
41. **a.** 362,880   **b.** 8640   **43.** 10,080   **45.** 735,471   **47.** 35   **49.** 3136   **51.** 12   **53.** 9   **55.** 48,620
57. 24   **59.** 45,360   **61.** 252   **63.** 1728   **65.** 48   **67.** 4512   **69.** 103,776   **71.** Answers will vary.
73. $x^5 + 5x^4y + 10x^3y^2 + 10x^2y^3 + 5xy^4 + y^5$

EXERCISE SET 12.3 *page 764*

1. {HHH, HHT, HTH, HTT, THH, THT, TTH, TTT}
3. {Nov. 1, Nov. 2, Nov. 3, Nov. 4, Nov. 5, Nov. 6, Nov. 7, Nov. 8, Nov. 9, Nov. 10, Nov. 11, Nov. 12, Nov. 13, Nov. 14}
5. {Alaska, Alabama, Arizona, Arkansas}   **7.** {BBB, BBG, BGB, BGG, GBB, GBG, GGB, GGG}   **9.** {BGG, GBG, GGB, GGG}

11. {BBG, BGB, BGG, GBB, GBG, GGB, GGG} **13.** $\dfrac{1}{2}$ **15.** $\dfrac{7}{8}$ **17.** $\dfrac{1}{4}$ **19.** $\dfrac{11}{16}$ **21.** $\dfrac{1}{12}$ **23.** $\dfrac{1}{4}$ **25.** $\dfrac{5}{36}$

27. $\dfrac{1}{36}$ **29.** 0 **31.** $\dfrac{1}{6}$ **33.** $\dfrac{1}{2}$ **35.** $\dfrac{1}{6}$ **37.** $\dfrac{1}{13}$ **39.** $\dfrac{3}{13}$ **41.** $\dfrac{1267}{3228}$ **43.** $\dfrac{804}{3228}$ **45.** $\dfrac{150}{3228}$ **47.** $\dfrac{26}{425}$

49. $\dfrac{18}{425}$ **51.** $\dfrac{58}{293}$ **53.** $\dfrac{36}{293}$ **55.** $\dfrac{1}{4}$ **57.** 0 **59.** Answers will vary. **61.** $\dfrac{1}{3}$ **63.** $\dfrac{3}{10}$ **65.** $\dfrac{8}{13}$ **67.** 1 to 4

69. 3 to 5 **71.** 11 to 9 **73.** 1 to 14 **75.** 1 to 1 **77.** 15 to 1 **79.** $\dfrac{49}{97}$ **81.** $\dfrac{3}{11}$ **83.** $\dfrac{3}{8}$

85. Answers will vary. **87.** $\dfrac{1}{108,290}$

EXERCISE SET 12.4 *page 775*

1. Answers will vary. **3.** $\dfrac{2}{13}$ **5.** $\dfrac{1}{9}$ **7.** 0.6 **9.** 0.2 **11.** $\dfrac{7}{10}$ **13.** $\dfrac{4}{5}$ **15.** $\dfrac{5}{18}$ **17.** $\dfrac{1}{2}$ **19.** $\dfrac{11}{18}$

21. $\dfrac{4}{13}$ **23.** $\dfrac{3}{13}$ **25.** $\dfrac{3}{4}$ **27.** $\dfrac{1150}{3179}$ **29.** $\dfrac{1170}{3179}$ **31.** 0.96 **33.** $\dfrac{74}{75}$, or about 98.7% **35.** $\dfrac{5}{6}$ **37.** $\dfrac{11}{12}$

39. $\dfrac{12}{13}$ **41.** $\dfrac{15}{16}$ **43.** 42.1% **45.** 36.1% **47.** 54.5% **49.** 88.8% **51.** 20.4% **53.** No **55.** 100%

57. Answers will vary.

EXERCISE SET 12.5 *page 785*

1. Answers will vary. **3.** $P(A\,|\,B) = 0.625$; $P(B\,|\,A) \approx 0.357$ **5.** $P(A\,|\,B) \approx 0.389$; $P(B\,|\,A) \approx 0.115$ **7.** $\dfrac{179}{864}$ **9.** $\dfrac{557}{921}$

11. 0.30 **13.** 0.15 **15.** $\dfrac{5}{18}$ **17.** $\dfrac{1}{5}$ **19.** 0.050 **21.** 0.127 **23.** $\dfrac{6}{1045}$ **25.** $\dfrac{3}{1045}$ **27.** $\dfrac{13}{102}$ **29.** $\dfrac{8}{5525}$

31. 0.000484 **33.** 0.001424 **35.** Independent **37.** Not independent **39.** $\dfrac{25}{1296}$ **41.** $\dfrac{1}{72}$ **43.** $\dfrac{1}{4}$ **45.** $\dfrac{1}{16}$

47. $\dfrac{1}{216}$ **49.** $\dfrac{1}{169}$ **51.** $\dfrac{1}{16}$ **53.** $\dfrac{1}{16}$ **55. a.** $\dfrac{1}{32}$ **b.** $\dfrac{13}{425}$ **57. a.** $\dfrac{100}{4913}$ **b.** $\dfrac{1}{51}$ **59.** 0.46 **61.** 0.11

63. a. $\dfrac{1}{8}$ **b.** $\dfrac{1}{4}$ **c.** $\dfrac{3}{8}$

EXERCISE SET 12.6 *page 792*

1. 49.5 **3.** -5 cents **5.** -24 cents **7.** -22 cents **9.** \$62.83 **11.** $-\$209.71$ **13.** More than \$42.18 **15.** \$39,100

17. \$20,250 **19.** -1.7 cents **21.** 7 **23.** 7 **25.** Red dice

CHAPTER 12 REVIEW EXERCISES *page 798*

1. {11, 12, 13, 21, 22, 23, 31, 32, 33} [Sec. 12.1] **2.** {26, 28, 62, 68, 82, 86} [Sec. 12.1] **3.** {HHHH, HHHT, HHTH,
HHTT, HTHH, HTHT, HTTH, HTTT, THHH, THHT, THTH, THTT, TTHH, TTHT, TTTH, TTTT} [Sec. 12.1]
4. {7A, 8A, 9A, 7B, 8B, 9B} [Sec. 12.1] **5.** 72 [Sec. 12.1] **6.** 10,000 [Sec. 12.1] **7.** 2400 [Sec. 12.1] **8.** 64 [Sec. 12.1]
9. 5040 [Sec. 12.2] **10.** 40,296 [Sec. 12.2] **11.** 1260 [Sec. 12.2] **12.** 151,200 [Sec. 12.2] **13.** 336 [Sec. 12.2]
14. $\dfrac{60}{143}$ [Sec. 12.2] **15.** 5040 [Sec. 12.2] **16.** 5040 [Sec. 12.2] **17.** 2520 [Sec. 12.2] **18.** 180 [Sec. 12.2]
19. 495 [Sec. 12.2] **20.** 60 [Sec. 12.2] **21.** 3,268,760 [Sec. 12.2] **22.** 165 [Sec. 12.2] **23.** 660 [Sec. 12.2]
24. 282,240 [Sec. 12.2] **25.** 624 [Sec. 12.2] **26.** $\dfrac{1}{4}$ [Sec. 12.3] **27.** $\dfrac{3}{8}$ [Sec. 12.3] **28.** 0.56 [Sec. 12.3] **29.** 0.37 [Sec. 12.3]

30. 0.85 [Sec. 12.3] **31.** $\dfrac{1}{9}$ [Sec. 12.3] **32.** $\dfrac{17}{18}$ [Sec. 12.4] **33.** $\dfrac{1}{6}$ [Sec. 12.4] **34.** $\dfrac{5}{9}$ [Sec. 12.4] **35.** $\dfrac{2}{9}$ [Sec. 12.5]

36. $\dfrac{1}{6}$ [Sec. 12.5] **37.** $\dfrac{3}{4}$ [Sec. 12.4] **38.** $\dfrac{4}{13}$ [Sec. 12.4] **39.** $\dfrac{12}{13}$ [Sec. 12.4] **40.** $\dfrac{2}{3}$ [Sec. 12.5] **41.** 5 to 31 [Sec. 12.3]

42. 1 to 3 [Sec. 12.3] **43.** $\dfrac{5}{9}$ [Sec. 12.3] **44.** $\dfrac{1}{2}$ [Sec. 12.3] **45.** 0.036 [Sec. 12.4] **46.** $\dfrac{173}{1000}$ [Sec. 12.4] **47.** $\dfrac{5}{8}$ [Sec. 12.3]

48. 1 to 5 [Sec. 12.3] **49.** $\dfrac{7}{16}$ [Sec. 12.5] **50.** $\dfrac{646}{1771}$ [Sec. 12.5] **51.** $\dfrac{7}{253}$ [Sec. 12.4] **52.** 0.37 [Sec. 12.3]

53. 0.62 [Sec. 12.4] **54.** 0.07 [Sec. 12.3] **55.** 0.47 [Sec. 12.5] **56.** 0.29 [Sec. 12.5] **57.** $\dfrac{1}{216}$ [Sec. 12.5]

58. 0.60 [Sec. 12.4] **59.** 0.16 [Sec. 12.5] **60.** $\dfrac{175}{256}$ [Sec. 12.4] **61.** 0.648 [Sec. 12.5] **62.** 0.029 [Sec. 12.5]

63. 0.135 [Sec. 12.4] **64.** −50 cents [Sec. 12.6] **65.** 25 cents [Sec. 12.6] **66.** −37.5 cents [Sec. 12.6]

67. About 5.4 [Sec. 12.6] **68.** $296.20 [Sec. 12.6] **69.** $765.30 [Sec. 12.6] **70.** $11,000 [Sec. 12.6]

CHAPTER 12 TEST *page 801*

1. {A2, D2, G2, K2, A3, D3, G3, K3, A4, D4, G4, K4} [Sec. 12.1, Example 3 and Check Your Progress 4] **2.** 72 [Sec. 12.1, Example 5]
3. 5040 [Sec. 12.1, Example 5; Sec 12.2, Example 2] **4.** About 1.09×10^{10} [Sec. 12.2, Example 2] **5.** 28 [Sec. 12.2, Example 6]
6. $\dfrac{7}{12}$ [Sec. 12.3, Example 3] **7.** 27.4% [Sec. 12.4, Example 4] **8.** $\dfrac{1}{17}$ [Sec. 12.3, Example 3] **9.** 0.72 [Sec. 12.4, Check Your Progress 3]
10. 1 to 7 [Sec. 12.3, Example 6] **11.** 0.635 [Sec. 12.5, Example 4] **12.** 0.466 [Sec. 12.5, Example 1] **13.** $\dfrac{1}{2}$ [Sec. 12.3, Example 5]
14. $38,350 [Sec. 12.6, Example 3]

Math Through the Ages

10,000 BC–0

10,000 BC Agricultural villages are evident in many parts of the world.

6300 BC The techniques for creating pottery are known.

6200 BC The first evidence of writing appears.

3400 BC The first symbols for numbers are used.

3250 BC Construction of the great pyramid Cheops begins.

3000 BC Babylonians use a place-value numeration system.

2200 BC Approximate time of the Trojan War.

1850 BC The Pythagorean Theorem is known by Babylonians.

1700 BC The approximate year the Rhind papyrus is written. It shows early algorithms for multiplication and division.

776 BC The first Olympic games are held.

700 BC A symbol for zero is introduced as a place holder. It is not considered a number.

575 BC Thales of Miletus, sometimes known as the first mathematician, uses deductive reasoning to prove theorems in geometry.

475 BC Babylonians use geometry to study astronomy.

450 BC The Greeks begin to use written numerals.

387 BC Plato founds his Academy.

300 BC Euclid writes *Elements*, the first systematic development of geometry.

290 BC Aristarchus uses geometry to calculate the distance to the moon.

260 BC Archimedes establishes pi as approximately equal to 22/7.

259 BC The construction of the Great Wall of China begins.

235 BC Eratosthenes estimates the circumference of Earth and develops a method for finding prime numbers.

225 BC Apollonius writes a treatise on conics and first uses the words *parabola*, *ellipse*, and *hyperbola*.

69 BC Cleopatra is born.

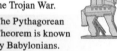

0–1499

79 Vesuvius erupts, destroying Pompeii and Herculaneum.

240 Mayan civilization uses a base 20 numeration system.

250 Diophantus writes *Arithmetica*.

400 Hypatia is the first recorded female mathematician.

476 The fall of Rome.

594 Decimal notation is used in India.

628 Brahmagupta gives rules for using zero and negative numbers in computations.

700 Mayan mathematicians introduce zero into their numeration system.

800 Charlemagne is crowned Holy Roman Emperor.

810 Al-Khwarizmi, also known as the father of algebra, writes a treatise on solving equations. From that treatise, the word *algebra* is derived.

875 Approximate date the first printed book is produced in China.

1202 Fibonacci writes *The Book of the Abacus* and helps popularize the Hindu-Arabic number system in Europe.

1206 The mechanical clock is invented.

1275 Yang Hui gives the first account of what becomes known as "Pascal's Triangle."

1290 Marco Polo travels the world.

1445 The approximate year in which the Gutenberg printing press is invented.

1482 Euclid's *Elements* becomes the first mathematics book to be printed.

1492 Christopher Columbus discovers America.

1492 Leonardo Da Vinci paints the Mona Lisa.

1500–1699

1545 Cardan publishes *Ars Magna*, which gives the first formula to solve any cubic equation.

1564 Shakespeare is born.

1585 Stevin writes *De Thiende*, which gives a treatment of decimal fractions.

1591 Letters are first used to represent variables.

1614 Johannes Kepler publishes his laws on elliptical orbits of the planets.

1614 John Napier publishes his work on logarithms.

1617 Henry Briggs introduces logarithms base 10.

1620 Pilgrims land on Plymouth Rock.

1620 First American Indian reservation established in Connecticut.

1620 Gunter makes the first primitive slide rule.

1647 Fermat states that $x^n + y^n = z^n$ has no integer solutions for $n > 2$. This becomes known as Fermat's Last Theorem.

1665 Isaac Newton develops his infinitesimal calculus.

1687 Isaac Newton publishes *Philosophiæ Naturalis Principia Mathematica*, which is regarded as one of the best scientific papers ever written.

1692 Salem witch trials are held.

1698 The first steam engine is patented.

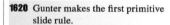

1700–1799

1706 William Jones introduces the symbol π to represent the ratio of the circumference to the diameter of a circle.

1707 Abraham De Moivre uses trigonometric functions to represent complex numbers.

1709 Bartolomeo Christofori invents the piano.

1727 Leonhard Euler introduces the symbol e for the base of the natural logarithms.

1731 Benjamin Banneker, the first African American mathematician and scientist, is born. He helped survey Washington, D.C.

1735 Leonhard Euler introduces the notation $f(x)$.

1736 Leonhard Euler creates introductory graph theory and uses it to solve the Königsberg bridge problem.

1748 Maria Gaëtana Agnesi writes the most respected calculus text in Italy.

1755 Samuel Johnson publishes the first dictionary of the English language.

1761 Pi is proved to be an irrational number.

1776 The Declaration of Independence is signed.

1781 Caroline Herschel and her brother William discover Uranus.

1785 Marie Jean Condorcet publishes a treatise on voting theory.

1786 Caroline Herschel becomes the first woman to discover a comet.

1789 George Washington becomes the first president of the United States.

1796 Edward Jenner develops a smallpox vaccination.

1799 The Rosetta Stone is discovered.

1799 Carl Friedrich Gauss proves the Fundamental Theorem of Algebra.